Sociology of the Sciences Yearbook

Volume 30

More information about this series at http://www.springer.com/series/6566

Sabine Maasen • Sascha Dickel
Christoph Schneider
Editors

TechnoScienceSociety

Technological Reconfigurations of Science and Society

Springer

Editors
Sabine Maasen
Munich Center for Technology
in Society (MCTS)
Technical University Munich
Munich, Bavaria, Germany

Sascha Dickel
Institute of Sociology
Johannes Gutenberg University Mainz
Mainz, Germany

Christoph Schneider
Institute for Technology Assessment
and Systems Analysis
Karlsruhe Institute of Technology
Karlsruhe, Germany

ISSN 0167-2320 ISSN 2215-1796 (electronic)
Sociology of the Sciences Yearbook
ISBN 978-3-030-43964-4 ISBN 978-3-030-43965-1 (eBook)
https://doi.org/10.1007/978-3-030-43965-1

This Springer imprint is published by the registered company Springer Nature Switzerland AG
The registered company address is: Gewerbestrasse 11, 6330 Cham, Switzerland

Contents

Chapter 1
TechnoScienceSociety: Technological Reconfigurations of Science and Society – An Introduction

Sabine Maasen, Sascha Dickel, and Christoph Schneider

1.1 14th March 2020, a Normal Day

Dear diary,

I wake up, grabbing my smartphone beside me. Any new messages or notifications? Sure, there is always something new. Every morning is a morning full of technologically enabled updates – from my friends, from organizations, from publics that make themselves visible. Algorithmic networks are underpinning my interactions, shaping them, filtering them, tracking them – expanding others, themselves and me. And making us dependent on an opaque infrastructure.

Look at my organic breakfast – an amazing result of a complex technological process. Industries of agricultural production, life sciences, transportation and consumption are forming a vast apparatus of nutrition. This apparatus hides behind romanticized images of nature: green forests, untouched by human hands or the grapplers of machines; animals, living in small herds. The newspaper shows a politician eating an in-vitro-meat burger for which supposedly no animal was killed.

Next, I drive to my office. Roads, signs, and traffic lights show me the way. They enable me to reach my workplace. However, they also define the space of permitted car movements and myself in time and space. The task of getting to my office is highly distributed. My arm, the wheels, my navigation system, the urban

S. Maasen (✉)
Munich Center for Technology in Society (MCTS), Technical University Munich, Munich, Bavaria, Germany
e-mail: sabine.maasen@tum.de

S. Dickel
Institute of Sociology, Johannes Gutenberg University Mainz, Mainz, Germany
e-mail: dickel@uni-mainz.de

C. Schneider
Institute for Technology Assessment and Systems Analysis, Karlsruhe Institute of Technology, Karlsruhe, Germany
e-mail: christoph.schneider@kit.edu

S. Maasen et al. (eds.), *TechnoScienceSociety*, Sociology of the Sciences Yearbook 30, https://doi.org/10.1007/978-3-030-43965-1_1

infrastructure, all the rules, procedures and regulations governing traffic. We all work together and sometimes against or past each other. I share these roads with some prototypes of self-driving cars, operated by what they call artificial intelligence, but hardly autonomous, bound by the same rules as my car and me, the rules of the contemporary traffic apparatus. Still, it's a peculiar feeling seeing these driverless cars and the news of the big companies in charge of them.

It is stormy today. Is this storm a force of nature? A symptom of anthropogenic climate change? How can I know? By checking my smartphone again. All it takes – at first glance – is my desire to know, the move of my finger, the touchscreen. I usually ignore the infrastructural assemblages that tie all these things together: the inner workings of my black-boxed smartphone, the cell towers, the factories that have assembled my phone, the workers in these factories, the rare earth elements, the logistics of production and consumption, the data centers, international climate science, the power plants providing them with energy. I ignore all this because I am a user. I connect with technology via the interface of a touchscreen. The Interfaces make the black boxes invisible. Should they be visible?

I choose a parking space half a mile away from the office. I need to walk a bit in order to reach my goal of 8.000 steps per day. My fitness tracker tells me what I still need to accomplish today: Some walking in the morning, some running in the evening (with special runners' shoes, manufactured far away from my home and these new socks with nano-coating). My body is not in its best shape (how could it be?). I know this because of medical studies performed on fellow citizens, aggregated into scientific evidence, translated into recommendations (from my physician, from online crowds, from health apps). By allowing my fitness tracker to upload data, I may myself contribute to technoscientific knowledge production, improve my achievements at work as well as in my life at large. Thus, I may contribute to innovations that in turn might shape future bodies and minds (including my own). I have to think of a civil society organization I once saw that advocates for neurological implants – cybernetic feedback loops everywhere?

I look through the window and see my city that extends to the horizon, thousands of buildings, millions of people living their lives in this place. The factory fumes have a weird color today. I heard that humanity may have become the central force that shapes our planet earth (because of its technologies), that the economy keeps growing (and this strange plastic island in the sea is as well), that mobile Internet usage is skyrocketing in Africa and Asia. I heard that antibiotics may no longer work like they used to, that data is the new oil, that rich people could enhance themselves to live longer than others, that we may make the shift towards renewable energies, that our working days may get shorter, that technology is moving in, through and in-between us.

I know something about the technologies and networks in my life. I know that I do not know much. I know that my knowledge depends on the technologies I use. I know that these technologies are shaping who I am and who I can be.

Welcome to the world we live in: Welcome to TechnoScienceSociety!

Can science and technology studies (STS) contribute to a diagnosis of the present? Can it observe contemporary phenomena as a series of special disruptions that take place as events in which we participate and examine what that present involves for us? If STS can contribute to such a diagnosis, might it at the same time escape the typical traps of generalized societal descriptions? Might it resist overstating one particular feature of the present while understating others? Might it refrain to construct a sharp contrast between an idealized past and a dramatized present (Osrecki 2015)? Might it avoid the jumps into imagined technological futures? Can it nevertheless say something distinctive about our contemporary world? We think it can, because we choose to follow the Foucauldian idea of composing an emerging account of the technoscientific present that dives into its manifold political, bodily and material investments as well as into their relations to technologies of power-knowledge (Foucault 1977: 30f).

To define and describe contemporary societies has always been one of the main tasks of sociology. It was often fulfilled by choosing a name for a certain epoch. Recent sociological works have labelled society as "post-industrial society" (Bell 1976), "knowledge society" (Böhme und Stehr 1986), "risk society" (Beck 1992), "network society" (Castells 1996), "high-speed society" (Rosa 2003), or "innovation society" (Rammert et al. 2016). In these descriptions, technology is both present and absent. On the one hand, it very much is *present* because all these fairly contemporary (self-)descriptions of society share the assumption that technologies constitute societal formations: Post-industrial society is enabled by technological reconfigurations of industries and services. Knowledge society is shaped by technological knowledge. The most prominent risks of risk society are technological risks. The networks of the network society are technological information networks. High-speed society is defined by technological acceleration. Innovation society is characterized by the ubiquity and reflexivity of sociotechnical innovation. On the other hand, technology is strangely *absent* in these diagnoses. It mainly serves as an invisible infrastructure of the societies described (Star 1999; Edwards 2003). Upon this infrastructure, the stage of the social is constructed. The assumed "real" social forces and relations are taking place on this stage, while technology stays in the background. As Weiser puts it succinctly in his article on "The Computer for the 21st Century" (1991): "The most profound technologies today are those that disappear. They weave themselves into the fabric of everyday life until they are indistinguishable from it." Hence, technology does not explicitly appear in the wordings of the catchy descriptions of society.

This has not always been the case. The technological shaping of society has indeed been an important topic of classic sociological diagnosis of the twentieth century (Ellul 1964; Schelsky 1961; Marcuse 1985; Mumford 1970). In these descriptions, technology typically appears as a *problem* of society. Technological rationality figures as the iron cage of social relations, stabilizing power, reducing contingency, destroying meaning. Social studies of science and technology (STS) have contributed to deconstructing these fears of technocracy. By first following the analytical strategies of the sociology of scientific knowledge (Bloor 1976), STS argued that technology is a social construct like any other, quite flexible and open to various competing interpretations (Pinch und Bijker 1984). This approach – sometimes explicitly, more often implicitly – questioned the macro diagnosis of a "technological society" (Ellul 1964) by turning our attention to the micro level, towards the very specific and heterogeneous configurations and networks that shape the technical and the social. Actor-network-theory even more prominently redefined the meaning of the social itself by extending and generalizing the principle of symmetry which was constitutive for the sociology of scientific knowledge (Latour 1987; Callon 1986; Law 2008). The analytical strategies of STS were highly successful in demonstrating that science and technology "in the making" are no monolithic forces. They neither determine the fate of society, nor can they be conceived as a taken-for-granted infrastructure.

Another strand of research emphasized an interventionist perspective: technology assessment, ethical, legal and social implications of science as well as – most recently – responsible research and innovation inquire into the ways in which socio-technical developments can be actively shaped – in firm view of the heterogeneity of actors, values and interests involved. These practices yielded new ideas on aligning technological and societal visions of our lives – sustainability, inclusion, and responsibility being among the major normative cornerstones. Grand challenges for science are, since then, accompanied by grand concerns as regards democratizing science and heading for a technological democracy.

Moving forward from here, this book wants to show that we need to encourage a bigger picture in analyzing the transformations of contemporary societies, technologies and their rationalities. STS may "run out of steam" (Latour 2004) if it cannot be translated into a diagnosis of the present, a tool to answer the classical question of sociology: *in which society are we living today?*

In this vein, this Yearbook Sociology of the Sciences introduces the term of *TechnoScienceSociety* to focus on the ongoing technological reconfigurations of science and society – and vice versa how these are reconfiguring technologies. More particularly, it aspires to use the breadth of analytical STS tools to perform a critical diagnosis of our contemporary society. However, instead of the pessimistic diagnoses of the past that constructed technology as society's "other", our book sets out to highlight the both complex and ambivalent entanglements of technologies and selves, socialities, and society at large. We aim to provide some tentative steps towards a diagnosis of a society in which individuals and organizations address themselves, their pasts, presents, futures, hopes and problems. This is geared increasingly in light of positive and/or negative aspects of technologies and the possibilities of technological re-design – in technosocial environments evermore suffused, or infrastructured, by technologies. In the following, we lay out how we can conceptually frame the reconfigurations of "techno", "science" and "society" that constitute much of the present complexities.

1.1.1 Techno

There is turmoil within contemporary technologies. For some time now, technical objects with new qualities have emerged: digital objects. The proliferation of personal computers from the 1980s onwards, the internet and its wires and servers from the 1990s onwards, mobile phones and later smart phones, giant server farms connected to all kinds of sensors – the amount of digital data and digital devices are exponentially growing and robots are entering spaces beyond factories. Just like people in the nineteenth century had begun to live with motorized machines, so have people in contemporary technologized societies begun to live with digital machines and devices, thus giving rise to digitized ways of life.

But there is even more turbulence in today's world of objects. Artificial nanoparticles are in our food, our building materials and our rivers. Biotechnological artefacts, some of them "biofacts" (Karafyllis, Chap. 8 this volume), can be found on fields, in supermarkets and on farms. 3D printers are merging digital design and material production and are entwined with new ways of organizing technical knowledge. And all these "emerging technologies" – most of which have been addressed by science and technology studies during the past decades – are spreading through the world while technological artefacts and systems from former technological transformations still exist, their numbers in part even growing. Coal power plants and cars, concrete buildings, high-speed trains or factories have been part of many people's lives across the globe. All these technologies, along with their side effects and uncertainties, seem to prevail while novel technologies push in the direction of highly uncertain futures.

This book is partly about these expanding technological realities that merge, mix, and transform in complex ways. Like no society before, TechnoScienceSociety is co-constituted by vast varieties and numbers of technological artefacts and systems. To the field of STS, this may come as no surprise. It is most probably no coincidence that the global growth of the STS community and the attention the field receives run parallel to the vast extensions and problematizations of technologies and associated activities mentioned before. This book draws on these diverse traditions. It analyzes the complex ways in which TechnoScienceSociety is co-constituted by different technologics in various forms.

To understand the challenge related to this task, let's reconsider the notion of technology first. Starting with Aristotle there is a varied thought tradition that distinguishes a narrow meaning of technology as the making and using of technical objects, and a wider one, which encompasses the rationalities and logics that guide the making and using of artefacts. Some languages allow to distinguish these meanings. In German, one can distinguish "Technik" (narrow meaning) and "Technologie" (wider meaning). In English and in French one can find the word "technique" for the narrow meaning. In everyday life and in academic discourse, however, we often encounter the narrow meaning of the term used in conjunction with the notion "technology". In fact, the meanings of the notion "technology" have changed over time – and it may well be that we are in an ongoing transformation of what "technology" means. "A composite of Greek techne (skill) and logos (study of), "technology" in its earliest usage, back in the seventeenth century, meant the study of skilled craft. Only in the 1930s did the word begin to refer to objects produced through the application of techne" (Jasanoff 2016, p. 8). This book, however, argues that TechnoScienceSociety is not only co-constituted by the spread and transformation of technological objects but also by novel complexities of techno-*logies* that spread across society, thereby transforming it. This entails, on the one hand, the specificities by which, for example, nanotechno-*logies* or biotechno-*logies* render matter malleable. On the other hand, this also entails the ways in which such technologies are organized, politicized, contested, exchanged or imagined; the logics by which they are embedded in society and the logics by which social actors organize and govern different technologies. It is also in this way that much of social life has become techno-*logical* (cf. Hörl 2011). We suggest that the technological has become constitutional and that new arrangements of the technological, the social and society are emerging. The algorithm is simple: wherever we find versions of "techno", we should look for the assemblages it gives rise to or is part of.

1.1.2 TechnoScience

Technology has always been a defining feature of scientific practices and just like technologies, science has been undergoing vast changes. In contemporary society, however, some suggest that technology increasingly shapes the modes of knowledge production themselves in ways that move beyond the scientific enterprise and towards a regime of *technoscience*. Notions of technoscience gained particular prominence with emerging fields such as nanotechnology, biotechnology, computer- and neurosciences (Kastenhofer und Schmidt 2011; Schmidt 2011; Wajcman 2008). Common features of these fields are rationalities of engineering, which are introduced to other academic fields, social contexts and societies. A prototypical example of this is synthetic biology, which tries to apply an engineering approach to biology in order to design novel biological systems and, in turn, to radically alter societal relations to nature (Bennett et al. 2009; Presidential Commission for the Study of Bioethical Issues 2010). This book draws on the thriving scholarship on technoscience and interrogates different understandings and conceptions spanning from technoscientific practices in research laboratories to politics and the technoscientific underpinnings of infrastructures.

In this connection, different interpretations of technoscience have been put forward by STS scholars in order to describe contemporary strategies of research and innovation (Haraway 1997; Ihde und Selinger 2003; Kastenhofer 2010; Michael 2016; Schmidt 2011; Wajcman 2008; Weber 2011). This diversity notwithstanding, in most debates and reflections, technosciences are compared to and distinguished from "classical" sciences in one way or another. Some authors describe the present condition of research and science as an unfolding "age of technoscience" (Nordmann 2011: 468) with a social imaginary that centers around technological design and control of nature, individuals, organizations, and even society and its "grand challenges" (see Kaldewey et al., Chap. 11 in this volume).

The notion of technoscience as something distinctively *new and emerging* differs from another prominent use of the term in science and technology studies. This understanding is shaped by Latour's Actor-Network Theory (ANT). In Latour's view, science and technology are, *and have always been*, inseparable on an ontological level. For ANT, the latent effect of "science" has always been the creation of new ontologies by means of "technologies." It was only with the rise of modernity that extensive purification work separated the construction of facts and artefacts. What distinguishes our contemporary world from the world of "the moderns" is not the emergence of a new kind of science called technoscience. Instead, the difference is one of reflexivity. Contemporary society is increasingly able to question and deconstruct the purification work of modernity. This questioning allows technoscientific innovations to become explicit activities formerly separated by notions of "science" and "technology" (Latour 1993).

Others, however, *historicize* this notion of technoscience (Forman 2007). Nordmann argues that the claim that science was never separated from technology and has always been technoscience is indeed an expression of the contemporary technoscientific epistemic regime itself (Nordmann 2011). According to Nordmann, technosciences seem to share distinct epistemic strategies, research objects, and cultural values: Whereas the aspiration of the scientific enterprise was the discovery of truths, the aspiration of technoscience is the production of technological innovations. Thus, technosciences are not primarily constituted by an interest in better theories and certified knowledge, but by an interest in novel technological capabilities. Under the technoscience gaze, ontological distinctions between nature and technology collapse. The very concept of nature as a distinctive realm outside of human intervention is no longer informative and meaningful in a world, in which nature is conceived as an object of intervention, reconfiguration and construction (Haraway 1991).

These competing notions of technoscience notwithstanding, they both indicate that we are entering an age in which technoscientific ambitions come to the fore. An important part of the regime of contemporary technoscience is the existence of diverse sociotechnical imaginaries (Jasanoff und Kim 2009) that entangle promising technoscientific solutions with grand societal problems (Kuhlmann und Rip 2014). Technoscience promises to reconfigure the world at micro, meso and macro levels, and to reengineer life, matter and information, configured as genes, atoms and bits (Roco und Bainbridge 2003). From the perspective of technoscience, everything can and should be designed and transformed through technological inventions and interventions. And this imperative is being constructed and enacted through various futuristic discourses: much of the public appeal of mainstream technoscience is based on grand promises or fears about how new technical capabilities might change society.

It would be misleading to interpret the suggested age of technoscience as a renaissance of technocracy (Schelsky 1961) under a new guise. While the technocratic discourse was primary shaped by dystopian fears, technoscientific imaginaries compromise dystopian as well as utopian elements. TechnoScience is not only a continuation of "big science" and its established linkages to industrial and military applications (Weber 2003), it also takes place in heterogeneous experimental settings (Callon and Rabeharisoa 2003) and technologized personal lifestyles (Maasen and Duttweiler 2012). In TechnoScienceSociety a central analytical challenge is to trace the reconfigurations of power constellations and the emergence of new, complex assemblages that transcend institutional and intellectual frameworks of the twentieth century. The epistemological, ontological and socio-cultural transformations in an unfolding age of technoscience demand ambitious and inventive efforts for a timely diagnosis of contemporary societies and their technosciences. We aim to re-focus science and technology studies towards the analysis of the unfolding age of technoscience and its novel, systemic technologies.

1.1.3 TechnoScienceSociety

Apparently, emerging technosciences increasingly set the terms of research, influence political agendas and transform the everyday lives of people. Although technosciences aim at re-designing matter, life, self and society, they do not operate independently: Technoscientific practices, not least due to their public appeal and supposed solution offers, are deeply socially and culturally constituted. The emergence of a regime of technoscience can therefore be understood as a reconfiguration of the relationships of science, technology and culture in an increasingly technologized society. Its rationalities enter everyday life, politics and the public sphere.

With the perspective of TechnoScienceSociety we emphasize the co-constitutive relationships of technological, scientific and social phenomena. To describe these, the book promotes a mode of transversal analysis, inspired not only by science and technology studies, but also by philosophy as well as social and cultural theory. It builds upon established concepts like post-normal science (Funtowicz and Ravetz 1993), Mode-2 (Gibbons et al. 1994), transdisciplinarity (Bogner et al. 2010) and the democratization of expertise (Maasen and Weingart 2005), yet weaves them into innovative analyses that address the thoroughly techno-*logical* contemporary character of these phenomena (Hörl 2011; Nordmann 2011).

While STS has begun to investigate the various implications and effects of technoscience as well as their specific ontology and epistemology, our diagnosis of TechnoScienceSociety extends this program by using STS as an approach for the sociological interpretation of the present. In contrast to the pessimistic diagnoses of the mid-twentieth century such as Ellul's "technological society" (1964), the notion of TechnoScienceSociety draws attention to the different technological rationalities, their complexities and ambivalences that have emerged. We assume that contemporary society is co-constituted by the importance it attributes to technological rationalities.

To begin with, a central idea of this book is the hypothesis that contemporary society *has become techno-logical in an encompassing sense* (Hörl 2011): it forms socialities that are fused together with technologies: techno-socialities. Technological agency is moving away from humans as subjects in control of technologies towards partly self-acting and complexly interacting digital environments in which people participate under technologically predefined and often opaque rules (Hörl 2016). The notion of the "Anthropocene" captures these reconfigurations, although insufficiently. The prime reason for the diagnosis of the Anthropocene is not just human activity per se, but rather the emergence of a "technosphere" spanning the globe and entwining with the Earth's ecosystems and atmosphere (Haff 2014).

The notion of TechnoScienceSociety does not only draw attention to the technoscientific shaping of matter, but also to the techno-*logical* interfacing of various societal domains, be it politics, business, education, sports, healthcare, mass media or religion. Even our most intimate relationships, as well as our bodies and minds, have been reconfigured by technoscientific rationalities, designs and applications (van Est 2014). We do not assume, however, that TechnoScienceSociety is governed by one singular technological principle. Rather, its dynamics and turbulences seem to be shaped by a multiplicity of different, partly incompatible techno-*logies*. It is shaped by the pervasiveness of contested technological ideas and ideals as well as societal imaginations of technological futures and constellations of power with which they entangle (Jasanoff 2015; Nordmann 2016; Urry 2016).

The diagnosis of an emerging TechnoScienceSociety is also a diagnosis of a crisis. We are observing a society challenged by the uncertainties, dynamics and ambivalences of novel technological capabilities. These are novelties that emerge on top of the arrangements and ambivalences that older and deeply embedded technological systems, forms of governance or economies have been creating. The crisis thus results as much out of the novelty as out of the asynchrony of changes and stabilizations. TechnoScienceSociety is therefore a thoroughly transformational society that has to deal with tensions, uncertainties, incompatibilities, lock-ins, disruptions and complexities of technosocial transformations in different ways.

Technological transformations have thus become politicized in different ways. Besides the "innovation equals growth" narratives of mainstream politics we increasingly find voices in civil society that emphasize risks or hopes in relation to different novel technologics. From renewable energy campaigners to self-enhancing cyborgs and robot communists TechnoScienceSociety is also shaping novel political actors. Even organized religions are thinking about our technoscientific futures: the pope has adopted the notion of technoscience and has emphasized its relevance for human life. In his recent encyclical letter on ecology and inequality, he writes: "Technoscience, when well directed, can produce important means of improving the quality of human life [...]. Yet it must also be recognized that nuclear energy, biotechnology, information technology, knowledge of our DNA, and many other abilities which we have acquired, have given us tremendous power. More precisely, they have given those with the knowledge, and especially the economic resources to use them, an impressive dominance over the whole of humanity and the entire world." (Francis of Assisi 2015).

The innovation hubs of "big tech" companies and institutions of "big science" seem to be the most obvious places of technoscientific power. Discourses on technoscience have often focused on large-scale innovations in fields like nano-, bio-, and information sciences and their assumed disruptive effects (Roco und Bainbridge 2003). Recent debates underlined the importance of the digital industries in Silicon

Valley for emerging innovations. Framed this way, TechnoScienceSociety appears as an expression of high-tech capitalism and its associated visions of "technological solutionism" (Morozov 2013). In recent years, however, the entanglement of technoscience and civil society has also received much scholarly attention. Practices of fabbing, making, and hacking might even be conceived as prototypes of "civic technoscience" (Wylie et al. 2014). It is telling that these articulations of material critique also make use of techno-*logical* rationalities and designs to counter contemporary power structures and get access to alternative futures.

In TechnoScienceSociety, it is increasingly difficult to adopt a view from "outside" our technoscientific world. Therefore, our book provides multiple perspectives for the reconstruction and reflexive ("internal") critique of TechnoScienceSociety which sheds light onto the co-constitution of technosciences and the society of which they are part. The chapters in this book thus help the reader understand the complexities of TechnoScienceSociety, and also inquire into how the study of this society changes.

1.2 TechnoScienceSociety Assembled

The significance of technoscience and technologies, notwithstanding TechnoScienceSociety, is highly diverse and it can and should be inquired from different perspectives. The contributions to this volume do this by drawing on different disciplines, intellectual traditions and cases that highlight different aspects of TechnoScienceSociety. The chapters in this book are based upon a workshop dedicated to raising the problem of TechnoScienceSociety where all authors participated and discussed initial versions of their contributions. In fact, had it not been for the lively and inspiring atmosphere that was created and shared by all workshop participants, this book and its arguments would look much different. As editors, we would like to warmly say thank you to the contributors for their joint exploration of TechnoScienceSociety! While the authors were free to choose their cases and approaches, we have found some recurring themes that allow us to group the chapters into the book's four sections – rationalities, infrastructures, governance and publics.

1.2.1 TechnoScienceRationalities

The chapters in the first section delve into guiding principles, structures and practices of technosciences. Different technoscientific rationalities are emerging that differ from the rationalities of "classical" sciences and have profound effects on the relationships of science and society. While much of the ideas about "science" in society (and often also in STS) are based upon natural sciences such as physics or biology, these chapters show that technosciences need and deserve distinct considerations.

The section starts with *Alfred Nordmann's* chapter "The Advancement of Ignorance" which argues that technosciences advance ignorance as much as they advance technological capabilities. In this chapter, Nordmann shows that technoscientific progress is as much about learning to control novel technical capabilities as it is about coping with ignorance and uncertainties that are inherent to the process. This does not end in the laboratory but rather challenges the governance of technosciences as well. Whilst "EnlightenmentScienceSociety" sought to advance theoretical knowledge about the world, TechnoScienceSociety is challenged by its technosciences to deal with the ignorance they produce as they advance. It builds complex systems that may no longer be tractable by human understanding.

Jan C. Schmidt's chapter "Tracing the technoscientific core of late-modern technologies in TechnoScienceSocieties" takes up this thread. Schmidt argues that a new type of technology has been emerging together with novel technoscientific fields such as nano-, bio- or information technologies: "late-modern technology". In contrast to "modern" technologies that were based upon predictability and control, late-modern technologies are based upon different ontologies. They show signs of "self-organization" and complex feedbacks between technological system and its environment. An inherent aspect of late-modern technologies is a certain unpredictability and volatile dynamics. Despite the rhetoric of mastery and technological control, TechnoScienceSociety may need to see its novel technologies in a different light.

Carsten Reinhardt shows that a novel field of research emerged with the ascent of technosciences: research teams, organizations and industries focused on creating technological instruments for science and technoscience. In the chapter "The Development of Research Methods as Driving Force of Technoscience", Reinhardt shows how particular research fields were co-constituted through "method makers" who specialized in producing scientific instruments. Far from being mere methods, these instruments are part of the reconfigurations of science, industry and governance. Method making has profound impacts on the commercialization of science and its reach into TechnoScienceSociety. Research in TechnoScienceSociety has become thoroughly technologized.

Sabine Maasen inquires into the ways in which the neurosciences and neurotechnologies become entangled with selves and socialities. Equipped with new tools and technologies for investigating the brain and the wider neural system, scientists and engineers are currently expanding their understanding of the brain and produce new possibilities for treating disease and damage, as well as applications for use outside the medical arena, including those that interface for machines and computer technologies. TechnoScienceSociality emerges from manifold interactions between various natural and technological, human and non-human fields. Ontologically speaking, this accounts for a "mediated posthumanism" (Sharon). Epistemologically speaking, it calls for acknowledging and coping with the primacy of thoroughly entangled knowledges, practices, and values.

1.2.2 TechnoScienceInfrastructures

The chapters in this section look at the infrastructural underpinnings of TechnoScienceSociety and their unruly dynamics. *Andrew Barry's* chapter "The Material Politics of Infrastructure" provides a wide-ranging review of the literature of scholarship on infrastructures, and simultaneously adds the dimension of temporality and ongoing changes to the picture. Infrastructures are not simply the stable foundations of society but need constant maintenance, repair and modification. This ongoing politics of infrastructures needs careful consideration. Living in TechnoScienceSociety means being enmeshed in layers of different infrastructures that are in various ways dependent upon each other.

Noortje Marres inquires into one particularly lively infrastructure in her chapter "What if nothing happens? Street trials of intelligent cars as experiments in participation". Experimental trials of intelligent cars receive a lot of attention in light of the transformations of mobility. They are, however, more than that, as Marres argues. Such trials are essential to configure relationships between innovation processes and society and are a means through which legitimacy for innovation is negotiated as much as technical capabilities are tested. TechnoScienceSociety is also a beta-testing society in which public participation is reconfigured, enabled, and challenged.

Nicole Karafyllis looks into agriculture and its infrastructures in her chapter "Biofacts, Bioprospecting, Biobanking: A Reality Check of Seed Banks". On the one hand, Karafyllis draws attention to the technoscientific governance of one of the older technological domains of humanity and on the other, she provides a fascinating case study of the recent emergence of seed banks. These infrastructures are devised to protect biodiversity in light of ongoing challenges and changes in the usage of crops. The chapter shows how the materialities and semantics created in the complex process of shaping infrastructures for the preservation of seeds highlight the lively entanglements of technology, nature and society in the agriculture in TechnoScienceSociety.

Jan Passoth in his contribution "Computing Collectives. Digital Media and the Politics of Platforms, Data, Algorithms" looks at the changes of ordering music by making it available to listeners on digital music platforms. Passoth takes apart the strong claims that digitization leads to qualitative changes and improvements in infrastructures. What is really happening, he argues, is that digitization simply makes things and orders computational. It is on this process of computation that we should focus to better understand the politics of platforms and algorithms.

1.2.3 TechnoScienceGovernance

The second section of the book looks at the governance of technoscience within TechnoScienceSociety. In fact, as the chapters show, the emergence of special forms of governance for novel technologies is a prominent characteristic of

TechnoScienceSociety. Discourses of vast promises and fears, governmental strategies, technologized politics and power constellations are entwined with technologies and their transformations.

The section starts with *Erik Fisher's* contribution who uncovers the logics of technoscientific governance with a special focus on the US nanotechnology context. His chapter "Governing Technoscience in Society: Tracing the Dialectics of Enthusiasm, Ambivalence and Adjustment" shows how technoscience, public policy and political and civic actors entangle around particular technological paths to enact their governance. As Fisher shows, it is the strong belief in the power of technoscience that at once helps to popularize it in state agendas and to problematize it in society. The dynamics and tensions that constitute technoscientific governance are thus themselves ambivalent. They foster a kind of technoscientific hubris and open up areas for public engagement and democratization of science and technologies. Fisher ends with a particular focus on the role of engaged STS scholarship within these processes and its ambivalences – in fact, the complex governance processes can be seen as a huge success in favor of STS informed perspectives, although this is not widely reflected as such.

The section continues with a chapter that focuses on the dynamics and historical emergence of imaginaries and future-oriented policy discourses in relation to technological transformations. *David Kaldewey*, *Daniela Russ*, and *Julia Schubert* look at "The Politics of Technoscience: From National Visions to Global Problems" and ask which rationalities guide technoscientific knowledge production and its governance. The contribution traces the emergence of the discourse on "grand challenges" and its links to technological innovation. With detailed historical case studies on discourses of "energy security" and "climate engineering", the text shows how technoscience is made political by transforming visions into problems.

Michael Nagenborg and *Jutta Weber* look at the role of surveillance technologies and their politics in their chapter "TechnoSecuritySociety: Catastrophic Futures, Pre-emptive Security & Mass Surveillance". The authors show how political questions of security and policing have deeply sunk into technoscientific culture and everyday technologized infrastructures. "Technosecurity" is a culture within TechnoScienceSociety that entwines with technosciences and the governance of everyday life. Security and surveillance technologies have strongly grown and form the basis of a new security regime in our societies.

"Mobilizing the Emergence of Phronetic TechnoScienceSocieties: Low-Carbon E-Mobility in China" is the title of *David Tyfield's* contribution who dissects Chinese infrastructural transformations from a perspective on transformations in "complex power/knowledge systems". Tyfield argues that in light of such complex transformations it is necessary to move to an "onto-politics" of world-making and situated engagement that draws on practical wisdom. He looks at both the ontological sensibilities of STS and their limitations, the latter often evoked by anti-normative stances. Referring to the mobility transition in China, Tyfield shows how a power/knowledge sensitive perspective can uncover the political dimensions of such infrastructural changes and projects and asks about strategic intervention possibilities. Tyfield's contribution outlines the challenges of the politics of "innovations" in TechnoScienceSociety and to STS scholars alike.

1.2.4 TechnoSciencePublics

A special trait of TechnoScienceSociety concerns changes in the public sphere and civil society. On the one hand, we find many publics and civil society actors in TechnoScienceSociety that debate and contest technologies. But on the other hand, and as a more recent development, we find publics that are constituted in entanglement with particular technologies and sometimes even shape technologies as a public activity. The final section of this book turns toward these technologically constituted publics.

Sascha Dickel analyzes particular technologically mediated practices of citizen science in his text "Technoscientific Citizenship in Citizen Science. Assembling crowds for biomedical research". His paper asks how volunteers are included in digital citizen science. The analysis shows that the technologies used to accomplish citizen science with the help of digital platforms do not just extend participation. They also (re-)define the conditions and the meaning of participation itself. The construction of citizen science platforms, Dickel argues, is an installation of interfaces of participation. Included individuals do not enact citizenship by taking part in political discourses, but by becoming productive volunteers in a predefined techno-social order.

maxigas looks at the history of the European hacker tradition and its relations to cybernetics. The chapter "Anti-Modern Techno-Science: Cybernetics, Ontology, Practice in the European Hacker Tradition" traces significant practices of hacking information technologies. The analysis reveals the specific relations between technological artefacts, organizations and hackers and how they are put into practice. Drawing on an extensive range of ethnographic material, the chapter aims to make a conceptual point about different ontologies of technology. Maxigas argues that hackers practice an "anti-modern" ontology of technology that embraces emergence and symmetry between humans and artefacts, and from which TechnoScienceSociety could learn different strategies in coping with technological complexity.

Christoph Schneider analyzes the production and organization of knowledge and technology in open source approaches to 3D printing and other digital fabrication technologies. The chapter "The TechKnowledgy of open digital fabrication" draws attention to the plurality of ways in which technologies are being produced in TechnoScienceSociety. The entanglement of open source approaches and digital fabrication technologies and novel organizations, such as FabLabs, has gained particular prominence in the past decade and is the case that Schneider uses to analyze the specific logics by which technology and knowledge are produced.

Throughout the book, the reader may encounter transversal themes or trajectories indicating the emergence of TechnoScienceSociety which, as yet, are but hypotheses in need of further empirical studies. Pointedly summarized, we see, e.g., transformations from "technology" to techno-sociality, from the co-production of knowledge to the co-design of technologies, from science policy to the politics of technosciences, from the construction of technology to the production of techno-social worlds. In this vein, this book encourages STS to go for more: for empirically-based diagnostics of the present techno-social condition of society.

Dear diaries of the future,

a thought has come to my mind. It has literally come to me via technological channels. Is it my thought at all? Who owns thought anyway? The thought has approached me in between. I have been clicking, messaging, eating superfoods, watching a dam collapse on the news, scratching the nano coating off my socks, ordering my car to park, waiting for the service to repair my smart heating, freezing a little, listening to the radio.

And there it was, that strange noise behind the voices. The interview was quite fascinating. Someone spoke about how we are within a great transformation. How technological rationalities are meandering throughout society, reshaping its social landscapes. How our governments and modes of governance are at the same time highly stimulated by technological options and a high-tech economy but also overwhelmed. Our sociotechnical infrastructures are no longer stable and in search of new modes of living and doing things. Yet, public concern remains diffuse and what we thought of as the public is in itself changing with digital systems. That program caught me.

What they said resonated with that feeling others and I have had for quite a while. A feeling that established institutions do not care about us. They said that now we have started to live in TechnoScienceSociety on the radio show. But the noise did not stop. I turned off the radio and still there it was. The noise did not come from the radio – it was that thought approaching me. And then silence. I had to come to you and write this note to ask: If our future will be technologically co-designed, is there still space for a "me" – and which one? What will have remained of what is now and remembered by "you"? For now, and somewhat anachronistically, you can read and store books and diaries, online and offline, but not for much longer, right? What will be newly thinkable on your "pages"? Who will write the future? And who will write you in the future?

All technoscientification of selves, socialities, and society notwithstanding, I think, I will just lie down comfortably on my couch and read this book. Well, I know I could buy individual chapters online, and will surely tweet my friends and colleagues about it. Reading a book is a bit anachronistic, I know. I don't care.

References

Beck, Ulrich. 1992. *Risk Society: Towards a New Modernity*, Theory, Culture & Society. London/Newbury Park: Sage Publications.

Bell, Daniel. 1976. *The Coming of Post-Industrial Society: A Venture in Social Forecasting*, Harper Colophon Books. New York: Basic Books.

Bennett, Gaymon, Nils Gilman, Anthony Stavrianakis, und Paul Rabinow. 2009. From Synthetic Biology to Biohacking: Are We Prepared? *Nature Biotechnology* 27 (12): 1109–1111. https://doi.org/10.1038/nbt1209-1109 .

Bloor, David. 1976. *Knowledge and Social Imagery*. London: Routledge and Kegan Paul.

Bogner, Alexander, Karen Kastenhofer, and Helge Torgersen, eds. 2010. *Inter- und Transdisziplinarität im Wandel? neue Perspektiven auf problemorientierte Forschung und Politikberatung*, Schriftenreihe "Wissenschafts- und Technikforschung", Bd. Vol. 4. 1st ed. Nomos: Baden-Baden.

Böhme, Gernot, und Nico Stehr. 1986. *The Knowledge Society. The Growing Impact of Scientific Knowledge on Social Relations: Sociology of the Science – A Yearbook*, Unter Mitarbeit von G. Böhme and N. Stehr. Dordrecht/Boston/Lancaster/Tokyo: D. Reidel Publishing Company.

Callon, Michel. 1986. Some Elements of a Sociology of Translation: Domestication of the Scallops and the Fishermen Af St. Brieuc Bay. In *Power, Action, and Belief: A New Sociology of Knowledge?* hg. v. John Law, 196–233. London/Boston: Routledge & Kegan Paul.

Callon, M., and V. Rabeharisoa. 2003. Research "in the wild" and the shaping of new social identities. *Technology in Society, Studies in Science, Technology, and Society (STS) North and South* 25 (2): 193–204. https://doi.org/10.1016/S0160-791X(03)00021-6.

Castells, Manuel. 1996. *The Rise of the Network Society*, The Information Age 1. Malden: Blackwell Publishing.

Edwards, Paul N. 2003. Infrastructure and Modernity: Force, Time and Social Organizations on the History of Sociotechnical Systems. In *Modernity and Technology*, hg. v. Thomas Misa, Philip Brey, and Andrew Feenberg, 185–225. Cambridge, MA: MIT Press.

Ellul, Jacques. 1964. *Technological Society*. New York: Vintage Books.

Forman, P. 2007. The Primacy of Science in Modernity, of Technology in Postmodernity, and of Ideology in the History of Technology. *History and Technology* 23 (1–2): 1–152.

Foucault, Michel. 1977. *Discipline and Punish: The Birth of the Prison*. New York: Pantheon Books.

Francis of Assisi. 2015. Laudato Si: On Care of Our Common Home. *Encyclical Letter*, Rom, 24. Mai. Zugriff: 9. März 2018. http://w2.vatican.va/content/dam/francesco/pdf/encyclicals/documents/papa-francesco_20150524_enciclica-laudato-si_en.pdf

Funtowicz, Silvio O., and Jerome R. Ravetz. 1993. Science for the post-normal age. *Futures* 25 (7): 739–755.

Gibbons, Michael, Camille Limoges, Helga Nowotny, Simon Schwartzman, Peter Scott, and Martin Trow. 1994. *The new production of knowledge: The dynamics of science and research in contemporary societies*. London: Sage.

Haff, Peter. 2014. Humans and Technology in the Anthropocene: Six rules. *The Anthropocene Review* 1 (2): 126–136. https://doi.org/10.1177/2053019614530575.

Haraway, Donna, Hg. 1991. *Simians, Cyborgs, and Women: The Reinvention of Nature*. New York: Routledge.

———, Hg. 1997. Modest_Witness@Second_Millennium. FemaleMan©_Meets_OncoMouseTM: Feminism and technoscience. New York/London: Routledge.

Hörl, Erich, ed. 2011. *Die technologische Bedingung: Beiträge zur Beschreibung der technischen Welt*. Berlin: Suhrkamp.

Hörl, Erich. 2016. Other Beginnings of Participative Sense Culture. Wild Media, Speculative Ecologies, Transgressions of the Cybernetic Hypothesis. In *ReClaiming Participation. Technology – Mediation – Collectivity*, herausgegeben von Mathias Denecke, Anne Ganzert, Isabell Otto, und Robert Stock, 93–121. Bielefeld: transcript.

Ihde, Don, und Evan Selinger, Hg. 2003. *Chasing Technoscience: Matrix for Materiality*. Bloomington: Indiana University Press.

Jasanoff, Sheila. 2015. Future imperfect: Science, technology, and the imaginations of modernity. In *Dreamscapes of modernity: Sociotechnical imaginaries and the fabrication of power*, herausgegeben von Sheila Jasanoff und sang-Hyun Kim, 1–47. Chicago: University Of Chicago Press.

Jasanoff, Sheila. 2016. *The ethics of invention: Technology and the human future. The Norton global ethics series*. New York: W.W. Norton & Company.

Jasanoff, Sheila, und Sang-Hyun Kim. 2009. Containing the Atom: Sociotechnical Imaginaries and Nuclear Power in the United States and South Korea. *Minerva* 47 (2): 119–146.

Kastenhofer, Karen. 2010. *Do We Need a Specific Kind of Technoscience Assessment? Taking the Convergence of Science and Technology Seriously*. Zugriff: 19. Mai 2011. http://springerlink.metapress.com.proxy1.library.jhu.edu/content/m86628t53478q211/fulltext.pdf

Kastenhofer, Karen, und Jan C. Schmidt. 2011. Technoscientia Est Potentia? Contemplative, Interventionist, Constructionist and Creationist Idea(L)s in (Techno)science. *Poiesis & Praxis* 8 (2–3): 125–149. https://doi.org/10.1007/s10202-011-0101-2.

Kuhlmann, Stefan, und Arie Rip. 2014. The Challenge of Addressing Grand Challenges. In *The Future of Research and Innovation*, hg. v. R. v. Schomberg. Brussels: European Commission.

Latour, Bruno. 1987. *Science in Action: How to Follow Scientists and Engineers Through Society*. Cambridge, MA: Harvard University Press.

———. 1993. *We Have Never Been Modern*. Cambridge, MA: Harvard University Press.

———. 2004. Why Has Critique Run out of Steam? From Matters of Fact to Matters of Concern. *Critical Inquiry* 30 (2): 225–248. https://doi.org/10.1086/421123.

Law, John. 2008. Actor Network Theory and Material Semiotics. In *The New Blackwell Companion to Social Theory*, Blackwell Companions to Sociology, hg. v. Bryan S. Turner, 141–158. Chichester/Malden: Wiley-Blackwell.

Maasen, Sabine, and Peter Weingart, eds. 2005. *Democratization of expertise? Exploring novel forms of scientific advice in political decision-making*, Sociology of the sciences, v. Vol. 24. Dordrecht, The Netherlands: Springer.

Maasen, Sabine, und Stefanie Duttweiler. 2012. Neue Subjekte, neue Sozialitäten, neue Gesellschaften. In *Handbuch Wissenschaftssoziologie*, herausgegeben von Sabine Maasen, Mario Kaiser, Martin Reinhart, und Barbara Sutter, 417–28. Wiesbaden: Springer VS.

Marcuse, Herbert. 1985. Some Social Implications of Modern Technology. In *The Essential Frankfurt School Reader*, hg. v. Andrew Arato und Eike Gebhardt, 138–162. New York: Continuum..

Michael, Mike. 2016. Engaging the Mundane: Complexity and Speculation in Everyday Technoscience. In *Remaking Participation: Science, Environment and Emergent Publics*, hg. v. Jason Chilvers und Matthew Kearnes, 81–98. Abingdon/New York: Routledge Is an Imprint of the Taylor & Francis Group, an Informa Business.

Morozov, Evgeny. 2013. *To save everything, click Here: The folly of technological solutionism*. New York: PublicAffairs.

Mumford, Lewis. 1970. *Pentagon of Power.* 2 Bd. The Myth of the Machine 2. New York: Harcourt Brace Jovanovich.

Nordmann, Alfred. 2011. The age of technoscience. In *Science transformed?: debating claims of an epochal break*, herausgegeben von Alfred Nordmann, Hans Radder, und Gregor Schiemann, 19–30. Pittsburgh, Pa: University of Pittsburgh Press.

———. 2016. Enhancing Machine Nature. In *Perfecting Human Futures*, herausgegeben von J. Benjamin Hurlbut und Hava Tirosh-Samuelson, 195–214. Technikzukünfte, Wissenschaft Und Gesellschaft/Futures of Technology, Science and Society. Springer Fachmedien Wiesbaden. https://doi.org/10.1007/978-3-658-11044-410.

Osrecki, Fran. 2015. Constructing Epochs: The Argumentative Structures of Sociological Epochalisms. *Cultural Sociology* 9 (2): 131–146. https://doi.org/10.1177/1749975514566511

Pinch, Trevor J., und Wiebe E. Bijker. 1984. The Social Construction of Facts and Artefacts: or How the Sociology of Science and the Sociology of Technology Might Benefit Each Other. *Social Studies of Science* 14 (3): 399–441. Zugriff: 5. Mai 2015.

Presidential Commission for the Study of Bioethical Issues. 2010. *New Directions: The Ethics of Synthetic Biology and Emerging Technologies*. http://www.bioethics.gov/documents/synthetic-biology/PCSBI-Synthetic-Biology-Report-12.16.10.pdf.

Rammert, Werner, Arnold Windeler, Hubert Knoblauch, and Michael Hutter, eds. 2016. Innovationsgesellschaft heute: Perspektiven. *Felder und Fälle. VS Verlagfür Sozialwissenschaften*. https://doi.org/10.1007/978-3-658-10874-8.

Roco, Mihail C., und William S. Bainbridge. 2003. *Converging Technologies for Improving Human Performance: Nanotechnology, Biotechnology, Information Technology and Cognitive Science*. Dordrecht: Kluwer.

Rosa, Hartmut. 2003. Social Acceleration: Ethical and Political Consequences of a Desynchronized High-Speed Society. *Constellations* 10 (1): 3–33. https://doi.org/10.1111/1467-8675.00309/asset/1467-8675.00309.pdf?v=1&t=gedxdz8f&s=e5dfe789e7e29e0b69f33d1f4d3aa48a2d550c95.

Schelsky, Helmut, Hg. 1961. *Der Mensch in der wissenschaftlichen Zivilisation*. Wiesbaden: VS Verlag für Sozialwissenschaften.

Schmidt, Jan C. 2011. Toward an Epistemology of Nano-Technosciences. *Poiesis & Praxis* 8 (2–3): 103–124. https://doi.org/10.1007/s10202-011-0104-z.

Star, Susan L. 1999. The Ethnography of Infrastructure. *American Behavioral Scientist* 43 (3): 377–391. https://doi.org/10.1177/00027649921955326.

Urry, John. 2016. *What is the future? Cambridge, UK*. Malden, MA: Polity Press.

van Est, Rinie. 2014. Intimate Technology: The Battle for Our Body and Behaviour. Unter Mitarbeit von V. Rerimassie, van Keulen, Ira und G. Dorren. Den Haag: Rathenau Instituut. Zugriff: 12. Mai 2015.

Wajcman, Judy. 2008. Emergent Technosciences. In *The Handbook of Science and Technology Studies*, hg. v. Edward J. Hackett, 3rd ed., 813–816. Handbook Advisory Board. Cambridge, MA: MIT.

Weber, Jutta. 2003. *Umkämpfte Bedeutungen: Naturkonzepte im Zeitalter der Technoscience*. Campus Verlag: Frankfurt a. M.

Weber, Jutta. 2011. Technoscience as Popular Culture: On Pleasure, Consumer Technologies & the Economy of Attention. In *Science Transformed? Debating Claims of an Epochal Break*, hg. v. Alfred Nordmann, Hans Radder, and Gregor Schiemann, 159–176. Pittsburgh: University of Pittsburgh Press.

Weiser, Mark. 1991. The Computer for the 21st Century. *Scientific American* 265 (3): 66–78.

Wylie, Sara Ann, Kirk Jalbert, Shannon Dosemagen, and Matt Ratto. 2014. Institutions for civic Technoscience: How critical making is transforming environmental research. *The Information Society* 30 (2): 116–126. https://doi.org/10.1080/01972243.2014.875783.

Sabine Maasen holds the Friedrich Schiedel Endowed Chair in Sociology of Science at the Technical University of Munich (TUM), Germany. At the same time, she is Director of the Munich Center for Technology in Society (MCTS) at TUM. Her research focuses on the dynamics of innovation by research and legitimation of research in order to understand contemporary TechnoSocieties. She is Member of various committees and advisory boards, including the German Science Council.

Sascha Dickel is Assistant Professor for Sociology of Media. His research concerns social theories of science and technology, digital participation, and technoscientific futures. The edited volume *Personal Health Science* (with Nils B. Heyen and Anne Brüninghaus) came out with Springer (2019).

Christoph Schneider works as a Sociologist at the Institute for Technology Assessment and Systems Analysis of KIT, Karlsruhe. His research focuses on the possibilities of democratizing technoscientific research and innovation. As Public Sociologist, he is active in progressive technopolitics and consults organizations in search of more humane forms of digitalization.

Part I
TechnoScienceRationalities

Chapter 2
The Advancement of Ignorance

Alfred Nordmann

2.1 A Baconian Opening

Francis Bacon's *The Advancement of Learning* (1605) is a founding document not only of modern science but also of modern society. It forged a connection between scientific and social progress, between the accumulation of knowledge and of wealth (Krohn 1987). To Bacon we can trace the productive illusion as well as expectation that societies flourish through the continuous increase of knowledge. Even as the illusion is exposed and ignorance makes its appearance in every corner of today's so-called knowledge societies, the expectation persists in the form of anxiety: Ignorance is seen as a shortcoming in the process of modernization (Beck 1996). Even when it is acknowledged as inevitable and productive, it is considered a problem, if not a threat to the fabric and even the foundation of social life. In the face of ignorance, conceptions of human progress and general enlightenment become questionable.

Upon closer scrutiny, however, Bacon provides an instructive irritation. His advancement of learning cannot simply be equated with the advancement of scientific knowledge and a theoretical understanding of the world. In his *Novum Organon* Bacon develops a conception of learning that builds on the knowledge of artisans who labor in the making of works, who do not win battles in the court of opinion but attain power through works and deeds, who do not contemplate the order of things but participate in it (Bacon 1990: 41, 43, 79; cf. Bono 1995; Gaukroger 2001; Smith 2004). In other words, Bacon's advancement of learning is not tied to the intellect and its requirement of intelligibility or conceptual understanding. On the contrary, what the mind can learn is constrained by what the hand can do – manual operations suggest ideas of causation that cannot penetrate any deeper than these operations themselves (Bacon 1990: aphorisms 1–3):

A. Nordmann (✉)
Technical University of Darmstadt, Darmstadt, Germany
e-mail: nordmann@phil.tu-darmstadt.de

S. Maasen et al. (eds.), *TechnoScienceSociety*, Sociology of the Sciences Yearbook 30, https://doi.org/10.1007/978-3-030-43965-1_2

> Towards the effecting of works, all that man can do is to put together or put asunder natural bodies. The rest is done by nature working within. (Bacon 1990: aphorism 4)

On Bacon's account, then, the advancement of learning is the advancement of mastery in the order of things. In this lies his break with conceptions of science and theoretical knowledge that are rooted in antiquity. But *theoria* informs the conceptions also of modern science and the Enlightenment project. As opposed to Bacon, the criteria for theoretical knowledge and for the advancement of learning answer to the questions of Plato and Aristotle and their concern with true belief. Here, knowledge consists in the achievement of human understanding and in the intellectual tractability of the models and theories that represent the world of observable phenomena. Modern science aims for the agreement of mind and world, theory and reality, and for justified true belief – it is thus situated in the sphere of ideas and opinions where true understanding may produce a foundation for technical agency but technical agency as such does not constitute knowledge. In the world of modern science, intellectual understanding is viewed as a necessary condition for technical control, and inversely, technical control cannot by itself be a sufficient condition for the possession of knowledge (Bunge 1966).

This is Bacon's instructive irritation: From the received point of view of basic and applied science, Bacon's advancement of learning is tantamount to the tolerance of ignorance and, indeed, might involve the deliberate advancement of ignorance in respect to theoretical understanding, intellectual tractability and intelligibility by the human mind.

The following pages explore the productive tension that arises when we acknowledge that there is an *Advancement of Learning* which is in terms of the human intellect an *Advancement of Ignorance*. In the modern world of science and technology, of transparency, accountability and regulatory action it would appear inconceivable to consider progress as the deliberate production of ignorance - where ignorance is not an unintended consequence but a precondition for the achievement of technologically embodied knowledge that is not constrained by the requirements and limitations of the human mind. And yet, it appears ever more evident that citizens in the age of technoscience (Nordmann et al. 2011) are not only coming to terms with, but embracing the hitherto inconceivable surrender of an intellectual conception of knowledge.[1]

[1] Bacon began with a diagnosis of the pitiful state of theoretical knowledge which – as opposed to the knowledge of artisans – had stagnated for centuries. It came easy for him to jettison this ideal of knowledge. Today and before the backdrop of the amazing success story of modern science it appears far less likely and far more unpalatable to surrender the demand that there ought to be an intellectual grasp of the technological conditions of the world.

2.2 Scientific and Technoscientific Knowledge Societies

The ambitiously forward-looking book on *The Advancement of Ignorance* will never be written – it would require a kind of pathos which is lacking in *TechnoScienceSociety*. And yet, if one takes *TechnoScienceSociety* seriously and seeks to distinguish it from the scientific knowledge societies of first and second modernity, one cannot help but note that it is built around the accommodation of ignorance as a necessary evil and as a matter of course. It comes with the abdication of the intellect in favor of the reliable and robust working order of a sociotechnical machinery.

In order to substantiate this claim, what is needed first of all is a definition of technoscience and an account of how it gives rise to today's technoscientific condition. From there it is only a small step toward exhibiting or exemplifying the advancement of ignorance. Accordingly it will be first encountered neither in individual decision-making under conditions of uncertainty or ignorance, nor in regulatory institutions that are confronted with a lack of action-guiding scientific knowledge about, for example, toxicological thresholds. Instead, the production of ignorance as absence of theoretical understanding takes place first and foremost in academia's technoscientific research laboratories. It is technoscientific knowledge production that rejects the demands of the human intellect and links the promise of innovation to the advancement of ignorance.

Our familiar stories of science and society are rooted in the Enlightenment and the institution of scientific communities, academies, and journals. From Kant to Habermas, these are stories of methodological commitments in the republic of scholars or intellectual public sphere. These commitments produce the emancipation from dogma and prejudice and, thereby, clear the ground for political and economic emancipation as well. According to *EnlightenmentScienceSociety*, history and theory asymptotically progress towards the unattainable ideals of truth, eternal peace, and a well-ordered, just society. In this context, science provides a theoretical description of the world. It puts forward hypotheses or ideas that aspire to truth, and it is the unending quest of the scientific community to ascertain whether these claims to truth can be justified. Accordingly, criticism becomes an essential feature of science, Enlightenment, and modern societies alike.

If this conception of science and society is a mere conceit, it is nevertheless a powerful conceit as witnessed by the standing to this day of philosophers like Karl Popper or Jürgen Habermas. And even where the conceit is questioned or undermined, this is done in the idiom of modern science as producer of representations, descriptions, ideas. Constructivists, for example, suggest that claims to representational truth are not discovered and justified but socially produced and sanctioned. But irrespective of realist, empiricist, or social constructivist accounts of the success of modern science, its knowledge claims remain subject to revision, and on all these conceptions the job of science is to make knowledge available such that society can draw upon or expand it for technological or administrative purposes.

Similarly, when Ulrich Beck, Anthony Giddens, and Scott Lash began to speak of reflexive modernization in second modernity (1994), they spoke of modernity becoming reflexive in the sense of successful modernization turning upon itself and creating unanticipated problems of its own. These are problems that arise from knowing the world as envisioned by modern science. With the increasing complexity of socio-technical systems, for example, it becomes more difficult to assess risks or anticipate failure. And with increasingly probabilistic or statistical interpretations of data, rational decision-making routinely takes place under conditions of uncertainty or ignorance. These new sources of ignorance at the very core of *EnlightenmentScienceSociety* provide for the theorists of reflexive modernization the occasion to reflexively question forms of knowledge and control, and thereby to attain a kind of self-knowledge. So, even as they show that the advancement of learning produces new forms of ignorance, they treat this as a problem of knowledge with the inherent goal to overcome or sublimate ignorance.

Accordingly and for the most part, the recent slate of books on *Agnotology* (Proctor and Schiebinger 2008) or *Ignorance: How it Drives Science* (Firestein 2012) draw attention to systemic sources of ignorance so we might better address them – be it in the form of secrecy or the suppression of indigenous knowledge or the Socratic confession of ignorance as initiation to a search for truth. The slogan "Not Ignorance but Ignorance of Ignorance is the Death of Knowledge" graces the cover of the recent *Handbook on Ignorance Studies*. The slogan is indicative of the wide range of topics that is covered in this collection of essays. It deals with ignorance in the modernist mode of gesturing at the ineffable that lies beyond the limits of knowledge of which a long tradition of thinkers and writers were painfully aware (Franke 2015). It shows that ignorance is vital for knowing and acting as it becomes incorporated in advanced conceptions of rational choice, as it provides a strategic resource for claiming "needs to know" and as it serves to motivate science and science education (Gross and McGoey 2015).

Ignorance appears quite differently on the stage of *TechnoScienceSociety*. This other kind of knowledge society is not built around the advancement of learning for the purpose of empowerment and criticism and thus for the emancipation of individuals and societies from dogma and prejudice. Science is not considered the hallmark or vanguard of social and political progress. Instead, by tying technoscientific research into the procedures and processes of contemporary society, technoscience becomes embedded within a socially and technologically robust working order. This reversal was first discussed under the heading "mode 2" knowledge production with its attendant "new social contract of science and society" (Nowotny et al. 2001). When the machinery runs well and the sociotechnical working order is well-composed, it becomes easy to abandon intellectual claims to knowledge of truth and easy to accept ignorance as to how and why things work as they do.

One example of this provides a very different perspective on the familiar diagnosis of a second modernity which cannot provide properly science-based risk assessments. At first, the trend towards miniaturization and technical control over ever smaller constituents of matter provides a story *par excellence* of the advancement of learning. When reaching the nanoscale, however, this advancement of learning turns upon itself by producing a new kind of ignorance: The properties of nanoparticles are subject to such minute variations that it becomes exceedingly difficult to perform a scientific risk analysis on whole classes of nanoparticles, whereas it would be futile to assess one nanoparticle at a time. The most optimistic predictions speak of many years, if not decades until proper risk assessment becomes feasible. This, one might expect, would put nanotechnology on hold until an adequate knowledge base has been attained. Instead, however, we witness ways to accept and work around this deficit (Nordmann 2010): The focus shifts on ways of being precautionary, prudent, and productive, with risk assessment replaced by risk management, with hard regulation and monitoring of thresholds replaced by an institutional regime of permanent vigilance. The problem of a lack of scientific knowledge disappears in the construction of a robust socio-technical machinery. In its place emerges a managerial or technical knowledge of how to accommodate risk in real-time under conditions of presumably irreducible, presumably harmless ignorance.

All of this implicitly suggests a conceptual structure that needs to be elaborated. Evidently, it is not sufficient to speak of different science-society relations or different conceptions of knowledge in order to appreciate the shift from mode-1 Enlightenment science to mode-2 technoscience. This also requires a substantial conception of technoscientific research as consisting not in representing the world by means of theory, but in developing working knowledge of how things can work together and of how to make things work for us. This working knowledge, in turn, is not the opposite of ignorance: We can know how to make things work without knowing how they work. On the face of it, this may sound paradoxical, which is all the more reason to now turn to the technoscientific production of working knowledge and its attendant advancement of ignorance – and then, of course, to provide examples of this.

2.3 Accommodating Ignorance

More so perhaps than *EnlightenmentScienceSociety,* this volume's *TechnoScienceSociety* captures the peculiar character of today's "knowledge societies." If that term were to highlight that societal institutions, their technical, economic, administrative capabilities were based in knowledge of some kind or another, "knowledge society" would designate all societies at all times. The term gains specific meaning only in that it refers to the production and management of knowledge and ignorance in some but not all contemporary societies. These are knowledge societies because they appreciate the value of knowledge and are equipped to make knowledge available where profitable or necessary, and to offer work-around solutions where knowledge is not necessary or available. Indeed, citizens of the knowledge society are deeply aware that knowledge is scarce, that it is hard to come by and to sort out once things become sufficiently complex (Böschen et al. 2010; Janich et al. 2012: 7–20). They are well aware that knowledge as resource and commodity takes on a wholly different value and character as we transition from the world of the classical sciences to the world of biomedical and nanotechnological research, of information and communication technologies or synthetic biology.

The predicament of the technosciences and the so-called emerging technologies is this: On the one hand, there is plenty of useful knowledge, though it is quite general and not tailored as yet to the specific demands of the highly complex processes and phenomena that need to be modeled, manipulated, predicted and controlled. On the other hand, there is a lot that we can't know or at least won't be able to know for quite some time – due to that very same complexity and its attendant limits of knowledge. But finally and perhaps fortunately, one discovers that we really don't actually need to know what we can't and won't know.

This predicament owes to what might be the main epistemological difference between science and technoscience: Science seeks to reduce ignorance by reducing complexity. Technoscience accommodates ignorance in order to generate complexity. The search algorithm of science is to discover how things are, so as to make them technologically useful. The search algorithm of technoscience is to discover how far we can get technologically with what little or much we already know.

This may sound like a bold characterization of "technoscientific research," especially since one will be hard-pressed to find researchers who are willing to admit that they have abandoned the search for knowledge. That is why it is important to qualify the claim: It is not that curiosity is over and done with, that theoretical understanding is not valued where it can be had, or that the quest for theoretical knowledge and the reduction of ignorance is over. Instead, the argument is that technoscientific research proceeds on a substrate of ignorance, that it welcomes and, indeed, seeks out short-cuts, work-around solutions, black-boxed modules, and data-management routines. And indeed, there are plenty of indications that tolerance of ignorance is necessary to succeed in contemporary technoscientific research. Many of these are properly subject of a sociology of technoscientific knowledge. After mentioning these only briefly, somewhat more sustained attention will be given to the role of ignorance in knowledge generation and explanation, that is, to two questions for a philosophy of technoscience.[2]

While scientists are usually thought of as champions of criticism, the criticism of theories and hypotheses requires the framework of a discipline, of shared concepts, paradigms, methodologies, theoretical frameworks. In the world of technoscience, however, the most attractive projects involve multiple disciplines – and in order to recruit a multidisciplinary research team, mutual trust is necessary and some degree of credulity becomes a virtue. Moreover, since everyone knows just how difficult one's own piece of the work is, one would like to imagine that that of the others is comparatively easy. Indeed, this kind of wishful thinking might be a prerequisite for convincing oneself that real progress towards achievement of a project goal is possible at all.[3]

[2] The brief survey of features of contemporary research that require a tolerance of ignorance draws on Nordmann 2008.

[3] To be sure, the question of trust was at issue also in the gentlemanly science of the seventeenth and eighteenth centuries. In the twentieth century, it was mostly Karl Popper who maintained the ideal of the scientist who takes nothing for granted and questions everything, while Ludwig Wittgenstein or Thomas Kuhn showed that mutual critique requires a shared language or paradigm that must be taken on trust. Compare Shapin 1994, 2008.

Another reason why researchers need to accommodate ignorance as a normal part of their daily lives comes with the instruments they use. The insides of their instruments are as opaque to most researchers as are the insides of computers to most of their users. Indeed, some of these instruments are useful precisely because they work beyond the limits of what the human mind can manage – as witnessed by the current interest in "data-intensive science" and a somewhat morbid fascination in popular culture with the rule of "algorithms," "self-learning" and "A.I." When such instruments, moreover, render data sets in a pictorial way, they intimate immediate visual access to the objects of research. Highly processed data is thus presented as if the instrument merely opened a window to the world. This offers neither incentive nor point of entry to critical questions about instrumental artefacts or alternative ways of rendering the data. Indeed, some instruments render obsolete even the very idea of an observer's perspective, a perspective that serves as a reminder of the constructedness of representations and the epistemological difficulty of relating one's detached standpoint to a remote world "out there." Instead, some instruments beckon researchers to become fully immersed, conflating real and virtual worlds as one flies through and interacts with molecular or cellular systems, exploring them as if from within (Nordmann 2006). All of this suggests that technoscientific researchers must be tolerant of ignorance and willing to "black-box" bothersome detail. As Eran Tal has pointed out, this is not just a matter of reluctantly paying the prize for the increasing complexity of research technologies: Technoscientific researchers gratefully embrace and appreciate instruments, tools, or routines that allow them to be ignorant of the precarious construction and maintenance of these instruments. Standards and measures, for example, are maintained by an international community of metrologists who take care of the uncertainties that are associated with the various parameters that might influence a measurement. They thereby create the conditions that render measurements intelligible and comparable around the world – by "affording ignorance" metrologists render their own work invisible to those who perform measurements (Tal 2013).[4]

With Tal's analysis, we have moved from rather more sociological to rather more epistemological considerations. And it might now be worth considering how ignorance enters into the very process of generating knowledge. At first glance this appears to be paradoxical since knowledge is thought to be the opposite of ignorance such that each new item of knowledge displaces some blind spot of previous ignorance. The air of paradox vanishes just as soon as one conceives of different kinds of knowledge – intellectual and technical – and different kinds of problems – representational and effectual. A representational problem of accurately describing the world challenges us to fill a gap in our intellectual understanding. This problem signifies something for which we have no satisfactory account as yet. By supplying the proper theories, hypotheses, explanations one produces intellectual understanding and thus reduces ignorance. In contrast, to solve the effectual

[4] From a very different standpoint Mario Bunge noted that applied scientists "can afford to ignore" the details of theories (1966, 333, cf. also Bowker 2005).

problem of reliably bringing something about often takes the form of creating a thing that exhibits a desired property or behavior – for example, to write software that emulates expert judgement, or to model global weather patterns, or to increase the productivity of an ecosystem, or to build a more efficient engine or solar cell. Here, the solution of the problem will generate technical knowledge but frequently without a corresponding gain in intellectual understanding or theoretical knowledge. Indeed, it can be rational not to wait for theoretical understanding before setting out to expand and achieve technical control.

A rational strategy for establishing technical control in the absence of detailed knowledge of all the relevant processes is the design cycle that has been explicated in software engineering and that informs the construction of expert systems, of simulation and climate models, of biological entities and processes in synthetic biology, and implicitly the creation of many experimental systems and technical artefacts.[5] If the target or end-point of the design process is some entity or system with desired properties or behaviors, one begins, of course, with the assembly of a model that incorporates the relevant available knowledge. Based on an analysis of the task, a first prototype is constructed. So far, then, this conforms to the normal expectation that theoretical descriptions underwrite technical constructions: When we study the initial prototype we find our own knowledge reflected in the principles of its construction. In the next step along the design cycle, the performance of the prototype needs to be considered and compared to the desired behavior. Typically the prototype either doesn't work at all or falls short of desired performance goals. Here, then, the discrepancy needs to be bridged which requires an analysis in its own right, not of the original task but of the behavior of the prototype. We systematically probe the prototype and introduce corrections, patches, additional routines of various kinds. These corrections are also based on knowledge and experience, of course, but this is knowledge of the technical systems, of the design process, of tried and tested algorithms, of software modules or technical adjustments that usually get the job done. One tweaks the prototype, in other words, observes it again, tweaks it again, and thus the process of approximating the desired performance amounts to a process of optimization, of fine-tuning, or calibration. Iterating this cycle again and again, one keeps adding elements, writing new code, altering conditions in such a way that the prototype gains complexity: It incorporates more and more bits of knowledge and practice such that, as a whole, its is no longer a transparent instance of our theories and principles. The technical achievement now eludes intellectual tractability and we are to some and perhaps to a considerable extent ignorant of what we made.

It is by no means a novel phenomenon that our technical creations can get ahead of our theoretical understanding. In some cases – most famously, the steam-engine and perhaps the computer – science will catch up with the new devices. In other cases – most famously, nuclear weapons – it has been suggested that we will never

[5] Arguably, it plays a major role in metrology as well (Chang 2007). It is fair to say, perhaps, that it inhabits the vast middle ground between ab initio constructions that implement a theoretically derived blueprint and haphazard trial and error tinkering (Nordmann 2014).

be able to fully comprehend the potential effects and implications of our own technologies.[6] If there is something decidedly novel about optimization in a design cycle, it is that this production of technical knowledge occurs at the forefront of technoscientific research and that it purposefully drives a wedge between explanation and prediction as the two-pronged goals of science. According to standard accounts of scientific knowledge, one needs good explanatory theories in order to render the world increasingly predictable and subject to technical control. Inversely, the testing of predictions allows us to assess the quality of our explanatory theories. In the current technoscientific „culture of prediction“ (Johnson and Lenhard 2011) one demands that models and other technical constructions behave predictably and allow for the successful control of behaviors or events – and this functions as a technical criterion which does not presuppose intellectual understanding or theoretical explanation. By iterating the design cycle one can optimize the retrodictive and predictive performance of a model without requiring an articulated conception of how the predictions are generated or precisely what role the various theoretical components of the model play in the process. Indeed, computer models are mostly used to relieve researchers of having to know any of this – that is, as ignorance affordances.

In the long run, the dissociation of explanation and prediction in *TechnoScienceSociety* valorizes predictive control and the technological agenda of building the machines that do the knowing for us. Though we regulate and, through fine-tuning, maintain the capacity of these machines to accurately model and predict phenomena and processes, we do not know and need not know how an "e-scientific" computer model or expert system knows what will happen next, nor can we predict what it will predict.

This is not the place to discuss the long-term viability of this development. Though the process of iteration and incremental optimization is meant to overcome limits of complexity, it may well encounter limits of its own. To the extent, however, that the technical achievement of predictive control takes priority over the intellectual achievement of a satisfactory theory or tractable explanation, it is evidence for the tolerance of ignorance at the forefront of contemporary research.

Undoubtedly, some will object to this account by pointing out that explanations are constantly produced, that most research publications offer explanations of their findings, and that predictive and explanatory capabilities still seem to grow together. Modelling, they will contend, is at least as much about explanation as it is about prediction. However, though explanation has a place also in the culture of prediction, what an explanation is and what counts as an explanation is not the same in a culture of prediction as it was traditionally. Indeed, it is further testimony to the tolerance and promotion of ignorance in *TechnoScienceSociety* that even so-called

[6] Günther Anders speaks here of the discrepancy between what we can produce (*herstellen*) and what we can conceive, survey, take responsibility for (*vorstellen*), see Anders 1956. This may also hold, albeit along different lines, for formal systems or computer models that instantiate complex non-linear dynamics and that offer a new kind of technically produced entity for investigation, one that cannot be fitted easily to the requirements of human understanding.

explanations can be inscrutable and opaque: When pharmacologists, materials scientists, nanotechnology researchers, or synthetic chemists discover an interesting feature, promising behavior, or surprising property, they exhibit it in the laboratory with the help of their apparatus. They thereby arrive at measurements, perhaps a visual output, an observable effect, something that works better than placebo, or the like. Having found that this is a robust and repeatable finding, they may well be content to publicize it as is, foregoing explanation altogether. They might also walk down the hall to the theorists in their department and ask them to produce an explanation. More often than not, these theorists will seek to reproduce the finding once more. However, they do not reproduce it in a petri-dish, a population of cells, or an atomic-force microscope. Instead of materially reproducing it in an experimental system, they reproduce it in the physical system of a computer. In order to reproduce *in silico* what the experimenters have produced *in vivo* or *in vitro*, the theorists or modelers assemble bits of theory (*e.g.*, in the form of algorithms) to piece together a process that appears to be quite like the one that was observed in the lab.[7]

Once the process can be modeled *in silico*, one speaks of an "explanation" – which is now the ability to obtain a phenomenon by theoretical means in the computer model, that is, by building into the model bits and pieces of theory. Accordingly the model reflects technical rather than intellectual knowledge in that explanation now consists in the capability of producing a phenomenon with the means of theory. Technoscientific explanation is not constrained by the requirements of human understanding, that is, it does not trace the phenomenon back to general principles or laws, it does not exhibit entailment relations or show that the phenomenon was to be expected, nor does it allow the retracing step by step of a causal mechanism. Accordingly, the question of truth does not arise, with no one asking whether this *in silico* model provides the only explanation, the best explanation, or the one that should be preferred over others. Instead, all we get is a proof of principle, namely that one can use available knowledge to account for the phenomenon, with explanations showing only *that* a process or a phenomenon can be reconstituted with the help of theory. And given the computational complexity of these explanations, they tell us that the phenomenon might be inscrutable but nevertheless lies within the reach of human technical control. Moreover, these explanations allow us to explore technical alternatives: If we get the observed phenomenon by assembling such and such bits of theory into our model, perhaps we can get quite another phenomenon if we change this or that parameter.

[7]The iterations of the design cycle once again play a role here as one seeks to produce a close match between the experimental and the simulated versions of the process.

2.4 Technoscientific Rationality

TechnoScienceSociety advances a model of knowledge production that bypasses intellectual limitations. It ascertains technical control of systems that are too complex so as to be tractable by the human mind. It thereby bypasses also the established knowledge requirement for public deliberation, regulatory safeguards, or accountability for impacts on the environment, human health, or the social fabric. Here, tolerance for ignorance may result in toxicology by computer simulation or iterative processes of social learning from collective experiments. And instead of aspiring to a state of certainty of truth from which to make judgements or take regulatory action, *TechnoScienceSociety* cultivates a state of permanent vigilance – seeking to monitor potential hazards in real time and through a large network of informal institutions (Nordmann 2010).

Technoscientific research thus supports contemporary knowledge societies not only in their quest for innovation but also regarding their ability to deal with ignorance. Where we do not have knowledge, we can still have something like robustness. And when we cannot satisfy our intellect or achieve a reasonable degree of certainty about risks, we can develop new governance mechanisms and adaptive management tools to detect and respond to emerging risks.

To be sure, epistemic virtues are themselves subject to public debate and might be adjudicated in the terms of political theory and constitutional principles. For example, whether simulation toxicology will ever be allowed to underwrite regulatory actions does not depend on its ability to meet the technical challenge of managing the requisite complexity. It depends much rather how the ambivalence and political conflict between modern *EnlightenmentScienceSociety* and *TechnoScienceSociety* will be settled or resolved. What is commonsensical from the point of view of *TechnoScienceSociety*, may well appear unprincipled, if not cynical from an Enlightenment perspective. This holds for the delegation to machines of judgements about potential harm to humans, and it holds more generally for questions of knowledge and ignorance and whether one should let sleeping dogs lie.

There are things that we don't know, things that we can't know, things that we are not supposed to know, things that we don't need to know, things that we don't want to know. Curious seekers of enlightenment are bothered by what they do not know, technoscientific rationality requires only that we know just what we need to know, no more and no less. Technoscientific builders and makers might even go to some lengths to determine just how little knowledge or theoretical understanding is sufficient to get along just fine – even if it means surrendering here and there the notion that the intellect should always and everywhere be the arbiter of human affairs.

References

Anders, Günther. 1956. *Die Antiquiertheit des Menschen: Über die Seele im Zeitalter der zweiten industriellen Revolution*. Vol. 1. Munich: C. H. Beck.

Bacon, Francis. 1990. *Neues Organon*. Vol. 2. Hamburg: Felix Meiner Verlag.

Beck, Ulrich. 1996. Wissen oder Nicht-Wissen? Zwei Perspektiven "reflexiver Modernisierung". In *Reflexive Modernisierung; Eine Kontroverse*, ed. Ulrich Beck, Anthony Giddens, and Scott Lash, 289–315. Frankfurt: Suhrkamp.

Beck, Ulrich, Anthony Giddens, and Scott Lash. 1994. *Reflexive Modernization: Politics, Tradition and Aesthetics in the Modern Social Order*. Stanford: University Press.

Bono, James. 1995. The Reform of Language and Science: Sir Francis Bacon's Adamic Instauration and the Alphabet of Nature. In *The Word of God and the Languages of Man*, ed. James Bono, vol. 1, 199–246. Madison: University of Wisconsin Press.

Böschen, Stefan, Karen Kastenhofer, Ina Rust Jens Soentgen, and Peter Wehling. 2010. Scientific Nonknowledge and Its Political Dynamics: The Cases of Agri-Biotechnology and Mobile Phoning. *Science, Technology & Human Values* 35 (6): 783–811.

Bowker, Geoffrey. 2005. *Memory Practices in the Sciences*. Cambridge, MA: MIT Press.

Bunge, Mario. 1966. Technology as Applied Science. *Technology and Culture* 7 (3): 329–347.

Chang, Hasok. 2007. *Inventing Temperature: Measurement and Scientific Progress*. Oxford: Oxford University Press.

Firestein, Stuart. 2012. *Ignorance: How it Drives Science*. New York: Oxford University Press.

Franke, William. 2015. Learned Ignorance: The Apophatic Tradition of Cultivating the Virtue of Unknowing. In *Routledge International Handbook of Ignorance Studies*, ed. M. Gross and L. McGoey, 26–35. London/New York: Routledge.

Gaukroger, Stephen. 2001. *Francis Bacon and the Transformation of Early-Modern Philosophy*. Cambridge: Cambridge University Press.

Gross, Matthias, and Linsey McGoey. 2015. Introduction. In *Routledge International Handbook of Ignorance Studies*, ed. M. Gross and L. McGoey, 1–14. London/New York: Routledge.

Janich, Nina, Alfred Nordmann, and Liselotte Schebek. 2012. *Nichtwissenskommunikation in den Wissenschaften: Interdisziplinäre Zugänge*, 7–20. Frankfurt: Peter Lang.

Johnson, Ann, and Johannes Lenhard. 2011. Toward a New Culture of Prediction: Computational Modeling in the Era of Desktop Computing. In *Science Transformed? Debating Claims of an Epochal Break*, ed. A. Nordmann, H. Radder, and G. Schiemann, 189–199. Pittsburgh: Pittsburgh University Press.

Krohn, Wolf. 1987. *Francis Bacon*. München: Beck.

Nordmann, Alfred. 2006. Collapse of Distance: Epistemic Strategies of Science and Technoscience. *Danish Yearbook of Philosophy* 41: 7–34.

———. 2008. Ignorance at the Heart of Science? Incredible Narratives on Brain-Machine Interfaces. In *Nanobiotechnology, Nanomedicine and Human Enhancement*, ed. J. Ach and B. Lüttenberg, 113–132. Berlin: Lit-Verlag.

———. 2010. Philosophy of Technoscience in the Regime of Vigilance. In *International Handbook on Regulating Nanotechnologies*, ed. D. Bowman, G. Hodge, and A. Maynard, 25–45. Cheltenham: Edward Elgar.

———. 2014. Synthetic Biology at the Limits of Science. In *Synthetic Biology: Character and Impact*, ed. B. Giese, A. von Gleich, C. Pade, and H. Wigger, 31–58. Berlin: Springer.

Nordmann, Alfred, Hans Radder, and Gregor Schiemann, eds. 2011. *Science Transformed? Debating Claims of an Epochal Break*. Pittsburgh: Pittsburgh University Press.

Nowotny, Helga, Peter B. Scott, and Michael T. Gibbons. 2001. *Re-Thinking Science: Knowledge and the Public in an Age of Uncertainty*. Cambridge: Polity.

Proctor, Robert, and Londa Schiebinger, eds. 2008. *Agnotology: The Making and Unmaking of Ignorance*. Stanford: Stanford University Press.

Shapin, Steven. 1994. *A Social History of Truth*. Chicago: University of Chicago Press.
———. 2008. *The Scientific Life: A Moral History of a Late Modern Vocation*. Chicago: University of Chicago Press.
Smith, Pamela. 2004. *The Body of the Artisan: Art and Experience in the Scientific Revolution*. Chicago: University of Chicago Press.
Tal, Eran. 2013. Enhancing Knowledge, Affording Ignorance, Presentation at the Conference *What Affordances Afford*, Darmstadt, November 2013. https://www.philosophie.tu-darmstadt.de/media/philosophie___goto/text_1/Book_of_Abstracts.pdf. Accessed on 9 April 2020.

Alfred Nordmann is Professor of Philosophy of Science and of Technoscience at the Technische Universität Darmstadt. Investigating epistemological, metaphysical, and aesthetic dimensions of different cultures of research, he seeks to ground central concepts of philosophy of science in the philosophy of technology. He is Editor of the book series History and Philosophy of Technoscience (Routledge).

Chapter 3
Tracing the Technoscientific Core of Late-Modern Technologies in *TechnoScienceSocieties*

Jan Cornelius Schmidt

3.1 Introduction

The reflexivity of reflexive modernity[1] seems to be coming to a close. Francis Bacon[2] is back, at least in the power rhetoric, narratives, visions, and programs surrounding technoscience. A prominent characteristic of *TechnoScienceSocieties* is their striking ascription of power and control to the advancements of technosciences.[3] Based on this observation, the main thesis of this paper is that, by strongly pursuing Baconian power and control ambitions, *TechnoScienceSocieties* surrender power and control to a novel type of technoscientific objects; an ambivalent or even dialectic momentum seems to be emerging.

Ambitious power and control ideals abound in *TechnoScienceSocieties*. Well-known examples include the US-NSTC slogan "shaping the world atom by atom" and the US-NSF motto "converging technologies for improving human performance" by "changing the societal 'fabric' towards a new structure" (NSTC 1999; Roco and Bainbridge 2002). Nanotechnology, to give one example, is frequently defined as "the ability to understand and control matter with ultimate precision" and as "the most powerful and enabling technology humankind has ever developed." It "is used to create materials, devices and systems with fundamentally new properties and functions that will change the world as we know it." The NSF, in a similar vein,

[1] For a discussion of the concept of "reflexive modernity" in the 1980s and early 1990s, see Beck et al. (1996).

[2] See, for example, Bacon's Novum Organon (Bacon 2004).

[3] For example, Hottois (1984), Latour (1987), and Haraway (2003) forged the term "technoscience". Further clarification has been achieved by Ihde and Selinger (2003), Weber (2003), Nordmann (2008b) and Schmidt (2011b). For an overview, see Nordmann and Radder (2011).

J. C. Schmidt (✉)
Darmstadt University of Applied Sciences, Darmstadt, Germany
e-mail: jan.schmidt@h-da.de

S. Maasen et al. (eds.), *TechnoScienceSociety*, Sociology of the Sciences Yearbook 30, https://doi.org/10.1007/978-3-030-43965-1_3

announces: “If the Cognitive Scientists can think it, the Nano people can build it, the Bio people can implement it, and the IT people can monitor and control it” (Roco and Bainbridge 2002: 13).

Synthetic biology is yet another technoscientific research field that alludes to the power “to design and create bio-systems” having desired functionalities “that do not exist as such in nature.” In 2010 the technoscience entrepreneur Craig Venter announced the forthcoming advent of an epochal break toward “engineering biology” and envisioned a fundamental shift in our technical capabilities to undertake “rational design”: Synthetic organisms “are going to potentially create a new industrial revolution if we can really get cells to do the production we want; [...] they could help wean us off of oil, and reverse some of the damage to the environment like capturing back carbon dioxide.” (Venter 2010; see Gibson et al. 2010) Even the critics of the new technosciences adopt a rhetoric of powerful intervention and control: The Action Group on Erosion, Technology and Concentration, better known as ETC Group, published a statement suggesting that the use of the so-called BANG technologies—an acronym for *B*its, *A*toms, *N*eurons, *G*enes—will allow the “flexible manipulation of the bits of information, the atoms of matter, the neurons of the brain and the genes that code for life.” (ETC 2007).

Thus, Donna Haraway appears to be right in her observation that recent “technoscience is about worldly, materialized, signifying and significant power.” (Haraway 2003: 43)[4] The main objective of this paper is to analyze the ambivalence and ambiguity of a certain kind of power[5] that is characteristic of *technoscience.* The power referred to is inherently linked to a novel type of technology, as will be shown. It is central, for example, to soft machines, synthetic biosystems, nanorobots, artificial intelligence, software agents, and the like. The main point, as will be argued in this paper, is that *TechnoScienceSocieties* lose their ability to control these novel technoscientific systems; they surrender their power and control to them.[6]

To analyze how recent *TechnoScienceSocieties* surrender their power and control to new technoscientific systems, I will coin a search term and tentatively refer to the (envisioned and partly already realized) ontologically[7] novel type of technology as “late-modern”.[8] In their diagnosis of an epochal shift or break, other authors prefer

[4] Pointing to the Baconian program, Haraway’s thinking is largely in line with Rouse’s perception as far back as in the 1980s. Rouse argued that, in light of the power of science and engineering, a “political philosophy of science” needs to be established (Rouse 1987).

[5] In this paper, power is understood as the possibility of humans and the capacity of societies to design, create, construct, and control the natural, social, human and, thus, the societal world.

[6] The power becomes dissipated and appears to be spread symmetrically among the hybrid things of the techno-social worlds. As will be shown, the power rhetoric is one-sided: it is not based on sound arguments.

[7] The term “ontology” is not used in the sense it is used in the deep thought tradition of ancient and modern philosophy but in a way common in the philosophy of technology and in science and technology studies: as an attempt to characterize general aspects of the material design, construction, structure and principle of technical objects.

[8] The term “late-modern” also reflects the relation between humans and technical systems. It encompasses aspects of knowledge, expectation and predictability, and in particular non-knowledge

different, but cognate labels such as "transclassic technology" (Hubig 2007), "naturalized technology" (Nordmann 2008a),[9] "biofacts" (Karafyllis 2003, 2007), "alternative design paradigm" (Kroes 2009), "informal technology" (Kaminski and Gelhard 2014) or "autonomous technology" (Rammert and Schulz-Schäfer 2002).[10] The main difference between these cognate concepts and the approach pursued in this paper employing the notion of "late-modern" is that the latter aims to address the materialized/algorithmic *technoscientific core* in detail and strives to find a *common denominator* for many recently emerging technologies.[11] Drawing on the notion of "late-modern", this paper distinguishes the novel type of technology from "modern" technology: the technoscientific ontological core of the "modern" type of technology was, and remains, intentionally designed and constructed on the ideal of predictability, reproducibility, transparency, and controllability, whereas the "late-modern" type—by harnessing self-organization and instability for the purpose of "self-productivity"—exhibits limitations with respect to the "modern" epistemological and methodological ambitions.[12]

The above analysis may initially raise concerns, as it is an approach that, on the surface, does not seem to be societally or culturally oriented. It might even appear reductionist, since it aims to explore the underlying materialized/algorithmic structure, including the basic nomological principles implemented *in* the ontological fundaments, of late-modern technoscientific systems. However, as will be shown by the epistemological analysis undertaken here, the novel type of technology is, on account of its specific ontological structures and principles, far more entangled with the human and societal sphere than technology has ever been before. Indeed, the entanglement shows an ambivalence of power that can be regarded as a crucial characteristic of *TechnoScienceSociety*.

(Böschen et al. 2004); it is therefore an *epistemological* notion that refers to and reflects the ontological core.

[9] Gilbert Simondon argued along similar lines in the 1950s, as Barthélémy (2011: 103) points out: Simondon already spoke of a "naturalization of technology".

[10] See also: Christaller and Wehner (2003) and Suchman and Weber (2015).

[11] An important point to consider is that all late-modern technologies are based on (the emergence and advancement of) technosciences; but the reverse is not true. Technosciences also advance the traditional (modern) type of technology (see below and Table 3.1).

[12] The notion of "late-modern" can be justified by two arguments: First, it has been shown by reference to physics, biology, computer sciences, mathematics and some engineering sciences that we are entering a (new) era of "late-modern" science, or technoscience. The era of "modern" sciences is undergoing an epochal broadening and transformation due to a new emerging paradigm based on self-organization and instability. The latter two concepts put certain limitations on "modern" assumptions such as those relating to predictability, reproducibility, testability and describability/explainability (see Schmidt 2008, 2015a, b). The term "modern" in this context applies to the fields of "modern" physics and biology where the notion is well established. Second, the discourse on late-modernity or post-modernity also underlines the perception and acceptance of the above mentioned limitations which include the lack of controllability.

3.2 A New Trend

It is hard to frame emerging technologies such as soft machines, synthetic bio systems, nanorobots, software agents, neuro-enhancement drugs, and the like from the perspective that has been developed to analyze "modern" technology.[13] In its very nature, a "modern" technical system is regarded as a stable causal input-output rule based on a boundary or interface between the system and its (natural, technical, human, social, cultural) environment. The assumption that such an interface exists is of crucial significance. The interface seems to be necessary to any kind of technology as it simplifies the complexity of the real world and isolates the technical system from its environment, thereby enabling the system to realize its intended functionalities.

"Modern" technology is very much linked to the control ideal behind the design and construction of a technical system. Although societal conditions and the consequences of a certain technology might be hard to control, every technical system is, in its designed (technoscientific ontological) core, deemed to be controllable. Control in the sense used here means regulating and governing the flow of material, energy, and information across the interface between the technical system and its environment (see Ropohl 1999: 120f.).[14] A point of utmost relevance in this respect is that there is a precondition for the possibility of control that is seldom explicitly reflected upon—but is implicitly assumed to exist: stability.[15] The assumption of stability at the interface, which encompasses the stability of the flow (material, energy, information), is extremely important because only stability enables the regularity, reproducibility, predictability, transparency, and testability of the technical system's desired functionality. Stability can be considered the central backbone of any kind of control, as we will see later on in this paper.

At the same time, it is arguable whether technical systems are always designed and constructed in a way that provides stability. Understanding technology as a stable technical arrangement—a perspective which will be referred to as "modern" technology throughout this paper—is, indeed, restrictive. This view has been challenged by Ernst Bloch (1995), Hans Jonas (1985), Gilbert Simondon (2012), and others.[16] According to these authors, there are other types of technical systems

[13] The traditional concepts are described by Luhmann (2003: 93f.), Hubig (2007: 41f.) and Baecker (2011: 179f.). These concepts can be called "modern"; they refer to "modern" technical systems.

[14] Kroes (2009: 539) argues that (what he calls) the "traditional design paradigm" is based on "the assumption that it is possible to clearly separate the object of design from its environment". Furthermore, "it is assumed that the behaviour of the systems designed can be fully controlled by controlling the behaviour of its parts". This provides a very clear description of what is here termed "modern technology".

[15] The distinction between "stability" and "instability" will be clarified later on in this paper. For the moment, it is sufficient to regard a stable system as one that shows the above mentioned properties.

[16] In the most recent discussion, other authors argue along similar lines, including Rammert and Schulz-Schäfer (2002), Karafyllis (2003), Christaller and Wehner (2003), Hubig (2007), Nordmann (2008a, b), Hörl (2011), Kaminski and Gelhard (2014) and Schmidt (2012).

grounded in processes of self-organization: dynamical, processual, and evolutionary systems that can be associated with activity and, to an even greater extent, with the notion of "coming into being". Bloch, Jonas and Sismondo envision a new type of technology and, equally relevant, stress the need for a new thinking about technology: we should rethink our thinking. Their work was pioneering and is still very inspiring, in particular in light of the recent wave of emerging technologies. The present paper takes up their trailblazing thoughts. Its point of departure is that some of the most prominent emerging NBIC technologies[17] do not fit into the above mentioned "modern" conventional perspective.

Besides other NBIC technologies the need for a more comprehensive understanding of technology has been prompted by the attempts to further explore synthetic biology.[18] What does the trendy umbrella term "synthetic biology" mean? At first sight, the "onto-politics"[19] of synthetic biology – framed as a cutting edge of "modern" technology – encompasses three most prominent definitions (see Schmidt 2015b).[20] Above all, the *engineering definition* frames synthetic biology as being new, since it brings an engineering control approach to the scientific discipline of biology: "Synthetic biology is the engineering of biology", or in other terms, the "birth of a new engineering science." According to the *artificiality definition*, synthetic biology deals with "bio-systems that do not exist as such in nature" but are intentionally designed, constructed, and controlled by bioengineers (TESSY 2008). Finally, the *extreme gene/bio-technology vision* sees synthetic biology as an "expansion of biotechnology" (Endy 2005), and in particular as an expansion of its methods. The latter vision stands in the biotechnological tradition. – In general, the three definitions can be related to either one of or both of the cognate traditions of "modern" technology: to the tradition of molecular biology and genetic technology/engineering on the one hand and to the tradition of synthetic chemistry on the other. Both traditions share the engineering ideal of rational control that Drew Endy (2005) reframes as "standardization", "decoupling", and "abstraction" (see Giese et al. 2013).

Given the three non-disjunctive definitions, including the two traditions, throughout most research programs and practices, synthetic biology seems to be a paradigm of a control-oriented technoscience. In the field of synthetic biology, living entities are regarded as systems with modularized technical functionalities that can be controlled

---

[17] The NSF Roco/Bainbridge report coined the acronym 'NBIC' of nanotechnology, biotechnology, information technology, and cognitive sciences/technologies (Roco and Bainbridge 2002; see Schmidt 2004).

[18] Furthermore, the analysis made in this paper generally holds for other NBIC technologies as well (see below Sect. 3.3).

[19] See the introduction (Maasen et al., Chap. 1 in this volume) and in particular Tyler (in this volume).

[20] Schneider (Chap. 16, in this volume) argues conclusively that the types of visionary discourse on emerging technologies are a central characteristic of TechnoScienceSocieties. And, in line with Karafyllis (Chap. 8, in this volume), we could frame the discourse on "synthetic biology" using the concept of "bioprospecting".

by the bioengineer—an ambivalent point which is manifest in a technomorphic language: nucleosomes are characterized as data storage entities, membranes as electrical arcs, adenosine triphosphate synthases as generators, or polymerase chain reactions as copying machines. Evidently, a unifying arc of the three definitions can be seen in the general ideal of deeper control and power. Even so it remains debatable whether these definitions—framing synthetic biology as a "modern" type of technology—are sufficient to grasp the technoscientific core of synthetic biology.

Two blind spots hinder an in-depth analysis. *First,* the biology bias: The focus on biology alone as a discipline, entailing a discipline-oriented framing of synthetic biology, seems to prevent an exhaustive characterization of the new technoscientific wave. Synthetic biology is obviously far more *interdisciplinary* than disciplinary at its nucleus. This needs to be taken into account in the search for an adequate definition: Biologists, computer scientists, physicists, chemists, physicians, material scientists, and people from different engineering sciences are engaged in synthetic biology. *Second,* the bias towards a preunderstanding of technology as modern technology: In line with what is termed bionano or nanobio research, the three visions look at synthetic biology from the angle of control. Synthetic biology seems to extend and complement advancements in nanotechnology and hence spurs a control oriented position that can be called "technological reductionism" (Schmidt 2004). The control slogan promoted by technological reductionism is: Shaping, designing, constructing, and creating the world "atom-by-atom"; it is the ideal of "rational design/engineering". But whether that is all that can, or should, be said to characterize synthetic biology still needs to be clarified. In fact, synthetic biology claims to pursue an approach that is complementary to technological reductionism and has been called "systems approach" or, in a more visionary sense, "holistic".

Consequently, a more comprehensive characterization and in-depth analysis of synthetic biology should include further aspects that are not restricted to a view of technology as a "modern" type of technology and to biological technology alone. In addition to the two traditions of synthetic biology—the molecular biological tradition and the synthetic chemistry tradition[21]—we should also consider another path breaking tradition: that of bio-informatics or bio-systems which is central to systems biology. Because systems biology can be considered fundamental to synthetic biology, the systems approach is essential to synthetic biology, as well. In order to "be successful," Petra Schwille (2011: 1253) argues, "synthetic biology of any kind will have to join forces with systems biology". The systems paradigm—interlaced with the ideal of non-reductionism"[22]—appears to be a central distinguishing characteristic of synthetic biology.[23] Looking at synthetic biology in more detail, we find

[21] See above.

[22] It should be noted that, although this is currently no more than a guiding ideal, we find a considerable internal dialectic between non-reductionism ("holism") and reductionism in synthetic biology.

[23] Evidently, systems thinking constitutes the main difference between synthetic biology and classical biotechnology, a point also stressed by Kitano (2002: 1662): "A transition is occurring in biology from the molecular level to the system level that promises to revolutionize our understanding

a further widespread definition in relevant research programs: *the systems* or *self-organization definition*. Synthetic biology harnesses, or at least aims to harness, the self-organization power of nature for technological purposes. This paradigm is broadly present in papers and practices associated with the domain of synthetic biology: "Harnessing nature's toolbox" in order to "design biological systems" (Drubin et al. 2007). The aim is to "exploit the principles of automatic self-organization that are seen in nature" (Roco and Bainbridge 2002: 258). The basic idea of synthetic biology is to "turn organisms into manufactures" and to make them "self-organizing and self-productive" (Pottage and Sherman 2007).[24] The goal is to "break[e ...] the limits on design complexity," as the computer scientist Jordan Pollack argues: "We think that in order to design products 'of biological complexity' that could make use of the fantastic fabrication abilities [...], we must first liberate design by discovering and exploiting the principles of automatic self-organization that are seen in nature" (Pollack 2002: 161). In summary, these few quotations underline how widespread the notion of self-organization is throughout synthetic biology.[25]

3.3 Ontology and Epistemology of Late-Modern Technology

The paradigm of complex, self-organizing systems is advancing at an accelerated pace, both in science and in technology (see Dupuy 2004). From a methodological perspective, it is frequently labeled "systems engineering".[26] The systems approach

of complex biological regulatory systems and to provide major new opportunities for practical applications". Synthetic and systems biologists cite classic works from the field of systems and structural sciences, for example, Bertalanffy's General System Theory and Wiener's Cybernetics (see Kitano 2002). Today, systems thinking is not primarily directed at theoretical insight, but at technological progress. Advancements in synthetic biology require systems thinking. The approach that will be termed "late-modern technology" later in this paper is based on the advancement of systems thinking and systems sciences.

[24] In addition, the biologists Westerhoff and Palsson perceive a shift in biology from a molecular-biological approach towards a systems paradigm encompassing fundamental questions relating to "self-organization. [...] The problem of biological self-organization is to understand how structures, oscillations or waves arise in a steady and homogenous environment, a phenomenon called symmetry breaking. The Prigogine school and others developed the topic from the perspective of non-equilibrium thermodynamics in molecular contexts. [...] Contemporary systems biology has a historical root outside mainstream molecular biology, ranging from basic principles of self-organization in non-equilibrium thermodynamics, through large-scale flux and kinetic models to 'genetic circuit' thinking." (Westerhoff and Palsson 2004: 1249)

[25] See further cognate approaches and similar quotations in Buchli and Santini (2005) and Nolfi and Floreano (2000). An analysis and reflection on the subject of self-organization is also presented by Kroes (2009).

[26] See, for instance, (Sage 1992). The ACM *Transactions on Autonomous and Adaptive Systems Journal* (TAAS), a leading journal in the field, presents further examples of the systems approach and self-organization in technical systems. See also the European Conference on Complex Systems (No. 5, Paris, 2005) which hosted a workshop on "Engineering with Complexity and Emergence" (http://complexite.free.fr/ECCS/).

placing self-organization at the center of a new technoscientific design has indeed enjoyed an impressive history over the last three decades. Most prominent is the vision painted by Eric Drexler (1990) with his programmatic ideas of "self-assembly technology" and "engines of creation". Others perceive a general trend towards "self-organizing soft machines" (see Jones 2004). Christoph Hubig and Sebastian Harrach (2014: 41), also reflecting such concepts, further substantiate the perception of an epochal shift regarding the relation of the technical, natural, and the human sphere: "[O]n the basis of self-organization processes" a novel type of technology that can be called "transclassic or informal" is emerging. It "blurs the interface between the technical, the natural and the human."

The above statements show that synthetic biology does not stand alone; it can be seen as a spearhead of a very general trend in the technosciences and in technoscientific engineering.[27] In order to give further substance to the thesis of the relevance and prevalence of self-organization, let us look at another branch of the technosciences.[28] The application of self-organizing processes is very common in artificial intelligence, information and communications technologies. Take, for example, the so called *autonomous stock agents* in high frequency trading (HFT) at the stock exchange and in the global neo-capitalist finance market.[29] Autonomous stock agents function on a mathematical-numerical basis of artificial neuronal networks, genetic algorithms, or similar numerical structures of artificial intelligence and deep *learning machines* (see Kaminski and Gelhard 2014). Based on principles of self-organization, the algorithms are "trained" with real "big data" in the context of application; they "learn" in an open environment in order to make "decisions". However, depending on the "decisions" they make, self-organizing autonomous stock agents can induce cascade effects and cause a crash at the stock exchange. In other words, they are the underlying abstract machines that constitute the algorithmic "market makers" governing today's finance system.[30] Although intentionally designed and constructed by engineers and computer scientists, self-organizing learning algorithms at work implemented in artificial intelligence systems show limits with regard to transparency and comprehensibility (epistemological limita-

---

[27] The relevance of the systems approach and processes of self-organization in a general sense is also underlined by Kroes (2009: 539) in his reflections on recent developments towards "systems engineering" (see Sage 1992): "Given the growing complexity of the systems that engineers have to deal with, it is to be expected that systems engineering will become an ever more important branch of engineering."

[28] See also the various robotic examples presented in Christaller and Wehner (2003).

[29] An introduction to this "hot" topic can be found in: Aldridge (2013) and Gomber et al. (2011).

[30] In addition, self-organizing algorithms today evaluate and rate the credibility of national economies. Also, they are "gatekeepers" of search engines and mail-order companies, and they play an indispensable role in social media. Cognate approaches are embodied and implemented in the field of "evolutionary robotics", see Nolfi and Floreano (2000), and in cognitive enhancement technologies, as Maasen and Sutter (2007) and Maasen and Duttweiler (2012) underline. Many of these technologies also utilize processes of self-organization: artificial intelligence and algorithmic learning via optimization by artificial neuronal nets and genetic algorithms.

tions), and in particular limited predictability and reproducibility (methodological limitations).

Self-organizing autonomous stock agents are merely another example besides synthetic biology. In general, the concept of self-organization is widespread in emerging technologies and technoscientific practice. It plays a constitutive role in all NBIC fields such as

1. nano-technologies;
2. bio-technologies;
3. information technologies, robotics, AI, software agents; and
4. cognitive and neuro-technologies.

In addition, self-organization serves as a leitmotif in science and technology politics. "Unifying science and engineering" seems possible by "using the concept of self-organized systems" (Roco and Bainbridge 2002: 84).

If we take the visionary promises as serious claims, they announce the emergence of a new type of technology. We do not know whether the visions can be fully realized. However, were this to happen, we would encounter a different kind of technology: a *late-modern technology*—as an extension and complement of modern technology. How can such a technology that is so inherently based on the development of technosciences be characterized (see Table 3.1)?

Although the notion of "late-modern technology" refers ontologically to a novel material/algorithmic structure of technical systems, it also addresses the relation between humans and technical systems: It focuses on human knowledge, perceptions and expectations, actions and interactions with regard to technical systems. A late-modern technical system has other phenomenological properties[31] and is perceived differently to our established perception of technical systems. It displays nature-like or human-like characteristics; it bears no resemblance to a technical system at all; it seems to be un-technical or non-artificial; the signs and signals, tracks and traces are no longer visible (see Hubig 2007). Technical connotations have been peeled off, well-established demarcation lines are blurred. A clear input-output mechanism and a systems/environment interface do not exist or cannot be ascribed. Paradoxically, the environment of the technical system is not excludable, it is part of it. Furthermore, late-modern technology often seems to possess an intrinsic momentum of rest and movement within itself—not an extrinsic one. The described characteristics come close to the Aristotelian common life-world understanding of nature or of human agents. Late-modern technical systems appear to act by themselves, which often means that they are imperceptible as technology: They (a) create and produce; (b) select means to ends and seem to follow a means-ends rationality; and (c) make decisions and act in response to environmental conditions. Technology evidently presents as an agent or, at least, as having agency properties. Activity and, more fundamentally, autonomy—a term that is central to our thought

[31] The term "phenomenological" is understood here in the sense commonly used in the natural and technical sciences as well as in the philosophy of science. It should not be confused or equated with the philosophical school of "phenomenology" (Husserl et al.).

Table 3.1 Shows some of the main characteristics of late-modern technologies

	Classic-modern technology	Late-modern technology
"Technology" term	Search and reflection term that encompasses (a) ontological aspects that reflect the technoscientific core and the nature of technical systems and, based on this, (b) epistemological aspects, including phenomenological, cultural, societal aspects of the human perception *of* and interaction *with* technological systems	
Paradigmatic reference sciences	Physics (to some extent: biology, chemistry); disciplinary orientation, based on traditional engineering and/or on technosciences	Systems/structural sciences; computer sciences; life sciences; interdisciplinary and transdisciplinary orientation; based necessarily on technosciences
Background concepts	Newton, Kant, Luhmann	Bloch, Jonas, Simondon
Approach	Technological reductionism	Systems holism as well as technological reductionism
Convergence/ integration/ unification program	If any, ontological convergence in the tiny objects (e.g., via nanostructures, molecules, atoms)	Convergence as a leitmotif: phenomenological convergence and structural/nomological convergence
Engineering concept/ imperative/ action	Intentional construction, rational engineering, control: standardization, decoupling, abstraction. Imperative: Stabilize, dampen, reduce noise! ("elimination of instability")	Self-organization, self-productivity, evolutionary capacity. Imperative: Induce, stimulate, provoke, make use of instability! ("positivization of instability")
Mechanism/ algorithm/rules/ causation	Multi-linearity, sometimes very complicated; strong causation	Non-linearity, complexity, sensitivity, bifurcations, randomness, noise; weak causation or stochastic processes
Environment/ tech-world relation	Functional simplification, causal isolation: boundaries and interfaces between the technical, the natural and the human sphere	Open interaction, processuality, time-dependency, activity, interactivity, learning: loss of clear boundaries and interfaces between the technical, the natural and the human sphere; no clear agency
Human perception/ experience of technology	Anticipatable, predictable, reproducible, describable, explainable, comprehensible, controllable, shapable (in principle)	Non-predictable, unexpectable, non-reproducible, autonomous, own decision making, real-world experiments, loss of control
Problems/ challenges	Classic technology assessment (TA)/ societal shaping challenges, Collingridge dilemma	Shaping challenges: limits of design/creation/construction and control: Is TA possible at all?
Examples/cases	All traditional/classic and modern technical systems, including high technology such as nuclear energy	Synthetic/systems biology, biofacts, evolutionary robotics, smart environments, autonomous stock agents, nano-assemblers

tradition and a basic anthropological characteristic of humans—seems to be ascribable to these systems.[32] Widespread notions include hybrids, cyborgs, chimers, biofacts, and technical agency. A symmetric anthropology between the human sphere and the technical domains seems to be emerging. Thus, the technoscientific development towards a late-modern technology is inducing an entanglement, and moreover a *phenomenological convergence,* of the human, societal, technical, and natural spheres.[33]

A phenomenological approach of this kind is, indeed, necessary and indispensable for characterizing the "techno-logical conditions" of societal life-forms in *TechnoScienceSocieties* today (see Hörl 2011). However, that is not everything an analysis should encompass. We should go further and address the background of the phenomenological convergence. Let us examine the claims made by advocates of the systems and self-organization paradigm. The underlying basis for phenomenological convergence is what could be called *nomological* or *structural convergence.* Structures that enable self-organization in technical systems are said to converge with those in nature or with those in the social and human sphere. It is assumed that, although the objects under consideration differ, the underlying structures and basic principles governing their dynamical behavior and the perceivable phenomenology show a certain kind of structural similarity. Looking back to the history of science and technology, the ideal of nomological convergence is, to some extent, already present in the cyberneticist's technomorphic view of nature. Norbert Wiener (1968), for instance, argues that there are "fundamental structures that can be applied to and found in machines and, analogously, in metabolic systems". But the cyberneticists depict nature, mistakenly, in a more or less static way.[34] In contrast to the classic cybernetic approach, the crucial point in late-modern technology is the non-static capability of self-organization to which a technical productivity is ascribed.[35] Some aspects of this ideal of technology can be traced back to Ernst Bloch's "nature alliance technology" (see Bloch 1995).[36]

---

[32] See an in-depth analysis of the notion of "autonomy" in technical systems in Christaller and Wehner (2003) and Rammert and Schulz-Schäfer (2002).

[33] This type of convergence is an ongoing topic in fields such as *Science and Technology Studies* (STS), *Sociology of Science* and *Technology Assessment.* Notions such as hybrids, agency, and the like are very common and have been discussed since the 1980s (for example, see Hackett et al. 2008; Maasen et al. 2012).

[34] And, furthermore, they regarded nature from a technomorphic perspective.

[35] These two related types of convergence differ from the type of convergence we find in the realm of modern technology. Modern technology mainly refers to physics, including the molecular-atomist view of the world. In consequence, within the microcosm different kinds of technical systems and different engineering disciplines are regarded as being similar—they are said to converge. This kind of reductionist convergence has been called *technological reductionism* (Schmidt 2004).

[36] Bloch refers to Schelling's view of nature and his concept of *natura naturans,* an ideal which explicitly traces back to Spinoza; it was coined in the Middle Ages and can be found in Aristotle's natural philosophy.

The nomological-structural convergence that constitutes the core of late-modern technology has been facilitated by advancements in the so called *systems and structural sciences* (Weizsäcker 1974): Following the first wave of developments in systems and structural sciences, such as cybernetics, game theory, and information theory, the second wave of advancements can be viewed as the most relevant for the aforementioned convergence: nonlinear dynamics, complexity theories, chaos theory, catastrophe theory, synergetics, fractal geometry, dissipative structures, autopoiesis theory, and the like (Schmidt 2011a). The later concepts can be understood as self-organization theories conceptualizing nature and other things, such as technical objects, in a processual, evolutionary, emergent manner, whereas the initial concepts put forward a more or less static view of nature. Although the two waves of developments differ in terms of the characteristics of their objects of study and therefore pursue different paradigms, they have much in common. Both combine a high degree of mathematization with theory development and with the need for computer simulations, and they both emphasize application orientation. On account of their dual orientation—in classic terms: theory as well as application—they represent a prominent example of *technoscience*. In this vein the US National Science Foundation stresses that the systems and structural concepts of self-organization enable a unification of science and engineering (Roco and Bainbridge 2002: 84).

What does "self-organization" mean? Although the concept of self-organization has been in flux since Kant and Schelling,[37] the term itself has retained its central meaning. Self-organization refers to the creation and construction, and to the perception and experience, of

1. *novelty:* the emergence and appearance of novel systemic properties—e.g., new entities, patterns, structures, functionalities, or capacities.

   Beyond the ongoing philosophical dispute on the notion and characteristics of novelty, the following further criteria are commonly used to specify "self-organization":
2. *processuality and dynamics*, in particular, time-dependency and evolutionary capacity;
3. *irreducibility* of the description and *limits of describability* in a very condensed form of a mathematical equation;
4. *non-separability* and *non-decomposability*;
5. *unpredictability* of the self-organizing phenomena.

A major consequence is that "[t]he engineers of the future will be the ones who know that they are successful when they are surprised by their own creations" (Dupuy and Grinbaum 2006: 289). Thus, although the notion of self-organization can be deemed to be based on ontological aspects, it is mainly epistemological because it refers to knowledge.[38] Even "novelty" is epistemological in the sense that

[37] For an in-depth treatment, see Schmidt (2011a, 2015a, b)

[38] Of major relevance is that this includes non-knowledge (see Böschen et al. 2004).

it is relational and relative to the knowledge someone has. Bearing that in mind, "late-modern technology" can certainly be regarded as a reflexive term.

In addition to the epistemological aspects described above, there is also a central ontological constituent which enables and provokes self-organization: namely, instabilities.[39] Philosophically speaking, instabilities can be seen as the condition for the possibility of self-organization. According to Ebeling and Feistel (1994), "self-organization is induced by instability of the 'old' structure through small fluctuations. This is why studying instability is of major importance." Nicolis and Prigogine (1977) argue that "instabilities [are] necessary conditions for self-organization." Instabilities can generally be thought of as situations in which a system is on a razor's edge: criticalities, flip or turning points, thresholds, watersheds. They generate novelty and general phase transitions.[40] But such systems also show epistemological limitations.[41] Insofar as technoscientists aim to harness self-organization power, they have to provoke and stimulate instabilities: self-organization requires that a system's dynamics pass through unstable situations. To put it metaphorically: Late-modern technology can be seen as the technoscientific attempt to stimulate and initiate a dance on the razor's edge.

3.4 Two Sides of a Coin

From a societal perspective, late-modern (instability-based, self-organizing) technology is ambivalent as it carries an internal conflict or considerable dialectic that cannot be overcome by minor modifications of the technical system itself.[42] The internal ambivalence of late-modern technology stems from insights gained in systems and structure sciences.

First, instabilities constitute the core of self-organization and, hence, of technically relevant self-organization and self-productivity. Functionalities and properties based on self-organization are desirable from a certain engineering point of view, but the drawback is that instabilities induce the limitations described above that undermine any control oriented "rational engineering".[43] When instabilities are

---

[39] See, for instance, Schmidt (2008, 2011a, 2015a, b).

[40] These processes are not governed by the (modern) *strong* type of causation. Rather, it is the (late-modern) *weak* type of causation that enables feedback procedures and amplification processes. Instabilities can induce random-like behavior, deterministic chance, and law-based noise—which are inherently linked to uncertainty. In general, instability should not be equated with the collapse of a system.

[41] The epistemological aspects are: predictability, reproducibility, testability, and reductive describability/explainability. The latter are linked with the central five characteristics of self-organization listed above.

[42] The kind of ambivalence diagnosed here could add a further (ontological) dimension to Zygmunt Bauman's (1992) general ambivalence diagnosis.

[43] The notion of *rational engineering* is exemplified by Endy (2005). It encompasses: standardization, decoupling, and abstraction. See, from both a critical and empirical perspective, the discussion in Giese et al. (2013).

present, tiny details are of major relevance because they are amplified during time evolution. Minor changes in some circumstances can cause tremendous, unforeseeable effects. These tiny details are technically and sociotechnically hard to control, due to empirical-practical and fundamental principle-related uncertainties. Buchli and Santini argue in a similar tenor that "there is a trade-off between self-organization on one hand and specification or controllability on the other: if you increase the control over your system you will suppress self-organization capabilities" (Buchli and Santini 2005: 3; see Schmidt 2010).[44] *Second,* because instabilities represent the core of self-organization, it is unclear where the dividing line between the technical system under consideration and its environment is to be drawn. The system can hardly be separated or isolated from its environment. It is inherently entangled with the natural, human, societal, political sphere. A clear input-output mechanism—as in modern technical systems—cannot be identified.[45]

Reasonable concerns can, therefore, be raised as to whether, for example, "synthetic biology will enable the design of 'biological systems' in a rational way" (EC 2005). The same also holds for other areas of technosciences, such as nanotechnology or artificial intelligence (AI). Due to the above discussed epistemological aspects, technology on the one hand and instability-based self-organization on the other hand have traditionally been regarded as being like fire and ice. According to the modern view of technology, instabilities ought to be excluded from technical systems. Whenever instabilities occurred, the traditional objective was to eliminate them. Industrial and post-industrial productivity, requiring a clear technosystem/environment interface with a controllable and regular input-output relation, seemed only feasible if stability could be guaranteed. According to the traditional, modern view, technology was (implicitly) equated with and defined by stability: technology was seen as a stable or stabilized causal system that, by its very structure and form, enables reproducibility and predictability. If technoscientific design today aims to harness instability-based self-organization, then this would, as Peter Kroes (2009: 539) argues, "indeed constitute a significant break with the traditional design paradigm [… and] with its ideal of total design control."[46]

It is still common to restrict the notion of technology to the modern type of technical systems. Even such a profound thinker as Niklas Luhmann draws only on "modern" technology when addressing the nature of technical systems in their very core.[47] At the same time, Luhmann's argumentation is crucial to facilitating a clearer

[44] Quoted in Kroes (2009: 538 f.).

[45] These kinds of technical systems are not "trivial" machines, as Heinz von Foerster argues in another context. How they can be characterized as "non-trivial" machines remains an open point (see Kaminski and Gelhard 2014).

[46] See for similar thoughts, including an analysis of what has been called "rational engineering": Giese et al. (2013).

[47] Luhmann does not frame "technical systems" from mainly a social or societal perspective, but by aspects of causation, isolation, functionality, and others. In general, this approach is very fundamental and thought provoking in addressing and analyzing the ontological core of technical systems and its ambivalences.

distinction between the two types of technology. Luhmann would no doubt have been skeptical whether a late-modern type of technology is even feasible, and also whether, if that were the case, it should be called "technology". Based on his understanding of technology as "functional simplification" or "simplifying isolation", implying the design, construction, and control of causal flows across the technosystem/environment interface, Luhmann anticipates problems that can be regarded as inherent *to* and characteristic *of* late-modern technology. In fact, Luhmann develops his thinking by referring *not* to late-modern technology but to modern "high technology", e.g., nuclear energy. The associated problems are, however, the same. Luhmann argues, we are confronted with "the chaos problem, the interference problem and the problem of singular, random-like events." Here, he draws on features that are induced by instabilities (Luhmann 2003: 100). According to Luhmann, the problems originate from the "form of the technical systems itself that constitutes the border [= "interface", J.C.S.] between the included and excluded causalities." The "form" of high technical systems, as well as of late-modern technical systems, renders it impossible to keep these two kinds of causalities apart or to control them. The non-existence of such a clear border or well-defined interface is a consequence of instabilities. A controllable separation of the included and excluded causalities across the interface between the internal and external side of the technical system cannot be guaranteed.

Referring to Luhmann, one might be tempted to argue that there is in principle no difference between what Luhmann calls "high technology" on the one hand and what has been named late-modern technology in this paper on the other. But this would be misleading, because in late-modern technical systems instabilities and self-organizing properties are intentionally built in and embodied, whereas in modern high technology instabilities occur accidentally. Regarding the latter, the aim is to reduce or eliminate them. Instabilities and instability-based self-organization are therefore viewed in different ways. Those who advocate late-modern technology see instabilities in a positive light and follow a technoscientific paradigm that puts forward a *positivization of instabilities*. In the realm of the modern account of technology, as in nuclear energy, by contrast, instabilities are highly undesired. Luhmann adheres to the modern ideal of technology as a stable system. He would question whether late-modern technology can conclusively be called "technology" or, more fundamentally, whether a late-modern technology is "*as a technical system* technically possible at all." (ibid.) Generally speaking, "modern" technology as termed in this paper is therefore in alignment with Luhmann's concept of technology, and in particular with his conceptual framework based on "functional simplification" and "causal isolation". Such restrictive understanding of technology seems to be changing today. But Luhmann's reservations and concerns can, from a present-day perspective, be reframed as ambivalence rooted in technically produced and implemented instabilities.

Hans Jonas (1985) anticipated such ambivalence a few years before Luhmann. Contrary to Luhmann, Jonas does not exclude instability-based self-organization from the concept of "technology" in advance. His approach enables him to perceive central characteristics of what he calls "engineering biology". He diagnoses a historically new technoscientific era and draws a dividing line between the classic engineering type of technology ("engineering art") and the new type. Jonas' distinction concurs with the theme of this paper—the distinction between late-modern and modern technology. As Jonas argues, the new type of technology differs in a qualitative way from our established understanding of what technology is, could or ought to be. "In th[is latter ...] case of dead substances, the constructor is the one and only actor with respect to a passive material [= modern technology]. [In contrast, in the case of the] biological organism, activity meets activity: biological technology is collaborative with the self-activity of an active 'material'." (Jonas 1985: 165) Jonas lists some characteristics of this new type of technology: *first*, collaborativeness, including self-activity and a certain kind of autonomy of the technical system; *second*, complexity, irreversibility, historicity and the capacity of evolution that limits the horizon of prediction and foresight; *third*, individuality and non-experimentability, which obstructs reproducibility and planned producibility; and, *fourth*, which Jonas underlines, "interactive causation" as a different kind of causality. However, Jonas does not take his clear analysis any further—and he does not dig deeper to find the root and source of such phenomena. Based on the argumentation developed in this paper, it is not primarily and solely the organismic criteria that constitute the central difference, but instability-based self-organization. Jonas' philosophy of technology is much more fundamental than Jonas himself seems to have assumed. An explication and exploration of the points said about late-modern technology could substantiate Jonas' formulation of his *imperative of responsibility*, including the precautionary principle and the heuristics of fear (Jonas 1984).

3.5 Shifting Relations

The recent wave of emerging technologies urges us to rethink our thinking about technology. In the age of *TechnoScienceSociety* the relations between science, technology, humans, and society are shifting. The shift has been induced by, amongst other factors, the recent development of technosciences that has been largely triggered by the systems and structural sciences, including systems engineering.[48] In analyzing this recent development, we have coined a provisional search term: *late-modern technology*, which—from an epistemological perspective—aims to reflect the shifting relations, in particular those that are linked with knowledge, control and power.[49] In conclusion, let us highlight the central characteristics of late-modern technology[50]:

[48] See, f.i., Sage (1992).

[49] As underlined before, late-modern technologies are necessarily based on technosciences, but the reverse is not true.

[50] Most of these aspects are sketched in Table 3.1.

First, ontology: The guiding idea behind the development and design of late-modern technology is to harness self-organization—including self-productivity, self-adaptation, self-repair, and more generally, self-activity, autonomy, and the like—for technological purposes. As instabilities constitute the necessary condition for enabling self-organization, they have to be guaranteed by the design and construction of the technical system. Since instabilities are highly valued as the source of self-organization—contrary to the devaluation of instabilities in the realm of modern technology—, they are intentionally constructed and built into the ontological structure of technical systems. At this onto-technological core we find a certain kind of convergence of nature, technical systems and human action, namely a structural-nomological convergence.

Second, appearance and perception: Late-modern technical systems challenge our established perception of technology, as these systems appear to be un-technical and non-artificial. Phenomenologically, they show emergent features, autonomous behavior, and agency properties—everything that has traditionally been associated more with human action or, to some extent, with natural phenomena than with technology. In the late-modern regime, a technical system appears dynamical, processual, and evolutionary. Technology is associated with activity and, even more, with a "coming into being". Traditional technical signs and signals, tracks and traces are no longer visible and perceivable. Thus, a kind of phenomenological convergence of nature, technical systems, and human action seems to be emerging in the late-modern regime of technology.

Third, interface: The next two aspects reflect the shifting relations between humans and technical systems to an even greater extent than the first two. Because of instability-based self-organization—which enables features such as novelty and emergence, adaptivity and flexibility, sensitivity and learning—late-modern technical systems can hardly be separated from their environment and from the context of application. An interface between the inner and outer side is no longer given by the form and structure of the technical system. Consequently, the material-artefactual core of technology is no longer excludable and isolatable from the social, natural and human environment, and vice versa. Late-modern technical systems are, therefore, always socio-technical. In light of this entanglement, the acquisition of knowledge about a technical system also needs to encompass detail knowledge about the socio technical sphere in which a certain system is placed.

Fourth, control: Besides the non-existence of an interface and the lack of separability, late-modern technology shows further epistemological and methodological limitations that are inherently linked to the capability of self-organization and to the provocation of instabilities. That is to say, there is a trade-off between self-organization on the one hand, and predictability, reproducibility, describability, transparency, and controllability—in other words: rational engineering—on the other hand. Too much control will eliminate self-organization capabilities, whereas allowing too much instability and self-organization will reduce the possibility of control, and of prediction, reproduction and explanation. Considering these epistemological limitations, the move towards late-modern technology will constitute an epochal shift from the traditional ideal of rational design and rational engineering towards a novel, yet not fully captured and established design paradigm (see Kroes 2009: 539; Giese et al. 2013). The new design paradigm placing self-organization at its very center seems to be characteristic for the recent wave of technosciences and, thus, for *TechnoScienceSocieties* in general.

New technical systems, devices, things and objects that are based on instabilities and show self-organizing capabilities are beginning to broadly and deeply populate the life-world of our *TechnoScienceSocieties*. Although one can argue with Donna Haraway (2003: 43) that *TechnoScience* is inherently linked to "worldly, materialized, signifying and significant power," such a claim merely seems to hold with regard to the rhetoric, narratives and visions. The power rhetoric is generally not based on sound arguments. A problematic ambivalence or even dialectic of power is emerging in *TechnoScienceSocieties*, embodied in a novel type of technical system. It seems the more *TechnoScienceSocieties* aim to control the technologized world, the more they lose their ability to control it. Contrary to the visionary aims and claims, it appears that late-modern technology does not empower humans any further in *TechnoScienceSocieties*.

References

Aldridge, I. 2013. *High-Frequency Trading. A Practical Guide to Algorithmic Strategies and Trading Systems*. 2nd ed. Weinheim/New York: Wiley.

Bacon, F. 2004. In *The Instauration Magna Part II: Novum Organon and Associated Texts* (1620), ed. Graham Rees and Maria Wakely. New York: Clarendon Press.

Baecker, D. 2011. Technik und Entscheidung. In *Die technologische Bedingung. Beiträge zur Beschreibung der technischen Welt*, ed. E. Hörl, 179–192. Berlin: Suhrkamp.

Barthélémy, J.-H. 2011. Simondon – ein Denken der Technik im Dialog mit der Kybernetik. In *Die technologische Bedingung. Beiträge zur Beschreibung der technischen Welt*, ed. E. Hörl, 93–109. Berlin: Suhrkamp.

Bauman, Z. 1992. Moderne und Ambivalenz. In *Das Ende der Eindeutigkeit* (1991). Hamburg: Junius.

Beck, U., A. Giddens, and S. Lash. 1996. *Reflexive Modernisierung. Eine Kontroverse*. Frankfurt: Suhrkamp.

Bloch, E. 1995. *The Principle of Hope*. Cambridge: MIT Press.

Böschen, S., M. Schneider, and A. Lerf (eds.). 2004. *Handeln trotz Nichtwissen. Vom Umgang mit Chaos und Risiko in Politik, Industrie und Wissenschaft*. Frankfurt: Campus.

Buchli, J., and C.C. Santini. 2005. *Complexity Engineering: Harnessing Emergent Phenomena as Opportunities for Engineering*. Santa Fe: Santa Fe Institute.

Christaller, T., and J. Wehner (eds.). 2003. *Autonome Maschinen*. Wiesbaden: Westdeutscher Verlag.

Drexler, K.E. 1990. *Engines of Creation: The Coming Era of Nanotechnology*. Oxford: Oxford University Press.

Drubin, D.A., J.C. Way, and P.A. Silver. 2007. Designing Biological Systems. *Genes & Development* 21: 242–254.

Dupuy, J.P. 2004. Complexity and Uncertainty. A Prudential Approach to Nanotechnology, *European Commission*. www.europa.eu.int/comm/health/ph_risk/documents/ev_20040301_en.pdf. Accessed 21 May 2015.

Dupuy, J.P., and A. Grinbaum. 2006. Living with Uncertainty—Toward the Ongoing Normative Assessment of Nanotechnology. In *Nanotechnology Challenges: Implications for Philosophy, Ethics and Society*, ed. J. Schummer and D. Baird, 287–314. Singapore: World Scientific.

Ebeling, W., and R. Feistel. 1994. *Chaos und Kosmos. Prinzipien der Evolution*. Heidelberg: Spektrum Akademischer Verlag.

EC (European Commission). 2005. Synthetic Biology. Applying Engineering to Biology. Report of a NEST-High Level Expert Group EU 21796; Brussels.

Endy, D. 2005. Foundations for Engineering Biology. *Nature* 438: 449–453.

ETC (Action Group on Erosion, Technology and Concentration). 2007. *Extreme Genetic Engineering. An Introduction to Synthetic Biology*. http://www.etcgroup.org/en/materials/publications.html?pub_id=602. Accessed 21 May 2015.

Gibson, D.G., J.I. Glass, C. Lartigue, V.N. Noskov, R.-Y. Chuang, M.A. Algire, G.A. Benders, M.G. Montague, L. Ma, M.M. Moodie, C. Merryman, S. Vashee, R. Krishnakumar, N. Assad-Garcia, C. Andrews-Pfannkoch, E.A. Denisova, L. Young, Z.-Q. Qi, T.H. Segall-Shapiro, C.H. Calvey, P.P. Parmar, C.A. Hutchison, H.O. Smith, and J.C. Venter. 2010. Creation of a Bacterial Cell Controlled by a Chemically Synthesized Genome. *Science* 329 (5987): 52–56.

Giese, B., S. Koenigstein, H. Wigger, J.C. Schmidt, and A. von Gleich. 2013. Rational Engineering Principles in Synthetic Biology: A Framework for Quantitative Analysis and an Initial Assessment. *Biological Theory* 8 (4): 324–333.

Gomber, P., B. Arndt, M. Lutat, and T. E. Uhle. 2011. *High Frequency Trading* (White Paper). Frankfurt: Frankfurt University.

Hackett, E.J., O. Amsterdamska, M. Lynch, and J. Wajcman (eds.). 2008. *The Handbook of Science and Technology Studies*. 3rd ed. Cambridge: MIT Press.

Haraway, D. 2003. Modest_Witness @ Second_Millennium. In *The Social Shaping of Technology*, ed. D. MacKenzie and J. Wajcman, 41–49. Berkshire: Open University Press.

Hörl, E. (ed.) 2011. *Die technologische Bedingung. Beiträge zur Beschreibung der technischen Welt*. Berlin: Suhrkamp.

Hottois, G. 1984. *Le signe et la technique*. Paris: Aubier.

Hubig, C. 2007. *Die Kunst des Möglichen II. Ethik der Technik als provisorische Moral*. Bielefeld: Transcript.

Hubig, C., and S. Harrach. 2014. Transklassische Technik und Autonomie. In *Zur Philosophie informeller Technisierung*, ed. A. Kaminski and A. Gelhard, 41–57. Darmstadt: WGB.

Ihde, D., and E. Selinger (eds.). 2003. *Chasing Technoscience. Matrix for Materiality*. Bloomington: Indiana University Press.

Jonas, H. 1984. *The Imperative of Responsibility. In Search of an Ethics for the Technological Age*. Chicago: University of Chicago Press.

———. 1985. Laßt uns einen Menschen klonieren: Von der Eugenik zur Gentechnologie. In *Technik, Medizin und Ethik*, ed. H. Jonas, 162–203. Frankfurt: Suhrkamp.

Jones, R. 2004. *Soft Machines*. Oxford: Oxford University Press.

Kaminski, A., and A. Gelhard (eds.). 2014. *Zur Philosophie informeller Technisierung*. Darmstadt: WGB.

Karafyllis, N.C. (ed.) 2003. *Biofakte*. Paderborn: Mentis.

———. 2007. Growth of Biofacts: The Real Thing or Metaphor? In *Tensions and Convergences. Technological and Aesthetic (Trans-) Formations of Society*, ed. R. Heil, A. Kaminski, M. Stippack, A. Unger, and M. Ziegler, 141–152. Bielefeld: Transcript.

Kitano, H. 2002. Looking Beyond the Details: A Rise in System-Oriented Approaches in Genetics and Molecular Biology. *Current Genetics* 41: 1–10.

Kroes, P. 2009. Foundational Issues of Engineering Design. In *Philosophy of Technology and Engineering Sciences*, ed. A. Meijers, 513–541. Burlington/Amsterdam: Elsevier.

Latour, B. 1987. *Science in Action*. Cambridge, MA: Harvard University Press.

Luhmann, N. 2003. *Soziologie des Risikos* (1991). Berlin/New York: De Gruyter.

Maasen, S., and S. Duttweiler. 2012. Neue Subjekte, neue Sozialitäten, neue Gesellschaften. In *Handbuch Wissenschaftssoziologie*, ed. S. Maasen, M. Kaiser, M. Reinhart, and B. Sutter, 417–428. Wiesbaden: Springer VS.
Maasen, S., and B. Sutter (eds.). 2007. *On Willing Selves. Neoliberal Politics vis-à-vis the Neuroscientific Challenge*. Houndmills: Palgrave Macmillan.
Maasen, S., M. Kaiser, M. Reinhart, and B. Sutter (eds.). 2012. *Handbuch Wissenschaftssoziologie*. Wiesbaden: Springer VS.
Nicolis, G., and I. Prigogine. 1977. Self-Organization in Nonequilibrium Systems. In *From Dissipative Structures to Order through Fluctuations*. New York/London: Wiley.
Nolfi, S., and D. Floreano. 2000. *Evolutionary Robotics: The Biology, Intelligence, and Technology of Self-Organizing Machines*. Cambridge, MA: MIT Press.
Nordmann, A. 2008a. Technology Naturalized. A Challenge to Design for the Human Scale. In *Philosophy and Design*, ed. P.E. Vermaas, P. Kroes, A. Light, and S.A. Moore, 173–184. Heidelberg/New York: Springer.
———. 2008b. Philosophy of Nanotechnoscience. In *Nanotechnology, Vol. 1: Principles and Fundamentals*, ed. G. Schmid, 217–244. Weinheim: Wiley.
Nordmann, A., H. Radder, and G. Schiemann (eds.). 2011. *Science Transformed? Debating Claims of an Epochal Break*. Pittsburgh: University of Pittsburgh Press.
NSTC (National Science and Technology Council). 1999. *Nanotechnology: Shaping the World Atom by Atom*. Washington, DC: NSTC.
Pollack, J. 2002. Breaking the Limits on Design Complexity in M.C. In *Converging Technologies for Improving Human Performance*, ed. M.C. Roco and W.S. Bainbridge, 161–163. Arlington: National Science Foundation.
Pottage, A., and B. Sherman. 2007. Organisms and Manufactures: On the History of Plant Inventions. *Melbourne University Law Review* 31: 539–568.
Rammert, W., and I. Schulz-Schäfer (eds.). 2002. *Können Maschinen handeln? Soziologische Beiträge zum Verhältnis von Mensch und Technik*. New York/Frankfurt: Campus.
Roco, M.C., and W.S. Bainbridge (eds.). 2002. *Converging Technologies for Improving Human Performance*. Arlington: National Science Foundation.
Ropohl, G. 1999. *Allgemeine Technologie. Eine Systemtheorie der Technik*. Karlsruhe: Karlsruher Universitätsverlag.
Rouse, J. 1987. *Knowledge and Power. Toward a Political Philosophy of Science*. Ithaca: Cornell University Press.
Sage, A.P. 1992. *Systems Engineering*. New York: Wiley.
Schmidt, J.C. 2004. Unbounded Technologies: Working Through the Technological Reductionism of Nanotechnology. In *Discovering the Nanoscale*, ed. D. Baird, A. Nordmann, and J. Schummer, 35–50. Amsterdam: IOS Press.
———. 2008. *Instabilität in Natur und Wissenschaft*. Berlin: De Gruyter.
———. 2010. Emergence. In *Encyclopedia of Nanoscience and Society*, ed. D. Guston, 180–184. London: Sage Books.
———. 2011a. Challenged by Instability and Complexity. On the Methodological Discussion of Mathematical Models in Nonlinear Sciences and Complexity Theory. In *Philosophy of Complex Systems*, Series Philosophy of Sciences, ed. C. Hooker, 223–254. Amsterdam: Elsevier.
———. 2011b. Toward an Epistemology of Nano-Technosciences. *Poiesis & Praxis* 8 (2): 103–124.
———. 2012. Selbstorganisation als Kern der Synthetischen Biologie. Ein Beitrag zur Prospektiven Technikfolgenabschätzung. *Technikfolgenabschätzung. Theorie und Praxis* 21 (2): 29–35.
———. 2015a. *Das Andere der Natur*. Stuttgart: Hirzel.
———. 2015b. Synthetic Biology as Late-Modern Technology. Inquiring into the Rhetoric and Reality of a New Technoscientific Wave. In *Synthetic Biology. Character and Impact*, ed. B. Giese, C. Pade, H. Wigger, and A. von Gleich, 1–30. Heidelberg/New York: Springer.
Schwille, P. 2011. Bottom-Up Synthetic Biology: Engineering in a Tinkerer's World. *Science* 333: 1252–1254.

Simondon, G. 2012. *Die Existenzweise technischer Objekte*. Zürich: Diaphanes.

Suchman, L., and J. Weber. 2015. Human-Machine Autonomies. In *Autonomous Weapon Systems. Law, Ethics, Policy*, ed. N. Buta, C. Kress, S. Beck, R. Geiss, and H.-Y. Liu. Cambridge: Cambridge University Press (in press).

TESSY (Towards a European Strategy for Synthetic Biology). 2008. *Information Leaflet: Synthetic Biology in Europe*. http://www.tessy-europe.eu/public_docs/SyntheticBiology_TESSY-Information-Leaflet.pdf. Accessed 23 May 2013.

Venter, C. 2010. 'The Creation of "Synthia" – Synthetic Life', *Science Interview* May 23, 2010, the Naked Scientists. http://www.thenakedscientists.com/HTML/content/interviews/interview/1332/. Accessed 20 Jan 2016.

Weber, J. 2003. *Umkämpfte Bedeutungen. Naturkonzepte im Zeitalter der Technoscience*. Frankfurt am Main: Campus.

Weizsäcker, C.F.v. 1974. *Die Einheit der Natur*. München: dtv.

Westerhoff, H.V., and B.O. Palsson. 2004. The Evolution of Molecular Biology into Systems Biology. *Nature Biotechnology* 22 (10): 1249–1252.

Wiener, N. 1968. *Kybernetik. Regelung und Nachrichtenübertragung in Lebewesen und Maschine*. Reinbek bei Hamburg: Rowohlt.

Jan Cornelius Schmidt is a Physicist and a Philosopher. He received a Ph.D. in Theoretical Physics and a Habilitation in Philosophy. He was an Assistant Professor at the Institute of Physics, University of Mainz (1996–1999), as well as at the Institute of Philosophy, Darmstadt University of Technology (1999–2006), and was Associate Professor for Philosophy of Technology at Georgia Tech (2006–2008). Since 2008, he has been Professor of Philosophy of Science and Technology at Darmstadt University of Applied Sciences. He was a Visiting Professor at the University of Jena (2011–2012) and invited Guest Professor at the University of Klagenfurt (2015) and at the University of Natural Resources and Life Sciences (Boku), Vienna (2016).

Chapter 4
The Development of Research Methods as Driving Force of Technoscience

Carsten Reinhardt

4.1 Introduction

In mid-twentieth century, novel kinds of instrumentation, often based on physics and electronics, quickly became routine in the physical and life sciences, as well as in industrial manufacturing and medical diagnosis.[1] Scientific research became inextricably linked to the deployment of these technical tools and the development of related methods and techniques. In this sense, I speak of technoscience as science being undertaken by the means of high-tech instrumentation. On this basis, I will argue, a novel field of scientists focusing on the development of research methods came into being, connecting instrument manufacturers, academic disciplines, and governmental science funding agencies. I will call these scientists 'method makers' and claim that the making of research methods reconfigured the relations of academic science with instrument technology and governmental science policy. Furthermore, the commercialization of instruments and methods in science, medicine, and industry contributed to an expanded reach of science's impact on society. At this level, technical and medical ends of science come into play. Thus, methods making leads to societal impact, both directly and indirectly, and the broader issues of TechnoScienceSociety are addressed. In this paper, however, I will focus on the scientific field itself, and the changes in its regulation caused by the advent of the method makers.[2]

[1] This paper draws heavily on, and is partially translated from, work first published in Reinhardt (2012). An earlier English version has appeared in the proceedings of the International Workshop on the History of Chemistry 2015 (Reinhardt 2016).

[2] It is important to distinguish between instruments and the methods or techniques associated with them for their use. Sometimes, innovation of instruments and research into methods go hand in

C. Reinhardt (✉)
University of Bielefeld, Bielefeld, Germany
e-mail: carsten.reinhardt@uni-bielefeld.de

S. Maasen et al. (eds.), *TechnoScienceSociety*, Sociology of the Sciences Yearbook 30, https://doi.org/10.1007/978-3-030-43965-1_4

Three general relationships are best suited to characterize the field of the method makers. The first relationship concerns their standing inside academic disciplines and research fields. The basis for this is the research practice, most notably the influence of high-tech instrumentation on experiments and data management, and the adaptation of instruments to established concepts and theories. The second is the academic-industrial partnership, most importantly that of academic scientists with industrial instrument manufacturers. The third applies to the organization of science funding, and the special relations of method makers with the relevant governmental agencies. In historiographical and sociological research, these three layers, or strands, are often disconnected. For the first, it is the new experimentalism and laboratory studies; for the second studies of the so-called science-based industry; for the third works about policy and governance. Some notions, such as triple helix, and mode 2, are designed to connect these layers (Leydesdorff and Etzkowitz 1996; Nowotny et al. 2003). The historian and sociologist of science Terry Shinn connects them with his notion of a transversal regime (Shinn 2006; Shinn and Ragouet 2005). In building on Shinn's work, my paper aims to identify and to characterize social and epistemic structures that allow us to better understand these connections and their working modes. In focusing on the United States in the period between the 1950s and 1980s, I aim at the macro-level, and especially the regulation of method-based technoscience, and not so much on its laboratory practice.

Although, as I will argue, the making of methods is a crucial and widespread activity in twentieth-century science, it is not well represented in its historiography.[3] An important exception is, of course, the notion of technoscience itself, as brought forward by, among others, Bruno Latour (1987). We may well be blinded by one of the myths of science, according to which it is the solution of problems that counts as the true, and sometimes only, activity of scientists. According to this view, the development of methods is just an intermediate step in the process of scientific research. By and large, these intermediate steps are forgotten, or suppressed, in both scientific and historiographical accounts. Still today the development of methods is marginalized in science. A case in point is the editorial of a new journal, *Nature Methods*, in 2004. Titled "Methods for Methods' Sake", the editors lament the neglect of the coverage of methods development in 'regular' journals (Anonymous 2004). In stark contrast to this neglect, the impact of methods development on science is huge: method makers work with and on all kinds of research instrumentation, ranging from big science to table-top instruments. They interact with a multitude of groups of scientists in different disciplines, and they have close and symbiotic cooperations with industrial instrument manufacturers. Many of their activities depend on the standardization and commercialization of research instrumentation that can be used by large groups of scientists in academia, government, and industry.

---

hand, but often they are separated activities, and executed by different communities of specialists.

[3] But see Nevers et al. (2001).

4.2 Isolation, Identification, Imaging, Interpretation, and Intervention

In order to establish my case, I wish to briefly address a conglomerate of research methods that increasingly gained momentum in twentieth-century science. Many uses of instruments in physics, biology, chemistry, medicine and engineering can be differentiated into methods of isolation, identification, interpretation, and intervention. In addition, I wish to single out imaging as a special case of identification. The first class, or family, of instruments serves the separation and purification of materials. Noteworthy are the chromatographic techniques, but also electrophoresis and the ultracentrifuge (Chiang 2009; Creager 2002; Gerontas 2012; Widmalm 2006). Next come the detecting techniques, most notably the spectroscopies (Hentschel 2002; Reinhardt 2006b). Of course, isolation normally entails identification, and spectroscopy works without isolation in many cases. However, often these two classes are not competing, but coupled and work in tandem. Imaging techniques got more and more independent from other identification methods while they expanded their reach tremendously. Among them are diverse kinds of microscopy, including electron microscopy, and imaging techniques based on x-ray and magnetic resonance in medicine (Joyce 2006; Kevles 1997). The fourth class is that of interpretation, and I refer to the computer, and algorithms. Most clearly this is the case with the applications of artificial intelligence (November 2012; Reinhardt 2006b), but there are many more cases. The fifth, and arguably youngest, group is that of intervention, for example applications of scanning tunneling microscopy in the nascent nanotechnology (Mody 2011).

Access to all, or at least most, of these functions is of crucial importance for any researcher in the physical or life sciences, and often these methods establish a large part of the identity in a discipline (Nye 1994). On the epistemic level, changes in the five classes, or families, of research methods contributed to the expansion, and partial transformation, of the relevant scientific disciplines during the twentieth century. The thinking in molecular structures which is so characteristic for chemistry (Steinhauser 2012), expanded into biology, physics, and later into nanotechnology, materials science and others largely on the basis of these methods. My examples come from the physical and life sciences, and relate to neighboring fields in engineering and medicine. Certainly, one could trace similar developments in many other fields, such as mathematics, the humanities, and the social sciences. An interesting case would be cybernetics with its claims of being a "super method."[4] Alongside with these epistemic changes, the emergence of a new type of scientist, that of a method-based expert, or specialist, can be traced.

[4] I am thankful to Jan-Hendrik Passoth for this hint. For the history of cybernetics, see for example Hagner and Hörl (2008).

4.3 Method Makers in the Scientific Field

'Normally,' scientists search for and apply methods that can be used to solve problems, which are in turn established by theories or applications. Method makers, in contrast, search for problems that are ready to be solved with their newest method at hand. My thesis, in a nutshell, is that in mid-twentieth century a division of labor between these two intertwined parts of scientific practice took place. Specialists for the development of instrument-based research methods complemented problem-solving researchers. Previously, I have analyzed their relations inside scientific fields with the model of expertise (Reinhardt 2011a). To further scrutinize the field of method makers, I use Pierre Bourdieu's concepts of habitus and field (Bourdieu 1998; Reinhardt 2011b). Each scientific field generates social structures, and these are governed by the distribution of symbolic capital. The command over research methods, and the ruling over their accurate and justified use, is one of the possibilities for scientific actors to change the rules of the game for the acquisition of symbolic capital. Thus, method makers have unique access to power in a scientific field, and their work can lead directly to scientific reputation. However, their claims can easily be regarded as competitive, and even subversive by the establishment (Reinhardt 2010). With regard to Bourdieu's phrase that the clients of scientists are their competitors as well (because the use of scientific knowledge is in the further development of it), we note an interesting exception: method makers are only indirectly the competitors of their clients, because they work for the use and dissemination of their methods and they intrude into problem solving only insofar as this is needed to prove the usefulness and impact of the method in question. The users or clients, moreover, have an interest in receiving ready-to-use, routine research methods and welcome an occasional intrusion with regard to high-end problem solving. This may be regarded as a win-win situation. In consequence, we will need to ask if the method makers generate their own scientific-technical field, and to what extent they are able to establish their autonomy with regard to their users. Moreover, we may ask if this represents an internal periphery, because the constitutive industrial and governmental relations—normally seen as exogenous to science—now play an endogenous role, as they are included in the field. The term 'method makers' entails fundamental questions, not least through the hint at methodology involved. In the literature, however, such questions are mostly addressed with the single scientific method in mind (Yeo and Schuster 1986). Here, we focus on a pragmatic approach, and use the term methods always in the plural, implying direct and far reaching consequences for scientific practice, interdisciplinary work, and technological innovation.

Eric von Hippel's model of user-dominated innovation describes the academic partners of instrument manufacturers as users (von Hippel 1995). However, these users were not passive, but actively engaged and stood in a symbiotic relationship with industry. Out of self-interest, method makers depended on innovation of instruments, in order to grow their scientific capital in form of reputation. The scale and scope of such relationships has been described by me with the example of the symbiosis, and sometimes competition, between the industrial scientist James Shoolery and the academic scientist John D. Roberts, featuring the example of NMR methods and organic chemistry (Reinhardt 2006a).

It is to be expected that the introduction of a whole new set of experimental methods for isolation, identification, imaging, interpretation and intervention into so many disciplines and sub-disciplines would have resulted in an overthrowing of traditional hierarchies. However, the novel methods did not completely change the power structures in the respective scientific disciplines, though they crucially impacted their course and structure, and they institutionalized new relationships between scientists. An often-used term for the relevant interactions is scientific collaboration (Maienschein 1993; Chompalov et al. 2002), and we see that in the 1960s new collaborative modes between method makers and their clients were institutionalized. An important case is that of the *facilities* or *special research resources* of the U.S. National Institutes of Health (NIH) (Reinhardt 2006c). In 1977, 52 such centers were in use in the U.S., with a range from computers to spectroscopies and imaging technologies. NIH introduced the categories of service, training, cooperation, and core research to characterize the functions of the centers. This enabled the scientists in charge of these special research resources (almost always placed at major research universities) to establish their specialty while at the same time influencing, and catering for, a scientific field. At the same time, most chemistry departments at universities expanded their analytical laboratories into service laboratories in charge of acquiring, maintaining, and developing an ever-growing array of instruments and techniques (Laszlo 2002). It is not by chance that much of the discourse related to the topic can be found under the term of research infrastructure. However, it is important to note that in the perspective taken here research infrastructure is the necessary, but not sufficient, platform for method makers' activities.

For saving resources, instruments were often shared, and this cost-saving argument played a large role in the establishment, and maintenance, of regional and national centers. At the same time, however, some of the new methods were performing so well, and so efficiently, that their capacity exceeded the abilities of a single research group to come up with interesting problems. Thus, the technical performance in data acquisition and interpretation contributed to driving the establishment of centers, and it led to more and more collaborative research projects. At the same time, the sharing of the instrument was closely connected to teaching, and training.

For all concerned, centers of various sizes constituted a win-win situation. Method makers gained access to relevant communities of users/clients in scientific disciplines. Instrument manufacturers generated a market for their instruments. Science funding agencies could point to driving scientific progress while at the same time having economic impact and acting with cost efficiency in mind. For the clientele of scientists in academia and industry, this arrangement constituted the right distance to the development of novel methods. Interestingly, we can observe an analogous development in medicine at the same time.

As Stuart Blume has analyzed, radiologists formed a similar community standing in between the clinic and the medical device manufacturers. In this case, the different size of the industry (most of the companies were multinationals), and the different economics involved led Blume to coin the phrase of the *medical-industrial complex* (Blume 1992; Timmermann and Anderson 2006; Timmermans and Berg 2003). Rosenberg, Gelijns and Dawkins emphasize the importance of the cooperation between the industrial instrument manufacturers (with their expertise in electrical engineering) and the clinicians (thus, not the researchers). In this case, the users are a crucial part of innovation, at least in the stage of clinical development. At the same time, the authors distinguish the role of NIH in the field of medical instrument technology from such techniques that involve chemical, biological and medical know-how, especially in the pharmaceutical field (Rosenberg et al. 1995). In post-World War II biomedicine, Keating and Cambrosio recognize novel structures, hugely different from the nineteenth-century origins of biomedical research. Now, the authors argue, the principal difference between the normal and the pathological has given way to gradual differences. Based on new technologies and institutions, clinicians and researchers developed screening tests and procedures that became available for large parts of a population, creating hybrid forms of therapy and research. Keating and Cambrosio's example are tests for cancer employing automatic cell counters (Keating and Cambrosio 2003).

Thus, similar patterns of methods development evolved in science and medicine at the same time. We have focused here on the roles of users and method makers in their interplay with the instrument manufacturers. In all these cases, the question of funding and governing modes requires more attention, though.

4.4 Regulation of Research Methods: The Case of the United States, 1950s to 1980s

In January 1957, members of the *Divisional Committee of Mathematical, Physical and Engineering Sciences* of the National Science Foundation (NSF) opined that the new research technologies challenged the established means of science funding:[5]

> Effective as it is, the research grants program is able to supply only the most modest of the scientists' needs for research equipment. Much can still be done with test tubes, slide rule, paper and pencil, but there is now convincing evidence that great scientific discoveries are to be expected through the development and use of the expensive new tools of scientific research. The necessary funds to provide such equipment are not now available.

But money alone was not the problem. The question was the distribution of funds. The tools of big science, such as particle accelerators, research ships and observatories, required and received special programs, just because of their size and

[5] National Science Foundation (NSF), MPE Divisional Committee, Chairman Thomas K. Sherwood to Bronk and Waterman, 21 January 1957. National Archives Record Administration (NARA), RG 307, Office of the Director, General Records, 1949–63, 1960–61, Box 48, folder Division of M, P, and ES.

cost. "Research apparatus for infrared, ultraviolet, Raman, mass, and nuclear magnetic resonance spectra and for many other uses," however, was a category that normally had been integrated into research projects and their funding. This restricted the amounts available for instrument and methods development. To find a solution to this bottleneck was an urgent task:

> Practically all areas of the physical and biological sciences are characterized today by an accelerating development of instrumental techniques permitting types of measurements and precisions which were not known a few decades ago.... As the pace of scientific advancement quickens, the amount and quality of research that is performed in some areas is limited by the sheer expense of such instruments (NSF 1957: 42).

For chemistry, NSF in 1957 established a program to specifically support research instruments. In the 1960s, it had a percentage of between 5% and 20% of all chemistry-related research projects funded by the NSF.[6] But the difference between research and equipment was never solved completely, and academic chemists had to lobby for the special role that instruments now played for their daily laboratory routine (Stine 1992: 241–242). The situation was similar in the biomedical field, when scientists claimed the establishment of instrument centers and the introduction of a "biological engineer" degree. Paul E. Klopsteg, a staff member of the NSF, in the so-called Klopsteg-Report of 1956 even called for a union of physics and biology already in education (National Research Council and Biology Council 1956). In a similar way argued members of the Biophysics and Biophysical Chemistry Study Section of the National Institutes of Health (NIH), founded in 1955. The roots of the above-mentioned special research resources can be traced back to this period. However, until the mid 1960s the overall rich supply with funding through governmental agencies allowed a smoothing over the structural deficits. The economic crisis of the late 1960s then led to a cutting-back especially with regard to equipment. For example, the proportion of instruments budgets in the total funding amount of NSF and NIH decreased from 12% to 6% between 1966 and 1971 (Stine and Good 1986: 42).

By the early 1970s, the scientific establishment had identified the large-scale availability of high-performance instrumentation, and its continuous renewal, as the *sine qua non* for continued scientific achievement and regarded it to be the guarantor of the perceived leadership of the United States in many scientific and engineering fields. The funding crisis thus threatened the supremacy of science, American style. The physicist Philip H. Abelson, president of the Carnegie Institution and editor of *Science*, wrote in 1971:

> To a large extent, American leadership in science has been based on the widespread availability of excellent instrumentation. ... Current trends indicate that, in the future, leadership in science will be even more contingent on pioneering the use of new and increasingly powerful equipment. American scientists are fortunate in having the support of an innovative instrumentation industry that has been a by-product of federal support of research. The

[6] See also Appendix A of Roberts to Haworth, 13. August 1963, NARA, RG 307, NSF Office of the Director subject Files, MPS Chemistry folder, 307-75-051, box 3; NSF annual reports 1965, 66, 67, 72, 73.

> grants system placed considerable sums of money at the disposal of a large number of investigators who were a good market for effective apparatus (Abelson 1971).

The biochemist Philip Handler, president of the US National Academy of Sciences, shared this opinion:

> Similarly, the commercial development of the electrostatic accelerator, the mass spectrometer, the nuclear-resonance spectrometer, the electron microscope, high-pressure equipment, and hundreds of other instruments, initially handmade with great travail by laboratory scientists, has permitted researchers to concentrate on the scientific questions rather than on merely reproducing research technologies already pioneered by others. The rapid commercialization of laboratory techniques and instruments has generated a new style of research in which the United States has been in the lead. It has been made possible by the quality and scale of United States research activity, the magnitude of Federal development programs, and the entrepreneurship of our industry (Handler 1970: 24).

Here we recognize the triad of the involved institutions: Governmental science funding led to a boom in the instruments manufacturing industry, and on this the success of the scientists depended (Reinhardt and Steinhauser 2008; Steinhauser 2012). We can find this intertwinement also at the level of specific research instruments. In 1980, C.V. Shank of Bell Laboratories, one of the inventors of a dye laser applied in high-speed spectroscopy, wrote:

> It is apparent that with the freedom to develop a new dye laser instrument capable of generating very short pulses, we have been able to influence a broad range of scientific endeavors. … Many of the techniques which we have developed have become or are becoming commercial products. The field of the investigation of picosecond phenomena was one that began in the early 60s with a great deal of excitement and enthusiasm but interest began to wane because of the difficulty in making measurements with primitive equipment. We now see this area of picosecond phenomena beginning to show a great deal of growth with the availability of commercial equipment (Shank 1980: 31).

Thus, method makers construed their own scientific-technical fields, using the commercial impact of their instruments. In a sense, method makers relied twofold on governmental research funding: First, they needed funds to develop their methods, and the related instruments. Second, they depended on the availability of resources for their scientific clients to buy the necessary instrumentation. These resources had a substantial size. In 1982, the number of instruments at US universities in the range of 10,000 $ to 1,000,000 $ per piece was estimated to be 25,000 in the physical sciences, the computer sciences and engineering—totaling 1 billion US$. Chemistry had a share of 25%, physics (without big science) 22%, and engineering 33%. Approx. 50% of the instruments were older than five years, 30% older than ten. 57% of cost was paid by federal agencies, with the National Science Foundation and the Department of Defense being the biggest spenders (Burgdorf and White 1984: 20, 23, 28).

In the mid 1970s, when it became clear that continued budget cuts would lead to an erosion of US research capabilities, scientists and members of governmental agencies attempted to stem the tide. At a meeting in March 1976, scientists asked the National Academy of Sciences to lead a study addressing "the general problem of major instrumentation for chemistry and biology"[7] and in July 1976 two of the involved scientists summarized the situation as follows:

> One of the most important problems facing research in chemistry and biology today is the escalating sophistication, size, and associated costs of major instrumentation. … The performance level of these instruments is often orders of magnitude ahead of their previous state-of-the-art counterparts, and they offer unusual opportunities for major breakthroughs in many areas of research. Effective usage of some of this sophisticated instrumentation, and efficient use of instrumentation dollars may require new institutional mechanisms (i.e. cooperative efforts and facilities) for research.[8]

The speed of innovation of instrumentation, to a large degree accelerated by the very same scientists involved in the study, led to an increase in spending needs. But rarely the situation was described as bluntly as by a representative of the department of energy in 1982: "The instrumentation problem is somewhat like the balance between food supply and population in primitive societies. If a tribal group, living at the margin of survivability, discovers a means by which it can increase its food supply, then it begins to overpopulate and gradually finds itself once again living at the margin" (Trivelpiece 1982: 17).

The scientists, however, argued that this was a virtuous, and not a vicious, circle, and that the leading international rankings of US science were based to a large extent on previous investments in instrumentation (Blout et al. 1978). The pleas based on this argument were twofold: "The first is that funds and opportunities have to be provided for the invention and development of new instruments. … The second point is that state of the art instrumentation has to be made accessible to research scientists who need to use it. And, considering the state of tight funding, this requires that instrumentation be shared."[9] Thus instrument-sharing became the miracle cure for the problems of the 1970s and early 1980s (Coulter 1978). Fewer means and an improved channeling of access worked hand in hand, and the latter worked for the benefit of the method makers. In 1984, a third of the instruments in the physical sciences were shared (Burgdorf and White 1984: 36, Fig. 15).

Regulating access to instrumentation, and especially steering the capability of instrument and methods innovation, constituted an instrument of science policy. For example, in 1981, the Biotechnology Resource Program of NIH held two workshops on the technical support of the life sciences. The 65 delegates from science, industry and government postulated a linear model of problem solving, which showed a convergence of the identification of problems and the development of new

---

[7] John D. Roberts et al. to Philip Handler, 15 March 1976, Oleg Jardetzky Papers, Ad hoc Committee, folder "Notes."

[8] John D. Roberts, Richard N. Zare, Appendix 3, 1. July 1976, "Report on Ad hoc Panel on Cooperative Efforts and Facilities for Research in Chemistry and Biological Chemistry," 23 October 1976, National Academy of Sciences. Oleg Jardetzky Papers, Ad hoc Committee, folder "Notes."

[9] Statement Oleg Jardetzky, Subcomittee on Science, Research and Technology of the Committee on Science and Technology, U.S. House of Representatives, protocol, p. 403, 26. February 1979. Oleg Jardetzky Papers, folder "Ad hoc Committee, Congressional Testimony."

technologies leading to the distribution of methods and their increased availability. For the panel, the problematic issue was the lack in the development of risky technologies, caused by the shortfall of physicists and engineers in commissions, and especially the lack of recognition of instrument and methods development as being part of science. Thus, they proposed "the interpretation of 'research' be classified and made explicit to include technologic innovation and discovery."[10] The duration and size of projects should be flexible, and the production of prototypes should be included. At the same time, the interaction of universities and industry should be strengthened.

The involved scientists and science managers had a much larger market in mind than just universities and governmental research institutions. By far the largest share was held by industry, in both research and manufacturing (Frost & Sullivan, Inc. 1972). Often, companies drove the innovation processes of instrumentation, especially and early on the petrochemical industry, and this gave rise to important spin offs of instrument manufacturing companies (Baird 2000; Rabkin 1987). The global market (the largest 22 nations, with the exclusion of the USA) in 1971 had a size of approx. 3.7 billion US$ (U.S. Department of Commerce 1972). For Germany alone the size was estimated to be more than one billion, with a prognosis of 1.5 billion in 1975 (U.S. Department of Commerce 1972: 7). Medical technology had an even larger size: "The industry that manufactures medical devices in the United States has grown … from less than $1 billion in 1958 (…) to more than $17 billion in 1983. Even after adjustment for inflation, industry sales increased sixfold during that period" (Office of Technology Assessment 1984: 3). Here, as well, similar mechanisms of research funding and regulation of innovation were in place as a report of the Office of Technology Assessment from 1984 demonstrates:

> A recent analysis of NIH, NSF, and Department of Energy grants and contracts active as of May 1983 revealed that almost $50 million was related to diagnostic imaging. This medical imaging R&D was scattered throughout the institutes and agencies and covered a wide assortment of subjects including not only development or refinement of new imaging devices, but the use of imaging techniques to enhance understanding of disease processes. A high proportion of these grants went to academic and other nonprofit institutions, and therefore supplemented the R&D on medical imaging conducted by industry. NIH funding in the medical imaging area has, in retrospect, had important impacts on the later development of commercial imaging devices (Office of Technology Assessment 1984: 80).

Thus, innovation of scientific and medical instruments showed a similar pattern. Funded by governmental agencies, the instrument industry supplied science with the necessary tools. At the same time, method makers enjoyed a key position as they were in charge for growing the market size. However, this led to problems in the funding of universities, as they could not keep pace with the speed of innovation. The question if this development is restricted to the US, and the Cold War, largely is still an open question. The cases of Japan and Germany, among others, provide

---

[10] Richard J. Johns, Ernest D. Cravalho, "Final Draft. Workshops on Technology in Support of Biomedical Research," 5. November 1981, p. 5. Oleg Jardetzky Papers, folder "Workshops on Technology in Support of Biomedical Research."

strong arguments that the symbiosis of science with a thriving instrument industry is not confined to the American case. Nevertheless, the United States excelled during the second half of the twentieth century and in many areas set the pace and the standards. Thus, we may well add the science, instrument industry and government nexus to the factors brought up by Jon Agar for explaining American supremacy in science (Agar 2012: 500–507).

4.5 Conclusion

The emergence of method-based technoscience in the middle of the twentieth century was the cradle for a novel type of scientist, with the major goal of development and dissemination of research methods. The support came from governmental science funding agencies and a well performing instrument industry. In the resulting triangle of science, industry and government, method makers had a central, but also precarious, role. Crucial for their long-term survival was their standing inside the scientific field, and with access to novel methods scientists obtained the means to gain reputation. However, method makers had the potential to threaten established power positions in a discipline. In the end, mostly a middle path was taken that led to the containment of the new methods in centers, but at the same time ensured change and innovation. As a result of this centralization and institutional separation, the transfer of methods through service, training, and collaboration impacted the further development styles of methods. As experts, the method makers could forward their research, but not completely rule disciplinary trends. Innovation processes and expert roles of method makers were thus dynamically intertwined. The more new methods became routine in certain disciplines, the more urgent it became for method makers to expand their reach. A large part of the relative autonomy of method makers was rooted in their ability to cater for different and growing clientele groups. Their alliances with instrument manufacturers and funding agencies stabilized their strong standing further.

The US style of scientific research, as portrayed by Abelson and Handler, relied in large parts on the commercial development of research instruments. This was the basis for the method makers' strategy to distribute their methods as widely as possible. At the same time it changed the definition of research projects in such a way that it included the innovation of instruments and the development of methods. The use of instruments outside the academy, for example in industrial and environmental applications, had some stabilizing effects, because it decreased the dependence on one sphere of application only and could smooth over times of financial crisis. On the other hand, science came to depend on the commercialization of its main research tools. This is not to say that method makers came to rely on the profits of their methods and associated instruments. On the contrary, method makers, as we have seen, became even more dependent on governmental funding, as they required the financial means to develop novel methods and their clients needed funds to acquire and update costly research instrumentation in order to apply these very same methods.

In this contribution I have analyzed the changes inside the scientific field, focusing on the US system in late-twentieth century. The emergence of method making in science led to new types of interactions with industry and medicine. Method makers typically "sweep" through many applications of their methods, ranging from the physical to the life sciences and to medicine and technology. In doing so, they serve as the hotbed of novel techniques that are adapted to a diverse array of applications. An example is the emergence of Magnetic Resonance Imaging techniques (MRI) from the field of NMR spectroscopy in the 1970s and 1980s. My hypothesis is that many of the techniques at play in TechnoScienceSociety are developed in social environments that are similar to that of the method makers in science, and may well overlap with these communities. Tracing some of their roots to the shifts and rearrangements of the US system of science in the second half of the twentieth century, and analyzing some of the social forces involved in this, can guide us in understanding better more recent developments as well.

References

Abelson, P.A. 1971. The Role of Scientific Instrumentation. *Science* 174 (4014): 1081.

Agar, J. 2012. *Science in the Twentieth Century and Beyond*. Cambridge: Polity Press.

Anonymous. 2004. Methods for Methods Sake. *Nature Methods* 1: 1.

Baird, D. 2000. Encapsulating Knowledge: The Direct Reading Spectrometer. *Foundations of Chemistry* 2: 5–46.

Blout, E.R., D.M. Grant, O. Jardetzky, W.D. Phillips, and K.R. Porter. 1978. Instrumentation Funding. *Science* 202 (4366): 381.

Blume, S.S. 1992. *Insight and Industry. On the Dynamics of Technological Change in Medicine*. Cambridge, MA: MIT Press.

Bourdieu, P. 1998. *Vom Gebrauch der Wissenschaft. Für eine klinische Soziologie des wissenschaftlichen Feldes*. Konstanz: UVK Universitätsverlag.

Burgdorf, K., and K. White. 1984. *Academic Research Equipment in the Physical and Computer Sciences and Engineering. An Analysis of Findings from Phase I of the National Survey of Academic Research Instruments and Instrumentation Needs*. Rockville: Westat Research Inc.

Chiang, H.H.-H. 2009. The Laboratory Technology of Discrete Molecular Separation: The Historical Development of Gel Electrophoresis and the Material Epistemology of Biomolecular Science, 1945–1970. *Journal of the History of Biology* 42: 495–527.

Chompalov, I., J. Genuth, and W. Shrum. 2002. The Organization of Scientific Collaborations. *Research Policy* 31: 749–767.

Coulter, C.C. 1978. Research Instrument Sharing. *Science* 201 (4354): 415–420.

Creager, A.N. 2002. *The Life of a Virus, Tobacco Mosaic Virus as an Experimental Model, 1930–1965*. Chicago: University of Chicago Press.

Frost & Sullivan, Inc. 1972. *The Analytical Instruments Market*. New York: Frost & Sullivan.

Gerontas, A. 2012. Chromatographie in chemischer Praxis vor und nach big science. In *Zur Geschichte von Forschungstechnologien. Generizität – Interstitialität – Transfer*, ed. K. Hentschel, 308–327. Diepholz: GNT-Verlag.

Hagner, M., and E. Hörl, eds. 2008. *Die Transformation des Humanen – Beiträge zur Kulturgeschichte der Kybernetik*. Suhrkamp: Frankfurt am Main.

Handler, P. 1970. *The Physical Sciences. Report of the National Science Board Submitted to the Congress*. Washington D.C.: Government Printing Office.

Hentschel, K. 2002. Spectroscopy or Spectroscopies? *Nuncius* 17: 589–614.

Hippel, E.v. 1995. *The Sources of Innovation*. New York: Oxford University Press.

Joyce, K. 2006. From Numbers to Pictures: The Development of Magnetic Resonance Imaging and the Visual Turn in Medicine. *Science as Culture* 15: 1–22.

Keating, P., and A. Cambrosio. 2003. *Biomedical Platforms. Realigning the Normal and the Pathological in Late-Twentieth-Century Medicine*. Cambridge, MA: MIT Press.

Kevles, B. 1997. *Naked to the Bone: Medical Imaging in the Twentieth Century*. New Brunswick: Rutgers University Press.

Laszlo, P. 2002. Tools, Instruments and Concepts. The Influence of the Second Chemical Revolution. In *From Classical to Modern Chemistry*, ed. P.J.T. Morris, 171–187. Cambridge: Royal Society of Chemistry.

Latour, B. 1987. *Science in Action. How to Follow Scientists and Engineers through Society*. Cambridge, MA: Harvard University Press.

Leydesdorff, L., and H. Etzkowitz. 1996. Emergence of a Triple Helix of University-industry-government Relations. *Science and Public Policy* 23: 279–286.

Maienschein, J. 1993. Why Collaborate? *Journal of the History of Biology* 26: 167–183.

Mody, C. 2011. *Instrumental Community: Probe Microscopy and the Path to Nanotechnology*. Cambridge, MA: MIT Press.

National Research Council and Biology Council. 1956. *Instrumentation in Bio-medical Research. Report of a Survey by Paul E. Klopsteg*, National Academy of Sciences, National Research Council Publication 472. Washington D.C.

National Science Foundation (NSF). 1957. *Federal Financial Support of Physical Facilities and Major Equipment for the Conduct of Scientific Research. A Report to the Bureau of Budget*. Washington D.C: National Science Foundation.

Nevers, P., R. Hasse, R. Hohlfeld, and W. Zimmerli. 2001. Mediating Between Plant Science and Plant Breeding: The Role of Research Technology. In *Instrumentation between science, State and Industry*, ed. B. Joerges and T. Shinn, 97–118. Dordrecht: Kluwer Academic Publishers.

November, J.A. 2012. *Biomedical Computing: Digitizing Life in the United States*. Baltimore: Johns Hopkins University Press.

Nowotny, H., P. Scott, and M. Gibbons. 2003. Mode 2' Revisited: The New Production of Knowledge. *Minerva* 41: 179–194.

Nye, M.-J. 1994. *From Chemical Philosophy to Theoretical Chemistry. Dynamics of Matter and Dynamics of Disciplines, 1800–1950*. Berkeley: University of California Press.

Office of Technology Assessment. 1984. *Federal Policies and the Medical Device Industry*. Washington, D.C.: U.S. Congress, Office of Technology Assessment.

Rabkin, Y.M. 1987. Technological Innovation in Science. The Adoption of Infrared Spectroscopy by Chemists. *Isis* 78: 31–54.

Reinhardt, C. 2006a. 'A Lead User of Instruments in Science. John D. Roberts and the Adaptation of Nuclear Magnetic Resonance to Organic Chemistry', 1955–1975. *Isis* 97: 205–236.

———. 2006b. *Shifting and Rearranging. Physical Methods and the Transformation of Modern Chemistry*. Sagamore Beach: Science History Publications.

———. 2006c. Wissenstransfer durch Zentrenbildung. Physikalische Methoden in der Chemie und den Biowissenschaften. *Berichte zur Wissenschaftsgeschichte* 29: 224–242.

———. 2010. Zentrale einer Wissenschaft. Methoden, Hierarchie und die Organisation der chemischen Institute. In *Geschichte der Universität Unter den Linden 1810–2010. Biographie einer Institution, Praxis ihrer Disziplinen, Volume 5: Transformation der Wissensordnung. Verwissenschaftlichung der Gesellschaft und Verstaatlichung der Wissenschaft*, ed. R.v. Bruch and H.-E. Tenorth, 575–603. Berlin: Akademie Verlag.

———. 2011a. Expertise in Methods, Methods of Expertise. In *Science in the Context of Application*, Boston Studies in the Philosophy of Science 274, ed. M. Carrier and A. Nordmann, 143–159. Dordrecht: Springer.

———. 2011b. Habitus, Hierarchien und Methoden. 'Feine Unterschiede' zwischen Physik und Chemie. *N.T.M* 19: 125–146.

———. 2012. Forschungstechnologien im 20. Jahrhundert. Transfer und Transformationen. In *Zur Geschichte von Forschungstechnologien. Generizität – Interstitialität – Transfer*, ed. K. Hentschel, 277–307. Diepholz: GNT-Verlag.

———. 2016. Physical Methods in the Twentieth Century Between Disciplines and Cultures. In *Transformation of Chemistry From the 1920s to the 1960s*, ed. M. Kaji, Y. Furukawa, H. Tanaka, and Y. Kikuchi, 144–152. Tokyo: Japanese Society for the History of Chemistry.

Reinhardt, C., and T. Steinhauser. 2008. Formierung einer wissenschaftlich-technischen Gemeinschaft. NMR-Spektroskopie in der Bundesrepublik Deutschland. *N.T.M* 16: 73–101.

Rosenberg, N., A.C. Gelijns, and H. Dawkins. 1995. *Sources of Medical Technology. Universities and Industry*. Washington D.C.: National Academy Press.

Shank, C.V. 1980. The History of Subpicosecond Optical Spectroscopy. In *The Scientific Instrumentation Needs of Research Universities. A Report to the National Science Foundation*, 30–31. Washington D.C: Association of American Universities.

Shinn, T. 2006. New Sources of Radical Innovation. Research Technologies, Transversality, and Distributed Learning in a Post-industrial Order. In *Innovation, Science, and Institutional Change*, ed. J. Hage and M. Meeus, 313–333. Oxford: Oxford University Press.

Shinn, T., and P. Ragouet. 2005. *Controverses sur la Science. Pour une sociologie transversaliste de l'activité scientifique*. Paris: Raisons d'Agir.

Steinhauser, T. 2012. *Zukunftsmaschinen in der Chemie. Kernmagnetische Resonanz bis 1980*. Frankfurt am Main: Lang Verlag.

Stine, J.K. 1992. Scientific Instrumentation as an Element of US Science Policy: National Science Foundation Support of Chemistry Instrumentation. In *Invisible Connections. Instruments, Institutions, and Science*, ed. R. Bud and S.E. Cozzens, 238–263. Bellingham: SPIE Optical Engineering Press.

Stine, J.K., and G.A. Good. 1986. Government Funding of Scientific Instrumentation. A Review of U.S. Policy Debates Since World War II. *Science, Technology and Human Values* 11: 34–46.

Timmermann, C., and J. Anderson, eds. 2006. *Devices and Designs. Medical Technologies in Historical Perspective*. Houndmills: Palgrave Macmillan.

Timmermans, S., and M. Berg. 2003. The Practice of Medical Technology. *Sociology of Health and Illness* 25: 97–114.

Trivelpiece, A. 1982. Untitled. In *Revitalizing Laboratory Instrumentation. The Report of a Workshop of the Ad hoc Working Group on Scientific Instrumentation, March 12–13, 1982*, ed. National Research Council, Office for Physical Sciences, 17–19. Washington D.C.: National Academy Press.

U.S. Department of Commerce. 1972. *Global Market Survey. Industrial and Scientific Instruments*. Washington D.C.. U.S. Government Printing Office.

Widmalm, S. 2006. A Machine to Work In. The Ultracentrifuge and the Modernist Ideal. In *Taking Place: The Spatial Contexts of Science, Technology, and Business*, ed. E. Baraldi, H. Fors, and A. Houltz, 59–82. Sagamore Beach, MA: Science History Publications.

Yeo, R., and J.A. Schuster, eds. 1986. *The Politics and Rhetoric of Scientific Method: Historical Studies*. Dordrecht: Reidel.

Carsten Reinhardt is Professor for Historical Studies of Science at the University of Bielefeld. Between 2013 and 2016, he was President and CEO of the Chemical Heritage Foundation in Philadelphia, USA. His research fields are the history of chemistry, industrial research, and scientific instrumentation and regulation. One of his books, *Shifting and Rearranging: Physical Methods and the Transformation of Modern Chemistry* (Science History Publications 2006), was awarded several prizes. Most recently, he coedited with Ursula Klein the volume *Objects of Chemical Inquiry* (Science History Publications 2014) and with Marianne Sommer and Staffan Müller-Wille the *Handbuch Wissenschaftsgeschichte* (Metzler 2017). His current work focuses on expert knowledge and regulation in its interaction with politics.

Chapter 5
On the Emergence of TechnoSociety by Way of Neuroculturality, for Example

Sabine Maasen

In 2010, the Dana Foundation invited seven directors of neuroscience-related institutes to the National Institute of Health to examine how brain research has progressed since 2000, the decade after *The Decade of the Brain*. To say the least, they all were highly enthusiastic. Dr. Nora D. Volkow of the National Institute on Drug Abuse for instance, held that „Neuroscience is at a historic turning point. Today, a full decade after the 'Decade of the Brain,' a continuous stream of advances is shattering long-held notions about how the human brain works and what happens when it doesn't. These advances are also reshaping the landscapes of other fields, from psychology to economics, education and the law "(Volkow 2010).

Indeed, it is safe to say that neuroscience is a rapidly developing area. Equipped with ever-new tools and technologies for investigating the brain and the wider neural system, scientists are steadily expanding their understandings of how the brain works. As this knowledge develops, so too come possibilities for new treatments for disease and damage, and applications for use outside the medical domain, including applications that interface for machines and computer technologies. Various communities of interest are involved in conversations about emerging possibilities. Philosophers, bioengineers, computer scientists, futurists, posthumanists, legal minds and policy makers all are beginning to consider the possible implications of the technological possibilities of neuroscience – if only rarely in inter-disciplinary settings reaching beyond the neuro-technical disciplines, applied ethics being an exception to this rule (for an overview, Leefmann et al. 2016).

Particularly as neurotechnologies gain more prominence, both invasive and non-invasive ones, associated social practices, such as self-experimentation with neuromodulation devices and the use of brain imaging in the courtroom, are raising new ethical, legal, social and regulatory questions. More recently, for example, questions

S. Maasen (✉)
Munich Center for Technology in Society (MCTS), Technical University Munich, Munich, Bavaria, Germany
e-mail: sabine.maasen@tum.de

S. Maasen et al. (eds.), *TechnoScienceSociety*, Sociology of the Sciences Yearbook 30, https://doi.org/10.1007/978-3-030-43965-1_5

arise, whether fundamentally new forms of governance might be required that help anticipate and shape technology as it develops, 'anticipatory governance' (Guston 2014; Aicardi et al. 2018a, b) being a prominent proposal. Responsible research and innovation (RRI) is another framework (e.g. OECD 2017) which has the capacity to address potential dysfunctional effects, namely by way of including stakeholder communication as well as collaboration with embedded researchers (Fisher et al. 2015) from the social sciences and humanities throughout the innovation process. Thus, almost 10 years after the Dana Conference, the neurosciences and their technologies have not lost momentum, at all.

However, why should scholars from the science and technology studies be so attentive to the neurosciences and neurotechnologies? On the one hand, given the social and economic burden that mental illnesses place on society, a number of governments – often in conjunction with foundations, private companies, and other parties – have established multi-year human brain research initiatives (Grillner et al. 2016), thereby creating nothing less but a "research infrastructure to decode the human brain" (see, for example, Amunts et al. 2016). The different national projects each have a particular focus, for example, gaining a more integrated understanding of brain structure and function through information and communications technology or advancing new technologies to engage brain function. At the same time, however, they share the goal of radically transforming the capacity of science and medicine to treat, cure and prevent brain-related condition, not least due to the convergence of neuroscience with information and communication technologies (ICTs). By developing new neurotechnologies such as brain imaging, neuromorphic computing, and brain-computer interfaces, neurotechnology becomes part and parcel of (trans-)national missions to improve public health, foster innovation, increase productivity, and promote economic growth. Neurotechnological research and innovation activities abound, already having an impact, or throwing off their shadows, on society.

On the other hand, neuroscience and their technologies not only affect societies in terms of neuromedical aspects, but they also affect selves and socialities (Gardner et al. 2018) on the level of everyday or professional practice and attitudes. By way of individuals and collectives interacting with neuroscience both co-produce what may be conceived as *neuro-culture*: „In general, we are witnessing the rise of a neuroculture (or neurocultures), in which neuroscience knowledge partakes in our daily lives, social practices and intellectual discourses. For instance, the dissemination of neuroscience theories, the availability of psychotropic medications and the latest neurotechnologies, such as fMRI (functional magnetic resonance imaging), are influencing healthcare strategies and legal policies as well as ways in which individuals think of themselves, their bodies or their mental disorders. For example, we relate aggressive and criminal behavior to dysfunctional firing in the pre-frontal cortex, brain images are used as evidence in court, it has become common to consider depression and sadness as a serotonin imbalance and to refer to the release of endorphins when talking about the sensation of pleasure from the consumption of chocolate or sex" (Frazzetto and Anker 2009: 815).

At second glance, however, there is more than just one single narrative about personhood or possible communities imposed on selves and society by the neurosciences. Rather, neuroscientific ideas that reach the public sphere do not encounter passive recipients of information, but active audiences who approach it through the lens of preexisting worldviews, assumptions, and agendas (Pickersgill et al. 2011). Neuroscience is thus open to a multiplicity of interpretations and uses in society, and has a corresponding multiplicity of effects (see O'Connor and Joffe 2013). In this perspective, we are confronted with a blend of imaginaries, styles of thought, and practices related to neuroscientific knowledge and technologies. In line with Dirk Baecker, I would argue that culturally we develop a new sense of complexity, that is, for mutual interdependence and incommensurability of body, consciousness, communication and technology (Baecker 2016).

From the angle of sociology of knowledge, culture, more generally, refers to the observation that cultural difference yields different patterns of perception, thought and action. And it does not matter at all, whether it concerns ethnic or epistemic culture, organizational or food culture. In any case, culture is a variable that creates differences. Reframed from a systems theoretical perspective, culture assumes the function of putting the human potential for change in a tension with society's potential for change and align it. In this view, culture indicates the capability or property of cultural beings, to knot these potentials of selves and society together and to consider it appropriate and desirable or inappropriate and thus criticizable. In contemporary societies, this is a continuous task (Baecker 2000).

This is particularly true for societies that consider themselves knowledge societies and thus are continuously affected and governed by scientific knowledge. They are characterized by cognitive patterns of expectations and hence by the willingness to question even long-held pattern of perception thought and actions, if challenged (Heidenreich 2003: 29). Knowledge thus equals a scheme of interpretation that implies the willingness to learn. Such schemes are inter-subjectively verifiable and, as a rule, are stabilized, because cognitive expectations are often confirmed. However, they can also be disappointed, whenever reality resists the expectation. In this case, one has to change one's scheme of interpretations, one has to learn. This resistance of reality can come in different varieties, for instance, caused by changing living conditions, be they natural or social. It can also be caused by new knowledge and technologies that prompt us to knot selves, socialities, and societies in a novel way. And that supply us with novel criteria to deem this knot appropriate or desirable.

Apparently, such a novel knot is just about to emerge: I would like to call it "neuroculturality". It not only refers to the brain that interlinks selves and society anew. Moreover, it does so constantly and in different forms and fashions (Dumit 2004; Vidal 2009; Rose and Abi-Rached 2013). Acknowledging the culturality of neurotechnological transformations of individual and collective selves means to acknowledge differences in processing these challenges, both over time and according to interests, pre-existing values, and so forth.[1] Generally speaking, neuroculturality results from the ongoing interaction of three dimensions: it entails ever-new forms of knowledge and practices based upon neuroscientific findings and theories. It arises from ever-new devices and procedures regulating the selves and collectives (neurotechnologies). Finally, neuroculturality results from specific values and identity politics interacting with neuroscientific challenges (e.g., neuro-ethical ones).

In the following, I would like to address prototypic examples for the ways in which neurotechnologies make up selves, communities, and rearrange values addressing these challenges. First, I will target the practice of neurofeedback. Based upon Jonna Brenningmejer's (2010) work, I will scrutinize the ways in which users of brain devices strive to understand how working on the brain operates on individuals, contributing to new ways on ways of *self-constitution*. Second, I will turn to the phenomenon called neurodiversity. Based upon Francisco Ortega's work (2009), I will show how the neurosciences are not only proactively employed to construct an autistic identity, but also to constitute a *neuro-community*. Third, based upon Michael Chorost's (2005) and Jake Dunagan's (2010) arguments, I will show how up-to-date hearing technologies challenge criteria for identity politics and indeed constitute a new *neuropolitics*. By way of conclusion, I would like to elaborate briefly on the idea, that it is not so much NeuroSociety we are confronted with. Rather, it is a society suffused by all kinds of ever-new technologies and a long-standing culture of working at oneself, individually and collectively. The interaction of those emerging technologies and the will to shape oneself is by no means deterministic. On the contrary, it is rooted in different motives and yields different results. Interestingly enough, technoscience and technologies of selves and socialities today meet in their joint indifference as to ontological issues and their appreciation of entangled knowledges and values.

By closer inspection, neuroculturality not only points to a 'political economy of brains' (Kwint and Wingate 2012: 20)[2] but also to a 'technical biopolitics': never before has life been so thoroughly technicized. Nor was it so dependent on and entangled with technology. With that, not only the old naturalism of life is gone. Both the epistemological claim of the objective determinacy of "life itself" and the political claims of biological essentialism can no longer be upheld (see Folkers 2017). Now it is about "technical biopolitics" that is the call to new, alternative experiments in a neuro-technologized art of living (and no "back to nature"). Politics of life beyond itself do not intervene at the molecular level of the genetic code or

[1] For the idea of culturality as designating difference, see Wimmer (2004), for the case of philosophy.

[2] "Brains are displayed, pictured, scanned, analysed and treated thus transforming them by scientific industry into academic papers, media spectacles of discovery, education and information, healthier citizens, more compliant workers and consumers – all to enhance national prowess, prosperity, security and so on. A multitude of spaces, aesthetic predilections, protocols, technologies and tools supports and enables this range of practices. There is, so to speak, a political economy of brains, by which the brain becomes invested with a level of hope and expectation that has usually exceeded the ability of science to make sense or use of it" (Kwint and Wingate 2012: 20)

brain physiology (Rose 2007), but rather focuses on the techno-environmental conditions of life. Government thus must be understood as regulating the complex ecology of humans and non-humans (Folkers 2017: 376). Life is hence characterized as an emergent phenomenon, and by its affectability through its (techno-cultural) ecology. In line with Gilles Deleuze and Félix Guattari (1987), this is articulated with the notion of "assemblages". Accordingly, life emerges by way of an assemblage of the natural and the cultural, the biological and the social – thereby avoiding both naturalistic and culturalistic fallacies. Especially under the condition of converging technological conditions[3], the mutual interdependence and incommensurability of body, consciousness, communication and technology (Baecker, see above) become ever-more visible, leading to rather open-ended, experimental ways of shaping selves and socialities. In order to disentangle this neuro-techno-cultural assemblage, let us look at instances of becoming neurotechnological selves and socialities as well as neurotechnically imprinted ways of identity politics, in turn.

5.1 Neurotechnologized Selves

Since about three decades ago, there is hope for individuals to become happier, healthier or more successful by non-invasive neurotechnological means. There are several devices available in clinics or on the internet which help people to modulate their brain rhythm so as to switch their brain state into meditation, hallucination or concentration, by using light-and-sound devices (see, for example, Heller 1991). Such 'brain machines' consist of a pair of glasses with shifting LED-lights and beeping headphones that operate at a specific frequency designed to 'train' the brain's frequency. These devices, which were used in esoteric art scenes in the 1960s, became popular as 'digital drugs' in the 1980s (Geiger 2003). Today, because of their stress-reducing and attention-increasing utilizations, they are also marketed for applications such as business training, as well as for driver distraction detection. At the same time, and predominantly, however, they are promoted for mental health problems and cognitive enhancement. Likewise, one may use electric or magnetic current to stimulate one's brain, the hope being to overcome depression, anxieties, physical pain or sleeping problems (see, for example, Naish 2007).

Neurofeedback, based on the theory of operant conditioning, is also designed to teach people to train their brainwaves. When used in clinics, the clients are first connected to a (quantitative) EEG device that measures their brainwaves on several points. Based on a brain map thus constructed and on the phenomenological

[3] "… the converging technologies promise/threaten to reverse the relationship between the natural and the artificial". Whether by genetic modification, metabolic pathway engineering, Biobricks, the addition of electronic implants, nanotechnological or neurotechnological engineering of organic life. Natural selection, which human culture has already completed with breeding, "is increasingly becoming an *un*natural selection of artificial elements. As a consequence, trans- and posthuman life will increasingly be '*natural by artifice*' (Mul 2014: 471).

complaints of the client, the practitioner determines the brain frequency and location that has to be trained and tunes the machine to the desired program. The user can, for example, watch a movie or play a race game: whenever her or his brain reaches the 'right' frequency on the defined location, she or he hears a tone and is rewarded, say, with an extended movie screen or an acceleration of the race car. In this manner, the user's brain is meant become conditioned to reach the desired brainwave state. After some exercise, this may be achieved more easily or even automatically in normal life as well, thus solving the real-life problems of the client. Again, whether depression, anxiety, chronic pain, brain damage, attention disorders, and hyperactivity, learning problems or to burn-out, neurofeedback can play a decisive role.

As brain scans, EEG devices and technical or pharmacological brain manipulators suffuse contemporary practices (medical and everyday one), they also have become important technologies to inscribe people. "People increasingly learn that their real-life problems are brain problems and thus some of them start to experiment with manipulating their brain states" (Brenninkmeijer 2016: 84). In this vein, brain devices are technologies of the neurobiological self, in the same way as psychotherapy was a technology of the psychological self (see, for example, Rose 2007). Users of brain devices emphasize that they themselves are not the cause of their problems, but their brains are. EEG devices make clear that people are troubled by their brains: accordingly, the self has to work on the brain to improve itself, notably by using a brain device.

In her study, all the users who Joanne Brenninkmeijer (2010) interviewed agree with the practitioners that their problems indeed are brain problems. Interestingly, they have begun to use the terminology of the practitioners and explain their problems in phrases like 'my 6 hertz mystery', 'my theta', 'those alpha and theta things', 'explosions in my brain norm', 'all 42 points in my head', 'the lack of a certain substance in the brain', 'brain tracks', 'neurons', 'my brain is out of balance'. Most plausibly, solutions of these brain problems are also to be found in the brain, neurofeedback being cherished as non-invasive, do-able, and empowering. Most tellingly, phrases like 'the computer rewires me', 'neurofeedback puts brakes on the race-car in my head' do not testify to a naïve dualism, though. Rather, in the perspective of selves articulating themselves in a neurotechnological ecology, it is but a short-cut terminology for a technologized way of seriously und effectively acting upon oneself. As regards the practitioners, they do not only teach their clients that their problems are brain problems, they also make this visible by showing and explaining brain maps to their clients. Again, these maps are not, however, mistaken to 'be' or 'represent' the clients' selves but rather they point at possible ways to act upon oneself within an ecology of techno-physiological elements.

At this point, Joanne Brenninkmeijer, following Ian Hacking, points out that neuroscience thus has created a 'neurologically nested triad' of 'mind, brain and body'. "Within the human organism of flesh and blood, one part, the brain, monitors the body, and another part, the mind (still flesh and blood), monitors the brain and its monitoring of the body" (Hacking 2005: 165). New technologies of the self give rise to new selves, as we know from Foucault, and in the case of the brain devices this is clearly the case. The act of manipulating the brain with a device makes people very aware of their thoroughly biological constitution. First, they learn that their problems are brain problems, then they see these problems visualized in their brain map, and next they interfere directly with their brains to fix these problems (Brenninkmeijer 2010: 121). In this process, the brain is separated from the self. But more importantly, users of brain devices make use of ever-more entities to explain their selves and advance their process of, e.g., modulation of attention – including computers, software, and a comfortable chair as well as, albeit optionally, medical doctors. They realize themselves as being articulated through heterogeneous entities.

In other words: neurotechnologized selves are not at all objects of reductionist practices (hence, nothing but their brain) as often stated. Rather, by closer inspection, they are characterized by an entire network of agentive entities. Whether human or non-human, they are performative in the sense that they induce self-regulating behavior (e.g. a 'yellow spot') but simultaneously become actors (as the cause of a depression, as the stimulation to do neurofeedback, as the spot that turns from yellow into green). It is by its affectability through its (techno-cultural) ecology, here illustrated by the practices of neurofeedback available and used in clinics and in everyday life, that selves adjust themselves without the need to be one and the same. While the immediate goal might be therapy or improvement, the overall mode of induced self-regulation is searching, experimenting, becoming, due to the perceivable interdependence and inevitable incommensurabilities of body, consciousness, communication and technology. Voilà Foucault's technologies of the self revisited: Neuroculturality allows for, or enforces, ongoing and variegated work at oneself, one's extended self, as Joanna Brenninkmeijer puts it (2015: 15). This particular *modus operandi*, an open-ended neuro-technologized way of living, is also true for the neurosciences transforming entire communities.

5.2 Neurotechnolized Socialities

Next, I would like to draw on Francisco Ortega's case study on the so-called neurodiversity movement, which so far has been dominated by autistic people who believe their condition is not a disease to be treated and, if possible, cured, but rather a human specificity (like sex or race) that must be equally respected (Ortega 2009). While situated within the tradition of disability movements, it shifts away from psychological and toward a genetic and neurobiological understanding of autism. Together with the availability of neurotechnological interventions, this poses nothing less but a challenge to the identity politics of an entire biosocial community, the deaf community. This identity politics, is not only about individual identity but also about collective culture. 'Neurodiversity' asserts that some features usually described as illnesses are actually only atypical or 'neuro-divergent', i.e. they result of a specific 'neurological wiring'. Therefore, neurodiversity advocates advance that proposing to cure a 'neuro-atypical' would be similar to curing a gay, a black, a left-handed. As opposed to this, neurodiversity provides the status of a biosocial group united by one particular feature: neuroatypicality.

This stance, however, is hotly debated within the deaf community itself, giving rise to pro-cure and anti-cure discourses: In the pro-cure faction, one finds parent and practitioner groups that favor the search for and use of behavioral and pharmacological therapies (Silverman 2008). The National Alliance for Autism Research (NAAR), founded in 1994, and the Cure Autism Now Foundation (CAN) are institutional supporters of each party. The latter foundation was created in 1995 and brings together families, medical doctors, and scientists devoted to accelerating the pace of biomedical research on autism, an additional goal being to raise funds for research and education in the area. By contrast, the anti-cure faction convenes so-called neurodiversity advocacy groups. They fight against the pro-cure faction for their alleged demonizing autistics and frightening their families. In their view, the anti-cure faction emphasizes narrow viewpoints on the disorder, and disregards real-life experiences of autistic adults. Put in a nutshell: While pro-cure representatives speak of themselves as *having* autism, anti-cure representatives consider themselves as *being* autistic (Ortega 2009: 433).

What is more, in particular in the US, the neurodiversity movement has been intensified by its enhanced exposure in the media. The internet has become the prime agent for forming advocacy and support networks, thus beginning to do "what was thought impossible, to bind autistics together into groups' (Singer 1999: 67). The internet is said to be for autistics what sign language is for the deaf or Braille is for the blind (Blume 1997). Prominent among the phenomena sustained by websites and blogs is the emergence of a specific self-advocate vocabulary for categorizing persons (Bagatell 2007): Aspie, NT, Autistic or Autie ('person with autism'), or Curebie (derogatory term for those who wish to cure autism). Websites also recommend fictional and specialized science literature as well as various online support organizations; blogs and chat rooms empower interaction among autistic individuals as well as provide information on symptoms; they assist in the sharing of experiences as well as in finding friends or partners (Silverman 2008). All of this contributes to a new ecology of human and non-human entities including the new media. Together they promote awareness and empower a biosocial culture that since 2005 has a Pride Day (18 July), an initiative of Aspies for Freedom, yet also provides spaces for contestation: among others, the debate between pro-cure and anti-cure has reached the courts. Parents fight for governmental support or health insurance companies pay for the expensive therapy. The advocates' claim that autism is not an illness and that attempts to cure it violate autistic rights may therefore provide a legal backing for refusing to subsidize the therapy.

Yet again, this comes with frictions and paradoxes, even. On the one hand, neuroscientific claims are mobilized in the construction of what counts as neurotypical and autistic experiences, so as to highlight their 'natural difference'. On the other hand, advocates of the neuro-atypical faction know how closely they are linked to the neurotypical 'faction' (Ortega 2009: 439) – e.g., by friendship or marriage. These ontological frictions are based on an outright paradox. It pertains to the ways in which autists refer to the neurosciences: for the most part, "neuroscience aims at locating the brain structures of circuits responsible for normal and pathological mental states. However, while neuroplasticity helps account for neurodiversity, neurodiversity advocates themselves tend to homogenize neuro-diverse brains and minimize their differences so as to support their claims for the existence of a brain-based autistic identity. Thus, the 'autistic brain' is displayed as ontological homogeneous and radically different from the also homogeneous 'NT brain'" (Ortega 2009: 441). Maybe it is on these grounds that the neurodiversity universe shows an, albeit uneasy, cohabitation of quite many ontologies while acting, thinking or speaking about themselves and interacting with others. They talk about 'my brain,' 'my mind', 'I' or 'me' synonymously, thereby smoothing contrasts they themselves help to build up within their biosocial community.

This example illustrates how the recent preference for cerebral explanations as regards autism can be understood within the context of neuroculturality: "the diffusion of neuroscientific claims beyond the laboratory into different domains of life in contemporary biomedicalized societies provides new possibilities for interdependence and incommensurabilities of body, consciousness, communication and technology "(Ortega 2009: 426). Within this ecology of neuroscientific theories, practices, technologies and therapies emerge new and never-stable ways of thinking and acting upon ourselves and relating to others, favoring forms of neurotechnological subjectivation and socialities. Neuroscientific claims are taken up not only in the formation of individual identities, but also in the form of community networks (Pitts-Taylor 2010; Pickersgill 2018).

The essential message is: Whether pro or con, all involved in discussing the 'true nature' of autism, relate to neuroscientific knowledge. While some autists are looking for a cure of autism and actively engage with neurotherapies, others believe their condition is not a disease to be treated, but rather a human specificity that must be equally respected – again relying on neuroscientific insights. For neurodiversity activists, the prestige of the neurosciences helps support their positive judgment about their condition; neuroscientific research, largely in the form of the 'wiring' metaphor, seems to legitimize their attitudes toward their diagnosis. From the perspective of neuroculturality, however, one observation is obvious: between these poles there is ample room for ambiguities, re-orientation and ongoing experimentation, advanced by an unruly assemblage of neurobiological, technological, new media and other cultural entities (such as law). In this so happening, the autistic community as a whole, however, recognizes a genetic and/or neurological origin to autism and uses (neuro/info) technologies to support their specific attitudes, if in specific ways.

5.3 Criteria to Work with in the Neuro-World

After having glimpsed into the making up of selves and socialities by referring to the neurosciences as a pool of knowledge and technologies to be used within an unruly ecology I call neuroculturality, we now turn to the ways in which they challenge the criteria that help to assess whether or not, neurotechnologies are appropriate or desirable. To this end, I will scrutinize the shifting evaluations implicit in, as it were, neuropolitical discourses, the example being advances in the realm of cochlear implants and their reception in the deaf community.

The question of whether or not deafness constitutes the basis of a culture is – again – an emotionally charged debate. On the one hand, deafness has historically been viewed as a physical impairment associated with such disabilities as blindness, cognitive, and motor impairments. On the other hand, views on deafness as a culture have recently emerged that consider deafness as a trait, not as a disability. Since its inception, the main goal of the deaf culture movement has been to remove the stigma and barriers to living a productive life in a hearing-centered world. It has strived to build respect for the specific qualities and accomplishments of the so-called deaf culture – including the invention of a sign-language based on visual communication. It was literally and metaphorically all about making the deaf "heard" in the public sphere and being acknowledged as political subjects. As so much work has been invested into making the non-hearing way of life a valued option in culture and the political terrain, deaf advocates do not welcome new technological implants, as they threaten their identity politics.

In this situation, an author entered the discussion who himself was born with severe hearing loss. Being an innovative science writer, Michael Chorost spent most of his life depending on hearing aids. Only after he received his implants, he wrote a book entitled "Rebuilt: How Becoming Part Computer Made Me More Human" (Chorost 2005). The title entailed a fascinating message: Chorost's forecast of the end of deafness was not a call to give up and join those with normal biological hearing. He rather challenged the deaf community at large, by emphasizing the many new possible interactions of sensation, biology, technology, and identity politics. In this vein, his book was also a plea for experimenting with hearing aids, individually and as a biosocial group. Not surprisingly, Chorost instigated enthusiasm and anger alike, not least within the deaf community.

As recent as two decades ago, there were little to no technological solutions available to assist a cochlear implant recipient, who presented with severe to profound hearing loss, with speech recognition in difficult listening situations – understanding speech in noisy and reverberant settings, over the telephone or television, and when spoken from a distance. Among the many things that happened ever since are advances in the domain of electroacoustic stimulation. As cochlear implant electrode arrays and surgical techniques have evolved, the potential to preserve residual hearing has greatly improved. For instance, hybrid cochlear implant systems feature shorter electrode arrays designed to facilitate hearing preservation and also feature sound processors capable of providing low-frequency acoustic amplification and high-frequency electrical stimulation. Future developments will focus on even better electrode-nerve interfaces by improving electrode technology. The robot-assisted surgery will allow for high precision and reliable hearing preservation. Biological therapies will support the bionic ear. Local drug delivery will focus on suppression of trauma reaction and local regeneration. Gene therapy by nanoparticles will hopefully lead to the preservation of residual hearing in patients being affected by genetic hearing loss. Due to converging available technologies, hearing preservation by cochlear implants will become routine (Lenarz 2018).

Many years before such advances, and being affected himself, Chorost proposed what only has gained in plausibility until today: instead of focusing questions of identity on deafness itself – and lamenting its extinction – the deaf community could and should rather become the vanguard of a new movement of sensory perception. While hearing remains important, it is not a dichotomy between hearing/non-hearing any more that is most relevant. Rather, thus his plea, what is at stake is a fascinating range of new and mediated ways of sensing the world. It is all about "new ways of using technology to communicate and build communities" (2005). For Chorost, this means the embrace of new neuro-sensory technologies. This embrace implies not only an outright indifference to any notion of 'mere' restitution of biological hearing, but rather a transformation of the very concept of hearing. He proposes that the deaf community should rethink the audible altogether, by making new sounds available to perception. Those with next generation implants that work directly on the auditory nerve could, for instance, program their hearing environment, filtering and amplifying certain sounds, and otherwise modulate their sensations and perceptions in ways unavailable to the 'normal' hearing community. Those with hearing modifications of the newest type can thus interact and 'use' their environment differently, changing their locus of cognition, and thereby gain a completely new realm of experience. In this perspective, hearing is thus characterized as an emergent phenomenon, and by its affectability through its techno-cultural ecology.

In other words, what is called for is identity politics framed as neuropolitics. From the angle of neuroculturality, it may be the accentuation, not the diminution, of embodied differences in sensibility through technological augmentation and modification that will define the politically qualified subject. What we can see here is a forceful example of challenging the criteria for maintaining or changing an identity politics. On this account, most recent cochlear implants prompt the deaf community to seriously reconsider whether their identity politics can be regarded as avant-garde any more. And regardless of whether the answer is 'yes' or 'no', given the neurotechnological ecology at hand, it has to be considered. Sheer ignorance in the name of the identity politics *quo ante* seems inappropriate.

5.4 (Neuro)TechnoSociety Ahead?

With new neurotechniques and neuroliteracies also come new expectations for interventions. These expectations and desires will encounter an increasing range of pharmaceuticals, brain training software, neurofeedback applications, and augmented cognition technologies … But what will modulating and experimenting with our brains do? In a competitive job market, better focus, better memory, better multi-tasking abilities, reduced need for sleep – all these effects are highly rewarded in this competitive landscape. A recent survey of scientists showed that over 20%

had admitted to taking some sort of pharmaceutical enhancement drug – be it Ritalin for concentration, or Provigil for attenuating the effects of fatigue (Maher 2008). Zack Lynch, President of the Neurotechnology Industry Organization, calls this the drive for "neurocompetitive advantage" (2010: 95–96).

Moreover, in an aging society, diseases such as Alzheimer's and Parkinson are looking for the neuro sciences to deliver (Gardner 2017). Likewise, in the case of depression. Somewhere on a continuum of therapeutic, modulating and enhancing functions the neuroscience not only deliver, but are also responsive to societal requests. Therefore, it may help to change the perspective and consider the society the neurosciences are part of: it is neoliberal society – a society demanding an active, enterprising, and responsible role of its members, both individual and collective. Hence, by debating the pros and cons of particular findings and technologies – the neurosciences also become a stage for public deliberation of modes and rules of being a self: individual agency, responsibility, and flexibility frame the debate on the possibility of sociality in neoliberal society, always challenged by new knowledge and technologies as well as new criteria to work with. As a consequence, we see a constant updating of selves, thereby smoothing the complexities evolving from the mutual interdependence and incommensurability of body, consciousness, communication, and technology.

In an analogous fashion, this applies to neuro-socialities as well: deaf communities and neurodiversity are biosocialities that not only define themselves by some typicality of the brain. In so doing, they also turn the typicality of their brains into a legitimating gateway to all kinds of interventions, be they medical, political, juridical or in terms of insurance. Whether this ultimately leads to a techno-progressive democracy or, by contrast, to a society of control "networked from synapse to street" (Dunagan), or both, remains to be seen. In any case, there are certainly new and opposite cognitive, social, and political forms taking shape before us: artificial intelligences, cyborgs, posthuman subjectivity, a breakdown of mind along with the climate collapse, and so on. The neurosciences and their technologies form part and parcel of this political development.

But there is more: science and technology, at large, have become thoroughly constitutive of modern life. From health to energy, transportation, information, communication, all the way to economic growth – science and technology shape the way we see, live in, and transform the world, both individually and collectively. While science and technology have led to tremendous gains in knowledge, prosperity and quality of life, they have also created new uncertainties, risks, and vulnerabilities for society, individuals and collectives including. The increasing scientization of social life requires new forms of public engagement, sense-making and trust-building, from citizen juries on GMOs to education on the use of heterogeneous information sources from the internet. A constitutive feature of contemporary life – and part of the reason why the relationship between science and the public is facing new challenges – is that technoscientific and social orders are becoming increasingly hard to separate.

In most general terms, technoscientific advances including neuroscientific ones should no longer be conceived simply as *prosthetics* designed to supplement insufficiently equipped selves, collectives, or and societies, at large, and then be analyzed with respect to their unintended side effects. Rather, following Simondon, they should be conceived of as open objects linked to, and extended through, the existence of socio-material networks, signifying today's technical-industrial world (Simondon 2009). As Weiser puts it somewhat provocatively: "The most profound technologies today are those that disappear. They weave themselves into the fabric of everyday life until they are indistinguishable from it" (Weiser 1991: 94).

In this view, neuroscientific knowledge and technologies can be revealed as co-producers (Jasanoff 2004) of culturality as they enable and come to be embedded in new relationships of selves and collectives such as self-help and advocacy groups, the health system and its professions, as well as the wider public, offline and online. Neurosciences, through their construction as novel epistemological practice, thus impact on society in various ways. In addition, both neuroscience and its various users, even when acting in parasitic or subversive ways, or against technoscientific evidence, share a certain ontological openness, if not indifference: if neuroscientific research is capable of measuring and visualizing anxiety in the brain and provides techniques to manipulate and control it, the question is not whether the result is natural or technical. Rather, it is an open-ended assemblage, co-produced by nature, social-cultural and political conditions suffused by the technological infrastructures and artifacts. Likewise, a user of neurofeedback systems can easily interact with ontological diversities such a minds, psyches, or brains without undue confusion or scrutiny. Analogously, neurodiversity groups can easily invert the aims and findings of brain research, and deaf culture can reject powerful hearing implants, as their ontological frame of reference is identity – the quest for identity functioning as a Deleuzian 'fold', as an ontology of becoming. More imminently, selves and socialities constantly maneuver within these unruly ecologies. Ultimately, unstable meanings of nature, technologies, and selves/socialities merge here, giving rise to new understandings of these foundational terms that imply a view of biology or nature that is at once given and given to control, that affects us and becomes affected by us. In so doing, we form assemblages or consistencies, that is, emergent unities that nonetheless respect the heterogeneity of their components.[4]

Note though, that respecting heterogeneity is not the same as essentializing it. As Katherine Hayles points out "the posthuman view configures human being so that it can be seamlessly articulated with intelligent machine" (Hayles 1999: 34). The removal of the line between machine, mind, and computer means there are no longer "essential differences or absolute demarcations between bodily existence and computer simulation, cybernetic mechanism and bodily organism, robot technology and human goals" (Hayles 1999: 3). For Merleau-Ponty, the body can adapt and extend itself via 'literal' or external devices and, after acquiring the ability to

[4] In this regard, I share Tamoar Sharon's view of technological mediation in that (bio-)technologies are "active mediators that help shape the relationship between humans and the world"– in a transformative, yet not deterministic way (Sharon 2014: 13). Such a mediated posthumanist perspective makes it possible to account for the mediated character of human existence and its originary technicity, and for the ambiguity of the emerging biotechnologies.

perceive the world through the device, it ceases to exist, having been absorbed, it now constitutes an essential part of what it means to be (1962: 440). While phenomenologically plausible, I would emphasize the process of becoming and, for certain periods of time, forming specific assemblages of technical and social, organic and non-organic entities, embracing the ever-new possibilities of neuro and information technologies shaping selves and socialities.

In summary, by the intertwining of different psychological, collective, and technical-medial milieus (Hörl 2015) we observe, among others, the becoming of neurotechnologized selves and socialities, of neuroculturality, at large. At the same time, they represent exemplary vehicles of the becoming of TechnoSociety. Posthuman subjects and communities come about and engage with neuro and information technologies in their capacity as "posthuman technologies" (Shaw and Shaw 2015: 6) as essential part of human-machine-information ecologies. Not surprisingly, we see constant challenges and changes in contemporary society prompted by an unruly neuro-techno-political assemblage. It seems as if TechnoSocieties, and NeuroTechnoSocieties, in particular, are societies that operate upon a new mode: the update-mode.[5]

5.4.1 … critique?

When trying to formulate an adequate way of critique – a critique on the level of these and similar socio-techno-medial assemblages, one is well advised to refrain from simple answers. Critique must operate on the very level of assemblages, thereby including and addressing the whole network of materialities, collectivities and subjectivities (as discussed in a variety of ways in Ong and Collier 2005; Tsing 2015; Rees 2018). What we can see here, in the age of the technosciences, is a deep, and in fact, regained, trust in engineering and re-engineering, also mirrored in stark protests against the rejection of novel technologies (Chorost 2005). When we see the astounding parallelism of info-, nano-, or biotechnical engineering *and* the increasing appraisal of neocybernetic approaches, this fits very well with a political rationality that tries, in a way, to ontologize ongoing flexibility as well as responsibility for reassembling the heterogeneous elements constituting selves and socialities today. Alongside energetic claims to, e.g., empowerment and democratization in TechnoSociety, ecologies of engineering can, simultaneously and easily, align with ongoing efforts at optimizing, enhancement, and/or increased responsibilization.

What happens to the participants of socio-technical arrangements also happens to their critics and their critique. As we have learnt not only from Foucault, critique, whatever it says, and whoever it expresses, is not simply and only an adversary, it is always also a „partner "of the arts of governing (Foucault 1997: 28). Moreover,

[5] A research program on this hypothesis is in preparation.

Foucault points out, "everything is dangerous is not exactly the same as (everything is) bad. If everything is dangerous, then we have always something to do (…) every day (we must) determine which is the main danger" (Foucault 1997: 256). Translate 'dangerous' with 'complexity' to be dealt with, it is up to us – all of us – to engage in critique, sometimes playfully, sometimes substantially. Whether this articulates within scholarly discourses or by publics in the social media, critique is always "embedded" and "procedural" – it co-evolves within the neuro-techno-political ecologies it addresses. This may be illustrated by a final example from the realm of prosthetic technology: One important barrier has always "been the general desire of patients to make their disabilities as inconspicuous as possible", sometimes at the cost of less-than-optimal performance". Interestingly, this is currently undergoing a sea change. "Consumers are becoming accustomed to wearing and even showing off visible technology such as neurofeedback systems for private use or 3D-manufactured prosthetics" (Loeb 2011: 172).

Thus, by implication, critique (here: as a new grass-roots identity politics of embracing prosthetics) is becoming a co-constitutive and productive element of the socio-techno-medial condition of TechnoSocieties – one prominent mode of its becoming being neuroculturality.

References

Aicardi, C., B.T. Fothergill, S. Rainey, B.C. Stahl, and E. Harris. 2018a. Accompanying Technology Development in the Human Brain Project: From Foresight to Ethics Management. *Futures* 102: 114–124. https://doi.org/10.1016/j.futures.2018.01.005.

Aicardi, C., M. Reinsborough, and N. Rose. 2018b. The Integrated Ethics and Society Programme of the Human Brain Project: Reflecting on an Ongoing Experience. *Journal of Responsible Innovation* 5 (1): 13–37. https://doi.org/10.1080/23299460.2017.1331101.

Amunts, K., C. Ebell, J. Muller, M. Telefont, A. Knoll, and T. Lippert. 2016. The Human Brain Project: Creating a European Research Infrastructure to Decode the Human Brain. *Neuron* 92: 574–581.

Baecker, D. 2000. *Wozu Kultur?* 1–24. Berlin: Kadmos.

———. 2016. *Die Währungen des Homo digitalis.* Presented at the Geldgipfel 2016 – Homo civilis et oeconomicus. Vom Fußabdruck zum Handabdruck. https://www.glsbankstiftung.de/media/pdfs/GeGi16/Dirk_Baecker.pdf. Accessed on 7 Oct 2019.

Bagatell, N. 2007. Orchestrating Voices. Autism, Identity and the Power of Discourse. *Disability & Society* 22 (4): 413–426.

Blume, H. 1997. *Autism & The Internet or 'It's The Wiring, Stupid'*. http://web.mit.edu/m-i-t/articles/index_blume.html. Accessed on 22 Aug 2019.

Brenninkmeijer, J. 2010. Taking Care of One's Brain: How Manipulating the Brain Changes People's Selves. *History of the Human Sciences* 23 (1): 107–126.

———. 2015. Brainwaves and Psyches: A Genealogy of an Extended Self. *History of the Human Sciences* 28 (3): 115–133.

———. 2016. *Neurotechnologies of the Self: Mind, Brain and Subjectivity*. London: Palgrave.

Chorost, M. 2005. *Rebuilt: How Becoming Part Computer Made Me More Human.* Boston: Houghton Mifflin Harcourt.

De Mul, J. 2014. Philosophical Anthropology 2.0. In *Plessner's Philosophical Anthropology. Perspectives and Prospects*, ed. J. de Mul, 457–475. Amsterdam/Chicago: Amsterdam University Press/Chicago University Press.
Deleuze, G., and F. Guattari. 1987. *A Thousand Plateaus: Capitalism and Schizophrenia*. Minneapolis: University of Minnesota Press.
Dumit, J. 2004. *Picturing Personhood: Brain Scans and Biomedical Identity*. Princeton: Princeton University Press.
Dunagan, J.F. 2010. Politics for the Neurocentric Age. *Journal of Futures Studies* 15 (2): 51–70.
Fisher, E., M. O'Rourke, R. Evans, E.B. Kennedy, M.E. Gorman, and T.P. Seager. 2015. Mapping the Integrative Field: Taking Stock of Socio-Technical Collaborations. *Journal of Responsible Innovation* 2 (1): 39–61.
Folkers, A. 2017. Politik des Lebens jenseits seiner selbst. Für eine ökologische Lebenssoziologie mit Deleuze und Guattari. *Soziale Welt* 68: 365–384.
Foucault, M. 1997. Ethics, Subjectivity, and Truth. In ed. P. Rabinow. New York: New York Press.
Frazzetto, G., and S. Anker. 2009. Neuroculture. *Nature Reviews Neuroscience* 10: 815–821.
Gardner, J. 2017. *Rethinking the Clinical Gaze Patient-Centred Innovation in Paediatric Neurology*. Cham: Palgrave Macmillan.
Gardner, J., N. Warren, P.H. Mason, D. Dominguez, and J. F. 2018. Neurosocialities: Anthropological Engagements with the Neurosciences. *Medical Anthropology* 37 (3): 189–193. https://doi.org/10.1080/01459740.2018.1439488.
Geiger, J. 2003. *Chapel of extreme experience: A short history of stroboscopic light and the dream machine*. Brooklyn, NY: Soft Skull Press.
Grillner, S., N. Ip, C. Koch, W. Koroshetz, H. Okano, M. Polachek, M. Poo, and T.J. Sejnowski. 2016. Worldwide Initiatives to Advance Brain Research. *Nature Neuroscience* 19 (9): 1118–1122.
Guston, D.H. 2014. Understanding 'Anticipatory Governance'. *Social Studies of Science* 44 (2): 218–242.
Hacking, I. 2005. The Cartesian vision fulfilled: Analogue bodies and digital minds. *Interdisciplinary Science Reviews* 30 (2): 153–166.
Hayles, N.K. 1999. *How We Became Posthuman: Virtual Bodies in Cybernetics, Literature and Informatics*. Chicago: University of Chicago Press.
Heller, Z. 1991. The Electric Kinesthetic Innerquest; this is the Ultimate Head Trip, The Independent. 15 December, p. 2.
Heidenreich, M. 2003. Die Debatte um die Wissensgesellschaft. In *Wissenschaft in der Wissensgesellschaft*, ed. S. Böschen and I. Schulz-Schaeffer, 25–51. Opladen: Westdeutscher Verlag.
Hörl, E. 2015. The Technological Condition. *Parrhesia* 22: 1–15.
Jasanoff, S. 2004. *States of Knowledge: The Co-Production of Science and the Social Order*. London/New York: Routledge.
Kwint, M., and R. Wingate. 2012. *Brains: Mind as Matter*. London: Wellcome Collection.
Leefmann, Jon, Clement Levallois, and Elisabeth Hildt. 2016. *Frontiers in Human Neuroscience* 10: 33.
Lenarz. 2018. Cochlear implant - state of the art. *GMS Curr Top Otorhinolaryngol Head Neck Surg* 16. https://doi.org/10.3205/cto000143.
Loeb, G.E. 2011. Neuroprosthetic Interfaces – The Reality Behind Bionics and Cyborgs. In *Human Nature and Self-Design*, ed. S. Schleidgen et al., 163–176. Paderborn: Mentis.
Lynch, Z. 2010. *The Neuro Revolution*. New York: St. Martin's Press.
Maher, B. 2008. Poll Results. Look Who's Doping. *Nature* 452: 674–675.
Merleau-Ponty, M. 1962. *Phenomenology of Perception*. Trans. C. Smith. London: Routledge & Kegan Paul.
Naish, J. 2007. Wake up – it's the Instant-Sleep Machine, The Times. 5 May, p. 3.
O'Connor, C., and H. Joffe. 2013. How has Neuroscience Affected Lay Understandings of Personhood. A Review of the Evidence. *Public Understanding of Science* 22 (3): 254–268.

OECD. 2017. *Neurotechnology and Society. Strengthening Responsible Innovation in Brain Science*, OECD Science, Technology and Innovation Policy Papers No. 46. Paris: OECD Publishing.

Ong, A., and S.J. Collier. 2005. *Global Assemblages Technology, Politics, and Ethics as Anthropological Problems*. Malden: Blackwell Publishing.

Ortega, F. 2009. The Cerebral Subject and the Challenge of Neurodiversity. *BioSocieties* 4 (4): 425–445. https://doi.org/10.1017/S1745855209990287.

Pickersgill, M. 2018. The Value of the Imagined Biological in Policy and Society: Somaticizing and Economizing British Subject(ivitie)s. In *Handbook of Genomics, Health and Society*, ed. S. Gibbon, B. Prainsack, S. Hilgartner, and J. Lamoreaux, 99–107. London: Routledge.

Pickersgill, M., S. Cunningham-Burley, and P. Martin. 2011. Constituting Neurologic Subjects: Neuroscience, Subjectivity and the Mundane Significance of the Brain. *Subjectivity* 4 (3): 346–365. https://doi.org/10.1057/sub.2011.10.

Pitts-Taylor, V. 2010. The Plastic Brain: Neoliberalism and the Neuronal Self. *Health: An Interdisciplinary Journal for the Social Study of Health, Illness and Medicine* 14 (6): 635–652. https://doi.org/10.1177/1363459309360796.

Rees, T. 2018. *After Ethnos*. Durham: Duke University Press.

Rose, N.S. 2007. *Politics of Life Itself: Biomedicine, Power, and Subjectivity in the Twenty-First Century*. Princeton: Princeton University Press.

Rose, N., and J.M. Abi-Rached. 2013. *Neuro: The New Brain Sciences and the Management of the Mind*. Princeton: Princeton University Press.

Sharon, T. 2014. *Human Nature in an Age of Biotechnology. The Case for Mediated Posthumanism*. Dordrecht: Springer.

Shaw, J.A., and H. Shaw. 2015. The Politics and Poetics of Spaces and Places: Mapping the Multiple Geographies of Identity in a Cultural Posthuman Era. *Journal of Organizational Transformation & Social Change* 12 (3): 234–256.

Silverman, C. 2008. Brains, Pedigrees, and Promises: Lessons from the Politics of Autism Genetics. In *Biosocialities, Genetics and the Social Sciences: Making Biologies and Identities*, ed. S. Gibbon and C. Novas, 38–55. London: Routledge.

Simondon, G. 2009. Technical Mentality. Trans. A. De Boever. *Parrhesia* 7: 17–27.

Singer, J. 1999. Why Can't You Be Normal for Once in Your Life?: From a 'Problem with No Name' to a New Category of Disability. In *Disability Discourse*, ed. M. Corker and S. French, 59–67. Buckingham/Philadelphia: Open University Press.

Tsing, A.L. 2015. *The Mushroom at the End of the World: On the Possibility of Life in Capitalist Ruins*. Princeton: Princeton University Press.

Vidal, F. 2009. Brainhood, Anthropological Figure of Modernity. *History of the Human Sciences* 22 (1): 5–36. https://doi.org/10.1177/0952695108099133.

Volkow, N. 2010. *Challenges and Opportunities in Drug Addiction Research*. https://dana.org/article/a-decade-after-the-decade-of-the-brain. Accessed on 22 Aug 2019.

Weiser, M. 1991, September. The Computer for the 21st Century. *Scientific American* 265: 94–104.

Wimmer, F.M. 2004. *Interkulturelle Philosophie. Eine Einführung*. Wien: Facultas.

Sabine Maasen holds the Friedrich Schiedel Endowed Chair in Sociology of Science at the Technical University of Munich (TUM), Germany. At the same time, she is Director of the Munich Center for Technology in Society (MCTS) at TUM. Her research focuses on the dynamics of innovation by research and legitimation of research in order to understand contemporary TechnoSocieties. She is Member of various committees and advisory boards, including the German Science Council.

Part II
TechnoScienceInfrastructures

Chapter 6
The Material Politics of Infrastructure

Andrew Barry

6.1 Introduction

In a survey of the anthropological literature on infrastructure, Brian Larkin opens with a definition: 'infrastructures are built networks that facilitate the flow of goods, people, ideas and allow for their exchange over space' (Larkin 2013: 328). In this view, infrastructures are fundamentally spatial formations; they facilitate connections, exchange, networks, flows and mobilities that cut across and reconfigure existing spatial borders and territories. This essentially spatial image of infrastructure as a series of 'built networks' reflects over 30 years of scholarship across the social sciences. In his monumental *Sources of Social Power*, for example, the historical sociologist Michael Mann placed great emphasis on the importance of road building in the consolidation of the Roman Empire (Mann 1986). The geographer David Harvey argued in turn that the development of the infrastructure of capitalism played a critical part in the process of what he called 'space-time compression' (Harvey 1989). And drawing on Foucault's analysis of biopolitics, Thomas Osborne demonstrated the importance of drains and sewers to the formation of what he termed the 'hydraulic city' (Osborne 1996: 114; see also Gandy 2014).

The wider interest of historians and social scientists in the spatiality of infrastructure over the last 30 years has also been reflected in STS. Historians and sociologists of the technosciences have long argued that the formation of enduring and spatially extended 'built networks' has come both to rely on and to foster the progressive consolidation of technoscientific standards. Ken Alder pointed to the vital role of metrological infrastructures in the political rationality of France after the revolution (Alder 1997; see also Foucault 1986), while historians of the physical sciences interrogated the critical links between electrical research, standardization and communication (e.g. Smith and Wise 1990). As Simon Schaffer reminded us,

A. Barry (✉)
University College London, London, UK
e-mail: a.barry@ucl.ac.uk

S. Maasen et al. (eds.), *TechnoScienceSociety*, Sociology of the Sciences Yearbook 30, https://doi.org/10.1007/978-3-030-43965-1_6

the integrity of the British imperial telegraphic network relied on the work of electrical laboratories and workshops dedicated to the values of precision, disciplined observation and standards (Schaffer 1992; Barry 1996: 134). And according to Susan Leigh Star, infrastructure has been integral to the spatial formation of 'large-scale systems' in general (Star 1999; see also Star and Ruhleder 1996). This line of interest in the critical relation between infrastructure and the formation of new spaces of political rule continues. More recent research has argued that infrastructural development is intimately associated with the history of state building (Guldi 2012; Rowland and Passoth 2015) and the reconfiguration of transnational political and economic spaces (Barry 2001, 2006).

In this chapter, I take the discussion of the relation between STS and the study of infrastructure in a different direction. Following Larkin I draw extensively on the work of anthropologists and urban geographers who, since the mid-2000s, have become increasingly interested in the study of infrastructures and their politics. However, whereas Larkin's brief definition emphasizes the networked spatiality of infrastructure, I note that recent work has also highlighted the temporality of infrastructure (Howe et al. 2016; Calkins and Rottenburg 2016: 254; Karsati et al. 2010): infrastructures acquire additions or 'accretions' (Anand 2015); they demand more or less continuous maintenance and repair; and they break down routinely (Barry 2015, see 2002). Moreover, whereas STS scholars once argued that infrastructures were often invisible, or only visible on breakdown (Star and Ruhleder 1996; Lampland and Star 2009: 207), the recent anthropological literature has tended to emphasize the mundane visibility of infrastructure in everyday life. Users of infrastructure have continually to be aware of at least some of infrastructure's instability, vulnerability, and even 'liveliness' (Edwards 2003: 188; Harvey and Knox 2010; Schwenkel 2013; Amin 2014; Björkman 2015). They may also recognize some of the 'sharp differences of access [to infrastructure] between the rich and the poor and the constant battle between them over public goods' (Amin 2014: 139; Graham et al. 2013). It emerges from these studies that a technoscientificsociety is one in which the urban citizen is enmeshed in multiple infrastructures, and their diverse and shifting materialities.

My focus on infrastructure in process raises two further issues, which lie at the heart of my argument, and return me to the concerns of STS. The first is that any focus on process and change must also address infrastructure's material composition. After all the development of infrastructures, such as power stations, cables and pipelines, has not only generated novel socio-technical assemblages, but also the emergence of new materials. At the same time infrastructural materials are subject to corrosion, fatigue, decay, and wear and tear, processes that may be exacerbated by neglect, theft, over use, sabotage, hacking, obsolescence or changes in the environments in which they are embedded. Whereas sociologists of technology once directed us to consider the ways in which the infrastructural systems were 'shaped' or constructed (e.g. Mackenzie and Wajcman 1985; Bijker et al. 1987), recent interest in the material politics of infrastructure draws our attention both to the progressive transformation of materials (Barry 2013), and technosocial worlds (Björkman 2015). Moreover, an analysis of the material politics of infrastructure should not equate infrastructure with a series of well-bounded objects or even a distinct infrastructural system (Star and Ruhleder 1996). Infrastructure is material but also always more than matter, and includes both scientific observations and plans as well as 'apparently immaterial phenomena' like imagination, emotion, pain and suffering (Latham and McCormack 2004: 705; Masquelier 2002; Petryna 2002; Sheller 2004; Elyachar 2010; Lockrem 2016). Any analysis of infrastructure's materiality must be concerned, then, *both* with the changing physico-chemical composition of infrastructure and with the apparently immaterial.

Second, once it is recognized that infrastructures are subject to transformation, the ongoing endurance of infrastructures becomes a question. Yet the endurance of infrastructures frequently depends not just on the development of standards, but also on the existence of bodies of expertise and regulation that are explicitly concerned with change, and managing 'the tension between long-term and short-term temporal scales' (Karsati et al. 2010: 401). Infrastructural 'zones of connectivity' such as telecommunication networks, rail tracks and oil pipes co-exist with scientific and legal infrastructures, or what I have termed 'metrological zones' and 'zones of qualification' (Barry 2006). These latter technoscientific and regulatory infrastructures play a vital and under-acknowledged part in the material politics of infrastructure. On the one hand, the on going state of an infrastructure maybe subject to more or less routine forms of inspection and monitoring. Infrastructures such as dams, rigs and power lines co-exist with technoscienfic infrastructures of, for example, flood risk management (Whatmore 2013), corrosion inspection (Barry 2013), subsea environmental monitoring (Parmiggiani and Monteiro 2016), and seismic early warning systems (Farías 2014). Thus routinised technoscientific practices are expected to detect or anticipate threats to the integrity or resilience of infrastructure and to pinpoint its vulnerability (Lakoff and Collier 2010). On the other hand, the potentiality social, economic and environmental impact of infrastructures are imagined, assessed, and anticipated in advance. Infrastructures take the form of technosocial *projects* (Ureta 2015) that maybe 'gestured' in advance (Weszkalnys 2016), even if the actualisation of the project is never realized or turns out to be quite different from its original projection (Khan 2006; Born 2007; Davies 2015; Pollock and Williams 2015).

In the paper that follows, I consider both of these issues. In the first part of the paper I focus on how recent research by anthropologists and urban geographers directs us to consider temporality of infrastructure; its endurance as well as its fragility and instability. In the second part of the paper, I address the ways in which infrastructures co-exist with further infrastructures of impact assessment, modeling, evidence gathering, monitoring, and public engagement. An analysis of the material politics and endurance of infrastructure needs to interrogate the relation between specific infrastructures and a second-order infrastructure of technoscientific knowledge production.

6.2 Infrastructure in Process

If the endurance or stability of infrastructure cannot be assumed then the question of how infrastructure endures, or fails to endure, comes to the fore. Endurance has to be understood not as given but as on going process of socio-material transformation. Endurance, one might say, is not the opposite of change, but its outcome (Whitehead 1920, 1985). In this context, one of the strengths of recent anthropological and geographical research on infrastructure, particularly in postcolonial contexts, is that it draws attention to the ways in which the 'variable endurance' of infrastructure is unevenly realized, often in precarious circumstances and harsh conditions (Howe et al. 2016: 6; see Lea and Pholeros 2010). Endurance is a situated achievement that may rely on the intervention of users, as well as government agencies, and professional scientists and engineers.

Whereas historians and sociologists have emphasized the spatial politics of infrastructure, there is also a politics to infrastructure's endurance. In part, this is because the endurance of infrastructure continues to raise questions of equality, justice, and geographically uneven development, and the relation between infrastructural provision and colonial and postcolonial history (Anand 2011; Appel 2012). After all, 'access to material systems of circulation – such as water, electricity, communication and transportation – is critical to participation in collective life' (Lakoff and Collier 2010: 243). In part the endurance of infrastructure is political because it poses the on-going (cosmopolitical) challenge of sustaining and transforming relations between persons and things that may be potentially dangerous or toxic (Hecht 2012), as well as enabling (Stengers 2005). Drawing on recent anthropological and geographical research, I highlight five ways in which infrastructures endure or fail to endure over time *Breakdown and Failure; Accretion; Repair; Remnants and After-Life;* and *Projection and Invention.* Each has its own politics and may, moreover, become explicitly politicised 'in concrete situations' (Stengers 2005: 994).

Historians and sociologists of technology have long been concerned with the significance of *breakdown and failure* to the social and political life of infrastructure. In her influential account of the 'ethnography of infrastructure', Susan Leigh Star observes that infrastructures establish connections between spatially or temporally separated sites or events. But she also makes a strong claim about the importance of attending to the moments at which they break down. 'The *normally invisible* quality of infrastructure', she argues, 'becomes visible when it breaks: the server is down, the bridge washes out, there is power blackout' (Star 1999: 382, my emphasis). This claim introduces two critical observations about the politics of infrastructure, both of which have been, in part, challenged. One observation, to which I will return, concerns the visibility or invisibility of infrastructure, and the 'normality' if its invisibility. How, when, where and in what way is infrastructure rendered either or invisible, and to whom? (Barry 2013: 116; Larkin 2013; Björkman 2015; Howe et al. 2016: 5; Harvey et al. 2016: 4). The other concerns the critical importance of breakdown, accidents and failures, including the risk of failures and accidents in both the past and future, in accounts of infrastructure's politics (Perrow 1999).

The fact that breakdowns, failures or accidents play an important part in infrastructure's public politics should come as no surprise. The occurrence of breakdowns is routinely publicized by news and social media, raising in this way as matters of concern the question of how well those infrastructures that have failed have been managed, designed and assembled. Indeed, when the construction of infrastructure is regarded as the responsibility of either the state or corporations, then infrastructural failures can readily become the catalyst for wider criticisms of the legitimacy of either state or corporate agencies (Humphrey 2003). If the users of infrastructure – whether a rail network or an electricity supply system – assume that political and economic interests have influenced its design and development (see Mackenzie and Wajcman 1985), then infrastructural failure maybe taken, rightly or wrongly, to reveal something about precisely how those interests shaped infrastructure's form (Barry 2002). Accidents or breakdowns, or the possibility of future breakdowns, can become indices of a more systematic failure in government. Infrastructural failure therefore opens up the potential for abductive inferences about the causes of failure; the particular failure tends to draw attention to the wider (political, economic, or cultural) context that is thought to explain why the particular failure occurred (Barry 2013).

An analysis of the relation between infrastructural breakdown and corporate and state interests is necessary but is not sufficient. Whatever we understand infrastructure to be, the properties and behavior of its nonhuman or material elements are surely irreducible to human interests and agency. This was, of course, one of the core arguments of actor network theory in its earliest formulations (Callon and Law 1982), and lies also at the heart of the political theorist Jane Bennett's analysis of what she understands to be the agency of more than human assemblages: 'the electrical grid is a volatile mix of coal, sweat, electromagnetic fields, computer programs, electron streams, profit motives, heat, lifestyles, nuclear fuel, plastic, fantasies of mastery, static, legislation, wire, water, economic theory, wood – to name just some of the actants' (Bennett 2005: 448, 2010). In Bennett's account, the failure of an electricity grid should *not* be taken to be simply a sign of the failure of the American state or political interests. It has to be understood as an index of the assemblages in which 'very active and powerful' nonhuman actants play a critical role. In this way, Bennett's 'political ecology of things' gives a vitalist twist to the relational materialism of actor-network theory. The failure of infrastructure becomes a sign not of the failure of the state but of the powers and the liveliness of materials that would otherwise be unrecognized.

Nonetheless, 'vital materialism' provides us with only a limited account of the material politics of infrastructure, for three reasons. First, as Bruce Braun and Sarah Whatmore ask, in ascribing liveliness to matter *in general*, do not vital materialist accounts like Bennett's side step the question of the specificity of particular material infrastructural forms, 'as opposed to a generic analogy to life that could be described as a metaphysics?' (Braun and Whatmore 2010: xxix–xx). An emphasis on what is variously termed the agency, activity or powers of materials in general tells us little about specific materials and their historically and situationally specific properties and forms. An account of material politics thus requires further analysis of the situated and relational properties of particular materials (Abrahamsson et al. 2015: 13; Barry 2013: 149), including the 'emergent relations between infrastructures, their human developers, and numerous other entities, such as animals and trees' (Brunn Jensen 2016: 228). Secondly, by focusing our attention on the unexpected liveliness of matter, a vital materialist account arguably downplays the degree to which the behavior of such materials is monitored, regulated and contained. In effect, vital materialism does not provide a developed account of the variable ways in which infrastructures are the objects of government, as well as the ways in which the properties of particular infrastructural materials are effectively or ineffectively governed in practice (Barry 2001; Braun 2011; Collier 2011). Thirdly, by highlighting the politics of socio-material assemblages in general, vital materialism fails to address the ways in which materials are *made* political, along with the specificity of politics as a set of practices, techniques, institutions, and forms of enunciation (Latour and Weibel 2005; Gregson 2011; Barry 2013; von Schnitzler 2016; see also Brown 2015).

Moreover, while accidents and breakdowns play a critical role in the politics of infrastructure, it is also important to consider other ways in which the condition of infrastructure is transformed over time. As anthropologists have increasingly argued, infrastructural failure or breakdown is not necessarily a rare occurrence, particularly in postcolonial contexts. Tess Lea and Paul Pholeros observe that the media in Australia tend to blame indigenous people for the poor state of their own housing infrastructure, rather than the practices of those 'policy makers, regulators, tradesmen and manufacturers' who provide housing of such poor quality that it is unlikely to function as housing (Lea and Pholeros 2010: 193). And, as Michele Lancione and Colin McFarlane argue, writing on urban sanitation, a focus on breakdown 'runs the risk of imposing a Western-centric conceptualization, given that for many urbanites infrastructures rarely "work" in reliable expected ways, and may be absent or denied entirely' (Lancione and McFarlane 2016: 48). In many postcolonial cities, urban residents need to be routinely concerned about the breakdown of transport, communication and electricity systems and the possibility of disruptions, fluctuations in service and interruptions in supply. In these circumstances, the user of infrastructure needs to be continuously updated, or may need to improvise in the context of changing conditions (Barry 2000). The user should not be understood as decoupled or detached from infrastructure, for the changing state of infrastructures immediately impacts on everyday life, pressing directly 'on the senses and the body' (Lancione and McFarlane 2016: 55; Latham and McCormack 2004: 719; Lockrem 2016).

Thus infrastructures change over time in diverse ways that do not necessarily lead to catastrophic breakdown, but are nonetheless significant. As urban geographers and anthropologists have recently argued, urban infrastructures acquire *accretions* (Anand 2015) or *increments*: 'inhabitation in urban postcolonial worlds has depended on practices of incrementalism, where residents seem to add on to built environments, livelihoods and social networks a "little bit at a time"' (Simone 2014: 24). As Anand observes, the users of infrastructure may add their own modifications, or divert or modify flows of materials, electricity and information by tapping into existing infrastructures. In turn, the practices of accretion and incrementalism and other modifications generate their own politics, as well as new informal markets. Graham et al. (2013), for example, document the occurrence of what they term 'water wars' in Mumbai, one aspect of which revolves around conflicts between the efforts of the urban poor to gain access to piped water supplies and attempts by the authorities to police and contain these efforts. In their attempts to access water, the urban poor continually add to the existence of infrastructure, diverting the flow of water along unauthorized channels. The same observation could be made about the tapping of electricity grids in the former Soviet Union or oil pipelines in Nigeria. In these circumstances, the transformation of infrastructure is not confined to moments of breakdown. The modification and incremental accretion of infrastructure is routine and does not necessarily lead to systemic failure; indeed it may contribute to the warding-off of potential failures, whether catastrophic or routine. Infrastructure is continually adjusted in response to temporary or localized interruptions or stoppages.

Anand's emphasis on accretion therefore draws our attention to the ways in which users transform infrastructures over time. As he reminds us, infrastructures 'are not smooth surfaces that perform as planned; instead they are flaky, falling-apart forms that constantly call out for projects of management, maintenance, and repair' (Anand 2015). Lisa Björkman (Björkman 2015), in her research on pipeline politics in Mumbai, also develops this argument. As she argues, from the point of view of the individual household, the failure of Mumbai's hydraulic infrastructure is common enough and the availability of water is routinely variable in time and space, yet the frequent interruptions to Mumbai's water supply cannot be understood simply as a manifestation of the chaotic growth of the city, nor as the product of privatization, of a prevalent corruption, or of the wider context of neoliberalism or 'planetary urbanisation' (Björkman 2015; see Brenner 2014). She argues that such interruptions require us to attend to the materiality of water and pipes and, above all, to the work of engineers and their assistants who, in the absence of reliable maps of the location of pipes and accounts of their state of repair, maintain (uneven and unequal) public access to the hydraulic network (see also Anand 2012). As Björkman's ethnography persuasively shows, the plumbers also engage in explicitly political work, which necessarily includes electioneering as well as engineering (see also Barnes 2016). In brief, a focus on the everyday work of repair 'highlights actors, sites, and moments', such as the work of Mumbai's engineers and plumbers, 'that have been absented or silenced by stories of design and origination, whether critical or heroic' (Jackson 2015; see also Graham and Thrift 2007).

Early research in STS was focused primarily on the design and development of new infrastructural systems, and had less to say about their ongoing evolution in a technoscientificsociety. But if many researchers have subsequently been concerned with the progressive transformation of infrastructure through both design and use (Pipek and Wulf 2009; Venters et al. 2014; see also Edgerton 1999), others have been increasingly interested in what we might call the *remnants and afterlife* of infrastructures, including their deconstruction and, in some cases, reconstruction in new forms. These include, for example, recent studies of the aftermath of the Chernobyl disaster and its radioactive residue (Petryna 2002; Kuchinskaya 2014), a growing body of research on the reverberations of violence as it is inscribed in destroyed, abandoned and reconstructed buildings (Navaro-Yashin 2012; Forensic Architecture 2014), and studies of urban ruination and decline (Mah 2012). Moreover, urban political ecologists and environmental justice activists have had a longstanding interest in the spatially uneven distribution of pollution in which may inhere the residue of earlier infrastructures, including ports, markets, power stations and pipelines (Heynen et al. 2005; Walker 2014). The legacy of past infrastructures is embodied not just in abandoned buildings and disused rail tracks, but also in the distribution of such chemicals and toxic materials as mercury, lead and asbestos. Infrastructure's material politics do not just revolve around the existence of distinct objects and functioning systems, then, but participate in controversies that foment around the geography of physical and chemical residues and waste and the epidemiology of the diseases they foster.

Infrastructures not only have after-lives in the form or ruins or residues, they also are *projections* in to the future, generating anticipations of future benefits or risks in the present (Kahn 2006; Born 2007; Howe et al. 2016; Hetherington 2016). Indeed, the engineer of infrastructure should be understood as the creator of fictions, 'since fiction, the projection of a state of technology from five or fifty years in the future to a time T, is part of their job' (Latour 1996: 24; see also Pollock and Williams 2015). In this light, ethnographers have been increasingly interested not just in the life of existing infrastructures but in the anticipation of future infrastructure. The anthropologist Gisa Weszkalnys, for example, highlights the ways in which future development of oil infrastructures may be gestured in the present in the form of agreements, plans and anticipatory laws (Weszkalnys 2016). Of course we should not expect that the projected form or anticipated impact of future infrastructural developments correspond to the ways in which they are imagined in the present; infrastructures may offer the unrecognized possibilities for new arrangements of persons and things. Indeed, the development of infrastructures can be *inventive*, opening up new possibilities or *anti-inventive*, reducing the space of future possibility, although we may not know in advance whether a particular infrastructure will prove to have primarily inventive or anti-inventive consequences (Barry 2001, 2007; see Zittrain 2006). In short, the development of infrastructures generates an ontological politics of both uncertain and future possibility (see also Mol 2002; Stengers 2005; Ribes and Polk 2015; Farías and Blok 2016).

6.3 Technoscientific Infrastructures

By adding these perspectives, recent anthropological and geographical research draws our attention to the ways in which infrastructures form critical elements of social and political life. The provision of infrastructure addresses human needs, often unequally and imperfectly, but it also generates the sights, smells, and sounds that animate everyday life, forming an essential part of the experience of the city in particular (Larkin 2008; Truitt 2008; Amin 2015). Moreover, as urban anthropologists and geographers have argued, infrastructural materials – from concrete and steel to copper and silicon – enter into diverse assemblies that make up the life of the city and contribute to the emergence and constitution of political movements (Amin and Thrift 2002). The politics of infrastructure thus entail what Ignacio Farías and Anders Blok view as a 'shifting and provisional articulation of human and non-human cohabitation' (Farías and Blok 2016: 8). In these mutating circumstances, urban residents have to manage with, add to and adjust the material infrastructures on which they depend (see de Laet and Mol 2000).

This anthropological and geographical emphasis on the fluidity of infrastructural assemblages needs, however, to be further supplemented in two ways. First, if it is critical to acknowledge the ways in which infrastructures include 'immaterial' elements (Latham and McCormack 2004; Elyachar 2010), it is equally important to consider the physico-chemical dynamics of specific materials, including their propensity to corrode, crumble and fracture. While much of the recent anthropological literature has been concerned with the materiality of infrastructure, it generally stops short of providing an analysis of the chemistry and physics of, for example, corrosion, contamination, fatigue or creep as they inhere in infrastructure (Bensaude-Vincent and Stengers 1996; Barry 2013; Harvey and Knox 2015). I label these diverse processes *corrosion*, observing that they do not necessarily lead to either partial or catastrophic failure or the demand for immediate repair, nor are they necessarily observable, except through their technoscientific mediation. A bridge or an underground pipe may, for example, accumulate rust and yet continue to function as a bridge or pipe; it is both the same infrastructure and yet the material composition is always differing. As AN Whitehead observed in a discussion of the life of a monument in London: 'a physicist who look on that part of the life of nature as a dance of electrons will tell you that daily [that the monument] has lost some of its molecules and gained others, and even the plain man can see that it gets dirtier and is occasionally washed' (Whitehead 1920: 167). One of the challenges that infrastructure poses to its users and to those who have responsibility for its repair and maintenance is to demand that they sense – either somatically or cognitively – when and where corrosion matters and when it can be ignored. The technosciences of corrosion need to not to generalize too hastily, but also take into account the ways in which the properties of materials are situated (Stengers 2010; Barry 2013: 138–139).

While the prevalent focus on the fluidity of infrastructure is valuable, it needs to be supplemented by attending to the ways in which infrastructures, whether railway networks, dams, canals, energy supply systems, telecommunication networks or pipes may co-exist with additional scientific, regulatory and informational infrastructures and devices which also demand our attention. It is these technoscientific or knowledge infrastructures that play a critical part in both the government and the politics of infrastructure (Calkins and Rottenburg 2016). In brief, infrastructures exist in conjunction with further infrastructures that, in turn, can be inventive or anti-inventive in practice, opening up or closing down the space of possibility. Drawing on the work of STS scholars, as well as my own research, I highlight seven distinct aspects of the politics of technoscientific infrastructures; *the government of infrastructural materials; monitoring infrastructures; informed materials; experts and publics; visibility; political situations; interference.*

Certainly, the material components of infrastructures maybe subject to a growing range of regulatory requirements, although the form and intensity of regulation is geographically uneven (see Smith 2010). The material components of a car are regulated in the EU, for example, in order to mitigate the risk of road accidents and injury to passengers, and to reduce the level of CO_2 emissions and other forms of atmospheric pollution. The car driver may or may not drive in a responsible way, but she nonetheless expects – rightly or wrongly – that regulation will act ethically on her behalf. Of course, as the recent scandals involving Volkswagen and other European motor manufactures demonstrate (Barry 2002), this does not mean that regulation enacts this ethical expectation in practice. Likewise, the circulation of solid, electrical and chemical waste is subject to monitoring and inspection in order to protect the environment, as well as on grounds of health and safety, security or effective resource use (Bulkeley et al. 2007; Kama 2015). In these contexts and others, a politics arises concerning the *government of infrastructural materials*, for what purposes, in what ways, and in relation to what locations and spaces (Barry 2006). The material politics of infrastructure necessarily revolve in part around the question of how its material composition is governed, while also attending to the ways in which materials exceed efforts to govern their behavior (Tironi and Calvillo 2016).

Thus in so far as the material elements of infrastructures are subject to regulatory requirements, this generates a regime of testing, modeling, monitoring and simulation. Such infrastructural entities as aircraft wings, electrical cables, train tracks, oil platforms and pipelines are maybe the subject of tests for corrosion, creep, fatigue and accidental or malicious damage more or less routinely and systematically throughout their operational life. In turn such mundane practices of testing and monitoring have their own regulatory infrastructures or 'metrological zones' associated with particular nation states or transnational actors or organisations like the European Union, multinational companies or industrial sectors (Barry 2001, 2006). At the same time, efforts to monitor and assess the performance of materials are delegated to corporations and consultancies, as well as some instances, NGOs and lay publics (Barry 2013), enabling the exercise of political power 'beyond the state' (Rose and Miller 1992). In these circumstances, regimes and practices of testing, simulation, evidence gathering, and monitoring have their own uneven and contested economy and organisation, which demands further interrogation (Ayre 2016; see Smith 2010). Yet just as 'zones of connectivity', such as wires and pipes are more or less fragile, unstable and prone to failure (Appel 2012; Mains 2012), so too are the technoscientific infrastructures that are expected to keep the fragility of infrastructure in check.

While the properties of infrastructural materials, such as glass, concrete or steel, maybe subject to regulation and inspection, the properties of material have been progressively enhanced or fine-tuned to specific tasks. In Marx's classical analysis, the *shape* of technical artefacts was determined by the demands of industrial capitalism: the specialist design of hammers, he suggested, mirrored the emerging specialist division of labour of the manufacturing system (Marx 1973; Mackenzie and Wajcman 1985). But the idea that infrastructural materials are shaped by external economic and physical forces provides us with only a limited understanding of the genesis of infrastructural materials today. For such materials are not produced or designed as bare objects isolated from the production of information about their properties. Rather, infrastructural materials such as steel and glass routinely co-exist with tests, samples, models and simulations that enable their properties and characteristics to be abstracted, circulated, traced, and translated into new material forms (see Mol 2002; Barry 2005). The engineer or materials scientist is not interested simply in shaping the design of material artefacts; instead, she is interested in creating a system of relations between the particular materials at issue and forms of testing, modeling and monitoring that enable the performance of those materials to be enhanced. In effect, the engineer translates between analyses of the internal microstructure of infrastructural materials and the wider economic and regulatory requirements that the development of this specific infrastructure is expected to address (Barry 2013: 142). As Katie Lloyd-Thomas observes, 'regimes of testing and standardisation make it possible for the quantification of both the properties of materials and [the] values associated with the structural and social uses they are to be put to….[I]n performance-engineered materials these goals become *internal* to their physical arrangement' (Lloyd-Thomas 2010: 161, my emphasis). In these ways the material component of infrastructure, whether a track, a window or a pipe, participates in an information-material assemblage – becoming what I have called, after Bernadette Bensaude-Vincent and Isabelle Stengers, an 'informed material' (Barry 2005; Bensaude-Vincent and Stengers 1996; Lloyd-Thomas 2010). The concept of informed material is historical; the accelerating production of information about the properties and performance of materials has led to the development of informed materials that are more or less 'richer and richer in information' (Bensaude-Vincent and Stengers 1996: 141).

In these circumstances, the question of which body has the responsibility, the legitimacy or the expertise to perform such monitoring of the performance or safety of a given infrastructure may become contested and politicized. STS scholars have rightly been concerned with the need to and democratize the practice of monitoring and assessment through fostering new forms of engagement between *experts and publics* and developing novel ways of organizing 'citizen science' (e.g. Callon et al. 2009; Maasen and Weingart 2006; Chilvers and Kearnes 2016; Haklay 2013; Whatmore 2013). They have also been increasingly interested in the use of material devices that animate participation (Barry 2001; Marres 2013). But arguably they have had less to say about the complex relations between state and commercial organisations in the conduct of monitoring (see Mitchell 1999), or the extent to which monitoring is delegated to corporations. In these circumstances, it is not only the failure or breakdown of infrastructural materials that is likely to be politically significant, leading to questions about the responsibility or failure of the national and local state. Those concerned with infrastructure's material politics have equally to be attentive to the practice and the failures of monitoring and assessment, or what we might term the 'modes of governing' infrastructure (Bulkeley et al. 2007; see also Barry 2002; Ureta 2014). Indeed, even in the absence of any demonstrable breakdowns in the functioning of infrastructure, critics of multinational oil corporations have pointed to failures in the conduct of assessment and monitoring – and taken these failures to be an index of both the weakness of transnational governance and the irresponsibility of the corporations themselves (Barry 2013).

A further argument follows, returning to the question raised by Star of the *visibility of infrastructure*. As noted earlier, Star contends that infrastructures are likely to be most visible at the moment of breakdown or failure. Yet although infrastructures are also increasingly visible through their technoscientific mediation, as well as through the proliferation of forms of monitoring and assessment, they are nonetheless unevenly so. Two political questions arise from this observation. One concerns what aspects of the evolving condition of infrastructure are rendered visible in practice. Any measurement of the condition of infrastructure, or of its performance or impact, necessarily abstracts from infrastructure's ongoing existence. The visibility of infrastructure is always both partial and mediated, and there is therefore a politics to the question of what is rendered visible, and what range of interventions is thus made possible, and for whom (see Mol 2002). A second inevitable question concerns the extent to which the technoscientific mediations of infrastructure are made public and open to public inspection and contestation. In recent years, there has been a growing emphasis on the importance of transparency as a principle of government (Hetherington 2011), yet the transparency is inevitably targeted and selective. The adoption of the principle of transparency by governments, transnational organisations, and corporations does not imply that all technoscientific information is made public; rather it establishes the conditions within which the question of the *visibility of infrastructure* can in principle become a political issue. It also raises the question for the researcher of what tends to be excluded or lies outside the realm of transparency, including the domains of state politics and finance (Brown 2015). Transparency has become what we can call, following Foucault, a 'diagram of a mechanism of power' (Foucault 1977: 205; see also Deleuze 1986), which has come to play a critical part in the politics of transnational infrastructure (Barry 2013).

The production of technoscientific knowledge about infrastructure can have a governmental function, yet also generates the possibility of dispute. Indeed, sociologists of scientific knowledge have long recognized that the production of scientific knowledge can lead to controversies that revolve around a wide of range of issues, including: the quality, relevance and origins of evidence; the competence, qualifications and trustworthiness of researchers; the relations between claims to knowledge and financial and political interests; the reliability of instruments; and the extent to which experts have been held publically accountable and their practices made transparent (Collins and Pinch 1993). All of these features of scientific knowledge controversies in general also apply to the controversies that revolve around the development and operation of technoscientific infrastructures of monitoring and assessment. Disputes may focus not just on the need for an infrastructure or the interests that have driven its development, but also whether its environmental impact was sufficiently assessed, or the degree to which interested publics have been adequately informed, consulted, or involved (Callon et al. 2009). Elsewhere I have argued that studies of the politics of infrastructure need to focus not so much on discrete issues or matters of concern, but on evolving *political situations* that are likely to contain multiple issues, which may have radically different and uneven historical trajectories and intensities (Barry 2013). For example, the construction of an oil pipeline or a nuclear power station may lead to disputes not just about the relation between energy infrastructures and longstanding geopolitical conflicts, but also to disputes about the reliability and trustworthiness of experts who work on behalf of the oil or nuclear industries, as well as quite specific questions about the material construction of pipelines, refineries and power stations.

As Larkin observes, infrastructures can be understood, in the first instance, as spatial networks. They enable persons, commodities and information to move along well-defined paths; yet in this way infrastructures inevitably cut across, reinforce or disrupt other territorial forms of political division. Moreover, if many infrastructures take a networked spatial form, these networks are not isomorphic with one another. Some infrastructures are more or less contained within the territorial boundaries of the nation-state, or within the borders of a geopolitical entity like the European Union. But others may cut across and reconfigure such national and transnational boundaries. Indeed, there is marked and significant disjuncture between the infrastructural networks of the nation state and those transnational networks forged by scientific institutions, transnational governments or multinational corporations (see, for example, Kama 2016). At the same time, infrastructures may both traverse and transform diverse environments.

In these circumstances, two observations follow. The first is that the immediate environment of an infrastructure should be thought of as part of the existence of that infrastructure. Larkin points towards this when he writes that the earth 'is the object that creates the ground on which other objects operate' (2013: 329). This is true, but the infrastructural qualities of the environment may need to be transformed, through cuttings, tunnels, dams and embankments, in order to render it a sufficiently stable 'ground' on which further specific infrastructures such as cables, pipes, canals, and rail lines can operate. As Ashley Carse observes, the 'natural' environment should not be considered to be external to infrastructure, but part of its constitution (Carse 2012). The second observation is that the dynamics of infrastructure and the dynamics of the earth interfere with each other contingently (Connolly 2011; Born 2015; Barry 2016). The earth forms part of the existence of infrastructure; yet at the same time, through the risks of landslides, earthquakes, tsunamis and floods, it potentially introduces instabilities into the existence of infrastructure. In turn, the potential instability of the earth generates the need for a further infrastructure of monitoring, which has its own infrastructure, materiality, politics and dynamics (Lakoff and Collier 2010; Whatmore 2013; Barry 2016). Infrastructures do not exist in isolation; their endurance, as well as their disintegration, is related recursively to the existence of further infrastructures.

6.4 Conclusions

Infrastructure has often been conceived of as a solid base for the construction of social and economic relations, forging connections across space and, in so doing, reconfiguring and reassembling the social and political. Here I've argued that any account of the material politics of infrastructure needs to supplement a focus on the spatiality of infrastructure in two related ways. One is to attend to the diverse ways in which infrastructures are in process. They do not form a static base on which social and political life can take place; they take enduring yet also shifting technosocial forms. The endurance of infrastructure therefore needs to be explained rather than assumed; infrastructures acquire accretions, demand repair, and their material components decay or corrode over time, and yet may remain operational and functional nonetheless. They also exist as residues and ruins, or as projections and gestures towards the future; their development both closes down and opens up possibilities, in ways that maybe both unpredictable and contested.

My second argument is that accounts of the material politics of infrastructure have to address the relation between infrastructures in general and the infrastructure of technoscientific knowledge production, including monitoring, inspection, modeling, simulation and environmental impact assessment. The endurance of infrastructures, as well as the sociomaterial worlds that they both generate and foreclose, have come to depend on their relation to further infrastructures that have their own materiality and spatiality. The study of these second-order infrastructures of knowledge production, and their own endurance, economy and visibility, has come to be critical for an analysis of the material politics of infrastructure.

References

Abrahamsson, S., F. Bertoni, A. Mol, and R.I. Martín. 2015. Living with Omega-3: New Materialism and Enduring Concerns. *Environment and Planning D: Society and Space* 33: 4–19.

Alder, K. 1997. *Engineering the Revolution: Arms and the Enlightment In France, 1763–1815*. Princeton: Princeton University Press.

Amin, A. 2014. Lively Infrastructure. *Theory, Culture and Society* 31 (7/8): 137–161.

———. 2015. Animated Space. *Public Culture* 37 (2): 239–258.

Amin, A., and N. Thrift. 2002. *Cities: Reimagining the Urban*. Cambridge: Polity.

Anand, N. 2011. Pressure: The Politechnics of Water Supply in Mumbai. *Cultural Anthropology* 26 (4): 542–564.

———. 2012. Municipal Disconnect: On Abject Water and Its Urban Infrastructures. *Ethnography* 4: 487–509.

———. 2015. 'Accretions', The Infrastructure Toolbox. *Cultural Anthropology*. http://culanth.org/fieldsights/715-accretion.

Appel, H. 2012. Walls and White Elephants: Oil Extraction, Responsibility, and Infrastructural Violence in Equatorial Guinea. *Ethnography* 13 (4): 439–465.

Ayre, B. 2016. Constructing an Arctic Laboratory. *Cultural Anthropology*. https://culanth.org/fieldsights/932-constructing-an-arctic-laboratory.

Barnes, J. 2016. States of Maintenance: Power, Politics and Egypt's Irrigation Infrastructure. *Environment and Planning D*: 1–19.

Barry, A. 1996. Lines of Communication and Spaces of Rule. In *Foucault and Political Reason: Liberalism, Neo-Liberalism and Rationalities of Government*, ed. A. Barry, T. Osborne, and N. Rose, 123–141. London: UCL press.

———. 2001. *Political machines: Governing a technological society*. London: Athlone.

———. 2000. Networks. In *City A/Z*, ed. S. Pile and N. Thrift, 160–163. London: Routledge.

———. 2002. The Anti-Political Economy. *Economy and Society* 31 (2): 268–284.

———. 2005. Pharmaceutical Matters: The Invention of Informed Materials. *Theory, Culture and Society* 22 (1): 51–69.

———. 2006. Technological Zones. *European Journal of Social Theory* 9 (2): 239–253.

———. 2007. Invention and Anti-Invention. In *Technoscience: The Politics of Interventions*, ed. K. Asdal, B. Brenna, and I. Moser, 287–308. Oslo: Unipub.

———. 2013. *Material Politics: Disputes Along the Pipeline*. Chichester: Wiley-Blackwell.

———. 2015. Infrastructural Times. *Cultural Anthropology*. https://culanth.org/fieldsights/724-discussion-infrastructural-times.

———. 2016. Infrastructure and the Earth. In *Infrastructure and Social Complexity*, ed. C. Bruun Jensen, P. Harvey, and A. Morito, 187–197. London: Routledge.

Bennett, J. 2005. The Agency of Assemblages and the North American Blackout. *Public Culture* 17 (3): 445–465.

———. 2010. *Vibrant Matter: A Political Ecology of Things*. Durham: Duke University Press.

Bensaude-Vincent, B., and I. Stengers. 1996. *The History of Chemistry*. Cambridge, MA: Harvard University Press.

Bijker, W., T. Hughes, and T. Pinch. 1987. *The Social Construction of Technological Systems*. Cambridge, MA: MIT press.

Björkman, L. 2015. *Pipe Politics, Contested Waters: Embedded Infrastructures in Millennial Mumbai*. Durham: Duke University Press.

Born, G. 2007. Future-Making: Corporate Performativity and the Temporal Politics of Markets. In *Cultural Politics in a Global Age*, ed. D. Held and H. Moore. London: Oneworld.

———. 2015. Making Time: Temporality, History, and the Cultural Object. *New Literary History* 46 (3): 361–386.

Braun, B. 2011. *Review of Jane Bennett, Vibrant Matter: A Political Ecology of Things*. Durham: Duke University Press, *Dialogues in Human Geography* 1(3): 390–405.

Braun, B., and S. Whatmore. 2010. *Political Matter: Technoscience, Democracy and Public Life*. Minneapolis: Minnesota University Press.

Brenner, N. 2014. *Implosions/Explosions*. Berlin: Jovis.

Brown, M. 2015. Politicizing Science: Conceptions of Politics in Science and Technology Studies. *Social Studies of Science* 45 (1): 3–30.

Brunn Jensen, C. 2016. Multinatural Infrastructure: Phnom Penh Sewage. In *Infrastructures and Social Complexity. A Companion*, ed. P. Harvey, C. Bruun Jensen, and A. Morita. London: Imprint Routledge.

Bulkeley, H., M. Watson, and R. Hudson. 2007. Modes of Governing Municipal Waste. *Environment and Planning A* 39: 2733–2753.

Calkins, S., and R. Rottenburg. 2016. Evidence, Infrastructure and Worth. In *Infrastructures and Social Complexity: A Companion*, ed. P. Harvey, C. Bruun Jensen, and A. Morita, 253–265. New York/London: Routledge.

Callon, M., and J. Law. 1982. On Interests and their Transformation: Enrolment and Counter-Enrolment. *Social Studies of Science* 2 (4): 615–625.

Callon, M., P. Lascoumes, and Y. Barthe. 2009. *Acting in an Uncertain World: An Essay on Technical Democracy*. Cambridge, MA: MIT press.

Carse, A. 2012. Nature as Infrastructure: Making and Managing the Panama Canal Watershed. *Social Studies of Science* 42 (4): 539–563.

Chilvers, J., and M. Kearnes. 2016. *Remaking Participation: Science, Environment and Emergent Publics*. London: Routledge.

Collier, S. 2011. *Post-Soviet Social: Neoliberalism, Social Modernity, Biopolitics*. Princeton: Princeton University Press.

Collins, H., and T. Pinch. 1993. *The Golem: What You Should Know About Science*. Cambridge: Cambridge University Press.

Connolly, W. 2011. *A World of Becoming*. Durham: Duke University Press.

Davies, G. 2015. Reading Alongside the Pipeline. *Dialogues in Human Geography* 5: 108–110.

de Laet, M., and A. Mol. 2000. The Zimbabwe Bush Pump: The Mechanics of a Fluid Technology. *Social Studies of Science* 30 (2): 225–263.

Deleuze, G. 1986. *Foucault*. Minneapolis: Minnesota University Press.

Edgerton, D. 1999. From Innovation to Use: Ten Eclectic Theses on the Historiography of Technology. *History and Technology* 16 (2): 111–136.

Edwards, P. 2003. Infrastructure and Modernity: Force, Time and Social Organisation in the History of Sociotechnical Systems. In *Modernity and Technology*, ed. T. Misa, P. Brey, and A. Feenberg, 185–225. Cambridge: MIT Press.

Elyachar, J. 2010. Phatic Labor, Infrastructure, and the Question of Empowerment in Cairo. *American Ethnologist* 37 (3): 452–464.

Farías, I. 2014. Misrecognizing Tsunamis: Ontological Politics and Cosmological Challenges in Early Warning Systems. In *Disasters and Politics: Materials, Experiments, Preparedness*, ed. M. Tironi, I. Rodriguez-Giralt, and M. Guggenheim, 61–88. Oxford: Basil Blackwell.

Farías, I., and A. Blok. 2016. Introducing Urban Cosmopolitics: Multiplicity and the Search for a Common World. In *Urban Cosmopolitics: Agencements, Assemblies, Atmospheres*, ed. A. Blok and I. Farías, 1–22. London: Routledge.

Forensic Architecture. 2014. *Forensis: The Architecture of Public Truth*. Berlin: Sternberg Press.

Foucault, M. 1977. *Discipline and Punish: The Birth of the Prison*. Harmondsworth: Allen Lane.

———. 1986. Space, Knowledge, Power. In *The Foucault Reader*. London: Penguin.

Gandy, M. 2014. *The Fabric of Space: Water, Modernity, and the Urban Imagination*. Cambridge, MA: MIT press.

Graham, S., and N. Thrift. 2007. Out of Order: Understanding Repair and Maintenance. *Theory, Culture and Society* 24 (3): 1–25.

Graham, S., R. Desai, and C. McFarlane. 2013. Water Wars in Mumbai. *Public Culture* 25 (1): 115–141.

Gregson, N. 2011. *Review of J. Bennett, Vibrant Matter*. Durham: Duke University Press, *Dialogues in Human Geography* 1 (3): 402–404.

Guldi, J. 2012. *Roads to Power: Britain Invents the Infrastructure State*. Oxford: Oxford University Press.

Haklay, M. 2013. Citizen Science and Volunteered Geographic Information: Overview and Typology of Participation. In *Crowdsourcing Geographic Knowledge*, ed. D. Sui, S. Elwood, and M. Goodchild, 105–122. Dordrecht: Springer.

Harvey, D. 1989. *The Condition of Postmodernity*. Oxford: Wiley-Blackwell.

Harvey, P., and H. Knox. 2010. The Enchantments of Infrastructure. *Mobilities* 7 (4): 521–536.

———. 2015. *Roads: An Anthropology of Infrastructure and Expertise*. Ithaca: Cornell University Press.

Harvey, P., C. Bruun Jensen, and A. Morita. 2016. *Infrastructures and Social Complexity*. London: Routledge.

Hecht, G. 2012. *Being Nuclear: Africans and the Global Uranium Trade*. Cambridge: MIT press.

Hetherington, K. 2011. *Guerilla Auditors: The Politics of Transparency in Neoliberal Paraguay*. Durham: Duke University Press.

———. 2016. Surveying the Future Perfect: Anthropology, Development and the Promise of Infrastructure. In *Infrastructures and Social Complexity: A Companion*, ed. P. Harvey, C. Bruun Jensen, and A. Morita, 40–50. London: Routledge.

Heynen, N., M. Kaika, and E. Swyngedouw. 2005. *In the Nature of Cities: Urban Political Ecology and the Politics of Urban Metabolism*. London: Routledge.
Howe, C., J. Lockrem, H. Appel, E. Hackett, D. Boyer, R. Hall, M. Schneider-Mayerson, A. Pope, A. Gupta, E. Rodwell, A. Ballestero, T. Durbin, Farès el-Dahdah, E. Long, and C. Mody. 2016. Paradoxical Infrastructures: Ruins, Retrofit, and Risk. *Science, Technology & Human Values* 41 (3): 547–565.
Humphrey, C. 2003. Rethinking Infrastructure: Siberian Cities and the Great Freeze of January, 2001. In *Wounded Cities: Construction and the Reconstruction in a Globalized World*, ed. J. Schneider and I. Susser, 91–107. Oxford: Berg.
Jackson, S. 2015. Repair. *Cultural Anthropology*. https://culanth.org/fieldsights/720-repair.
Kahn, N. 2006. Flaws in the Flow: Roads and their Modernity in Pakistan. *Social Text* 24 (4): 87–113.
Kama, K. 2015. Circling the Economy: Resource-Making and Marketization in EU Electronic Waste Policy. *Area* 47 (1): 16–23.
———. 2016. Contending Geo-Logics: Energy Security, Resource Ontologies, and the Politics of Expert Knowledge in Estonia. *Geopolitics* 21 (4): 851–856.
Karsati, H., K. Baker, and F. Millerand. 2010. Infrastructure Time: Long-term Matters in Collaborative Development. *Computer Supported Cooperative Work (CSCW)* 19 (3–4): 377–415.
Khan, N. 2006. Flaws in the Flow: Roads and their Modernity in Pakistan. *Social Text* 24 (4): 87–113.
Kuchinskaya, O. 2014. *The Politics of Invisibility: Public Knowledge About Radiation Health Effects After Chernobyl*. Cambridge, MA: MIT Press.
Lakoff, A., and S. Collier. 2010. Infrastructure and Event: The Political Technology of Preparedness. In *Political Matter: Technoscience, Democracy, and Public Life*, ed. Braun and Whatmore, 243–266. Minneapolis: University of Minnesota Press.
Lampland, M., and S.L. Star, eds. 2009. *Standards and their Stories: How quantifying, classifying, and formalizing practices shape everyday life*. Ithaca: Cornell University Press.
Lancione, M., and C. McFarlane. 2016. Infrastructural Becoming: Sanitation, Cosmopolitics and the (Un)Making of Urban Life at the Margins. In *Urban Cosmopolitics: Agencements, Assemblies, Atmospheres*, ed. A. Blok and I. Farías, 45–62. London: Routledge.
Larkin, B. 2008. *Signal and Noise: Media, Infrastructure, and Urban Culture in Nigeria*. Durham/London: Duke University Press.
———. 2013. The Politics and Poetics of Infrastructure. *Annual Review of Anthropology* 42: 347–343.
Latham, A., and D. McCormack. 2004. Moving Cities: Rethinking the Materialities of Urban Geographies. *Progress in Human Geography* 28 (6): 701–724.
Latour, B. 1996. *Aramis, or the Love of Technology*. Cambridge: Harvard University Press.
Latour, B., and P. Weibel. 2005. *Making Things Public: Atmospheres of Democracy*. Cambridge: MIT press.
Lea, T., and P. Pholeros. 2010. This Is Not a Pipe: The Treacheries of Indigenous Housing. *Public Culture* 22 (1): 187–209.
Lloyd-Thomas, K. 2010. *Building Materials: Conceptualising Materials via the Architectural Specification*, unpublished PhD thesis, University of Middlesex.
Lockrem, J. 2016. Bodies in Motion: Attending to Experience, Emotion, the Senses and Subjectivity in Studies of Transportation. *Mobility in History* 7 (1): 50–57.
Maasen, S., and P. Weingart, eds. 2006. *Democratization of Expertise? Exploring Novel Forms of Scientific Advice in Policy Making*. Dordrecht: Springer.
Mackenzie, D., and J. Wajcman. 1985. *The Social Shaping of Technology*. Milton Keynes: Open University Press.
Mah, A. 2012. *Industrial Ruination, Community, and Place*. Toronto: Toronto University Press.
Mains, D. 2012. Blackouts and Progress: Privatization, Infrastructure, and a Developmentalist State in Jimma, Ethiopia. *Cultural Anthropology* 27 (1): 3–27.

Mann, M. 1986. *The Sources of Social Power, Volume 1*. Cambridge: Cambridge University Press.
Marres, N. 2013. Why Political Ontology Must Be Experimentalised: On Eco-Homes as Devices of Participation. *Social Studies of Science* 43 (3): 417–413.
Marx, K. 1973 [1867]. *Capital, Volume 1*. London: Penguin.
Masquelier, A. 2002. Road Mythographies: Space, Mobility, and the Historical Imagination in Postcolonial Niger. *American Ethnologist* 29 (4): 829–856.
Mitchell, T. 1999. Society, Economy and the State Effect. In *State/Culture: State Formation after the Cultural Turn*, ed. G. Steinmetz, 76–97. Durham: Duke University Press.
Mol, A. 2002. *The Body Multiple: Ontology in Medical Practice*. Durham: Duke University Press.
Navaro-Yashin, Y. 2012. *The Make-Believe Space: Affective Geography in a Post-War Polity*. Durham: Duke University Press.
Osborne, T. 1996. Security and Vitality: Drains, Liberalism and Power in the Nineteenth Century. In *Foucault and Political Reason: Liberalism, Neo-Liberalism and Rationalities of Government*, ed. A. Barry, T. Osborne, and N. Rose, 99–121. London: UCL press.
Parmiggiani, E., and E. Monteiro. 2016. A Measure of 'Environmental Happiness': Infrastructuring Environmental Risk in Oil and Gas Offshore Operations. *Science and Technology Studies* 29 (1): 30–51.
Perrow, C. 1999. *Normal Accidents: Living with High Risk Technologies*. Princeton: Princeton University Press.
Petryna, A. 2002. *Life Exposed: Biological Citizens After Chernobyl*. Princeton: Princeton University Press.
Pipek, V., and V. Wulf. 2009. Infrastructuring: Toward an Integrated Perspective on the Design and Use of Information Technology. *Journal of the Association of Information Systems* 10 (5): article 6.
Pollock, N., and R. Williams. 2015. The Venues of High Tech Prediction: Presenting the Future at Industry Analyst Conferences. *Information and Organization* 25 (2): 115–136.
Ribes, D., and J. Polk. 2015. Organizing for Ontological Change: The Kernel of and AIDS Research Infrastructure. *Social Studies of Science* 45 (2): 214–241.
Rose, N., and P. Miller. 1992. Political Power Beyond the State: Problematics of Government. *British Journal of Sociology* 43 (2): 173–205.
Rowland, N., and J.H. Passoth. 2015. Infrastructure and the State in Science and Technology Studies. *Social Studies of Science* 45 (1): 137–145.
Schaffer, S. 1992. Victorian Metrology and Its Instrumentation: A Manufactory of Ohms. In *Invisible Connections*, ed. R. Bud and S. Cozzens. Bellingham/Washington: SPIE.
Schwenkel, C. 2013. Post/Socialist Affect: Ruination and Reconstruction of the Nation in Urban Vietnam. *Cultural Anthropology* 28 (2): 252–277.
Sheller, M. 2004. Automotive Emotions: Feeling the Car. *Theory, Culture and Society* 21 (4/5): 221–242.
Simone, A. 2014. Relational Infrastructures in Postcolonial Worlds. In *Urban Infrastructure in Context*, ed. S. Graham and C. McFarlane, 17–38. London: Routledge.
Smith, N. 2010. *Uneven Development: Nature, Capital, and the Production of Space*. London: Verso.
Smith, C., and M.N. Wise. 1990. *Energy and Empire: A Biographical Study of Lord Kelvin*. Cambridge: Cambridge University Press.
Star, S.L. 1999. The Ethnography of Infrastructure. *American Behavioral Scientist* 43 (3): 377–391.
Star, S.L., and K. Ruhleder. 1996. Steps Towards an Ecology of Infrastructure: Design and Access for Large Information Spaces. *Information Systems Research* 7 (1): 111–134.
Stengers, I. 2005. The Cosmopolitical Proposal. In *Making Things Public: Atmospheres of Democracy*, ed. B. Latour and P. Weibel. Cambridge: MIT press.
———. 2010. *Cosmopolitics Volume 1*. Minneapolis: Minnesota University Press.
Tironi, M., and N. Calvillo. 2016. Water and Air: Territory, Tactics, and the Elemental Textuality of Urban Cosmopolitics. In *Urban Cosmopolitics: Agencements, Assemblies, Atmospheres*, Questioning Cities Series, ed. A. Blok and I. Farías, 207–224. Abington/New York: Routledge.

Tironi, M., I. Rodríguez-Giralt, and M. Guggenheim. 2014. *Disasters and politics: Materials, Experiments, Preparedness*. Oxford: Basil Blackwell.
Truitt, A. 2008. On the Back of a Motorbike: Middle Class Mobility in Ho Chi Minh City, Vietnam. *American Ethnologist* 35 (1): 3–19.
Ureta, S. 2014. Normalizing Santiago: On the Challenges (and limits) of Repairing Infrastructures. *Social Studies of Science* 44 (3): 368–392.
———. 2015. *Assembling Policy: Transantiago, Human Devices, and the Dream of a World Class Society*. Cambridge: MIT press.
Venters, W., E. Oborn, and M. Barrett. 2014. A Trichordal Temporal Approach to Digital Coordination: The Sociomaterial Mangling of the CERN Grid. *MIS Quarterly: Management Information Systems* 38 (3): 927–949.
Von Schnitzler, A. 2016. *Democracy's Infrastructure: Techno-Politics and Protest After Apartheid*. Princeton: Princeton University Press.
Walker, G. 2014. *Environmental Justice: Concepts, Evidence, and Politics*. London: Routledge.
Weszkalnys, G. 2016. Infrastructure as Gesture. In *Infrastructure and Social Complexity*, ed. C. Bruun Jensen, P. Harvey, and A. Morito, 284–295. London: Routledge.
Whatmore, S.J. 2013. Earthly Powers and Affective Environments: An Ontological Politics of Flood Risk. *Theory, Culture and Society* 30 (7–8): 33–50.
Whitehead, A.N. 1920. *The Concept of Nature*. Cambridge: Cambridge University Press.
———. 1985. *Science and the Modern World*. London: FaB.
Zittrain, J. 2006. The Generative Internet. *Harvard Law Review* 119: 1974–2040.

Andrew Barry is Professor of Human Geography at the University College London. He was originally trained in Science and Technology Policy Studies at Sussex University and subsequently taught at Brunel University and Goldsmiths College, where he co-founded the Centre for the Study of Invention and Social Process. He has written extensively on the politics of technology and is the Author of *Political Machines: Governing a Technological Society* and *Material Politics: Disputes along the Pipeline* and Coeditor of *Interdisciplinarity: Reconfigurations of the Social and Natural Sciences* and *The Technological Economy*.

Chapter 7
What If Nothing Happens? Street Trials of Intelligent Cars as Experiments in Participation

Noortje Marres

7.1 Introduction

In recent years, industries and governments have made significant investment in so-called real-world testing of intelligent car technology in everyday environments, on public roads. These tests take different forms, from the large-scale roll-out of experimental driver-assist technology by tech companies like Tesla, to government-funded field experiments which demonstrate new transport concepts, such as the driverless buses called WePods, which introduced 'mobility-as-a-service' into urban streets in the Netherlands and Germany, and the Gateway trials in Greenwich (UK) which placed driverless pods and other intelligent concept vehicles on a pedestrian path along the Thames River water front (Fig. 7.1).

On-the-road testing has a long history in the automotive sector, going back to at least the 1930s (Dennis and Urry 2009). So do efforts to render cars intelligent: some components of automotive systems, like fuel injection, have been controlled by computers, not just mechanics, since the late 1960s, and today most cars run software to do anything from breaking to navigating to steering.[1] While some experts today claim that in 5–10 years it will be feasible for cars to "drive themselves," others have emphasised that some technologies that today are labeled as enabling 'autonomous' driving are not as new as is often thought: they have been customary in higher-end vehicles for some years already, as in the case of pedestrian detection systems of Volvo cars.[2] However, while the newness of on-the-road

[1] With thanks to reviewer 1 for pointing me to this history of the computerized car: https://itstillruns.com/car-computer-history-5082250.html

[2] Intelligent Vehicles research at WMG, Warwick Manufacturing Group, video, 23 February 2017, http://www2.warwick.ac.uk/fac/sci/wmg/research/smarter/ For a playful presentation of the Volvo

N. Marres (✉)
Centre for Interdisciplinary Methodologies, University of Warwick, Coventry, UK
e-mail: N.Marres@warwick.ac.uk

S. Maasen et al. (eds.), *TechnoScienceSociety*, Sociology of the Sciences Yearbook 30, https://doi.org/10.1007/978-3-030-43965-1_7

Fig. 7.1 Outside the O2 Centre, Oxbotica Trial, Greenwich, 19th August 2016

testing and of the computational enhancement of cars can be debated, there are also other reasons to take an interest in these trials: *a distinctive approach to the introduction of technology to society is detectable in today's street trials of intelligent cars,* and it is this I want to examine here.

In their study of an urban trial of an electric car-sharing service in St Quentin (France), Laurent and Tironi (2015) argue that such a street test is exemplary of a new, "experimental" mode of industrial innovation. They contrast this local experiment with an older, more centralized approach to the introduction of automotive technology to society. Instead of constructing "complete socio-technical systems in-house," car companies today increasingly enter into partnerships with a variety of other agencies in government, business and society in order not just to implement a new form of transport, but to configure "a whole ecosystem," in which the role of each of the above agencies is at stake (Laurent and Tironi: 211). In the test studied by Laurent and Tironi, for example, Renault envisioned itself in the role of no longer just a manufacturer of cars, but a "provider" of mobility solutions in partnership with IT companies and municipal government (on this point see also Wentland 2016). In what follows, I will query the extent to which such street trialling of automotive technologies qualifies as "experimental" in all respects, and indeed, whether the mode of innovation in evidence deserves to be called "experimental" at all, from the perspective of society. However, even if street pilots and urban demonstrations are not necessarily experimental, in the way that we would wish to understand the term, it does seem that these tests present a distinctive *innovation in the procedures of introducing technology to society*. Laboratory-based testing not on-the-road testing is today still considered the established paradigm in the automotive sector. In his article 'From the Road to the Lab to Math' (2010), the organization studies scholar

feature, see: Volvo S60 Pedestrian Detection System Test Failure, August 2010, https://www.youtube.com/watch?v=w2pwxv8rFkU

Paul Leonardi, on the basis of extended fieldwork in car companies, shows how over the course of the twentieth century car safety and other forms of automotive testing like societal impact modelling have been increasingly confined to dedicated test sites and lab-based computer simulations. How, then, should we account for the fact that today automotive testing is being explicitly, publicly, located in society, in everyday environments like the street?

In this chapter, I will then examine the ways in which street tests of so-called intelligent automotive technology configure relations between innovation and society, and develop the argument that, notwithstanding some of the revolutionary ambitions pronounced in relation to the 'real-world testing' of autonomous driving, these tests *cannot* be understood in terms of the displacement of technology testing from the laboratory to society. This becomes clear when we consider the 'networked' character of today's 'intelligent' car technology, which connects cars, if not exactly to the laboratory, then at least to data centres. Furthermore, efforts to relocate technology testing in 'real-world' environments must be considered incomplete, insofar as they facilitate only highly limited forms of social engagement and public participation in the evaluation of new technology. Through a discussion of recent street trials of intelligent automotive technology, I will propose that the testing of intelligent technology 'in society' today enable a double-edged operation upon existing arrangements for participation in the societal evaluation of new technology. On the one hand, testing technology in everyday environments, amidst social complexity, makes possible a more active engagement with users, audiences and participants as part of the testing process than would be feasible in laboratory settings. And, indeed, a significant consequence of recent streets tests of intelligent cars is the curation of "environments of participation" in infrastructural environments in society. However, these trials may also contribute to the dismantling of existing mechanisms of public accountability of innovation. For one, they more often than not involve the use of public spaces by private agencies (Laurent and Tironi 2015), and it is not self-evident what ensures the democratic legitimation of such initiatives. How then should we judge, and engage with, these double-edged, decidely ambivalent, implications of street trials of intelligent automotive technology for the relations between innovation and democracy? To answer this question, this chapter will discuss four different street tests of intelligent automotive technology: the roll-out of Autopilot by Tesla; an emission tests of a VW diesel car on the streets somewhere in Germany; the Gateway street trial in Greenwich (UK); and finally, an attempt by myself and colleagues to devise a street test of sorts of intelligent autonomotive technology in collaboration with colleagues, using social research methods to conduct a participatory evaluation of driverless cars at the University of Warwick. Before introducing these particular tests, however, I would like to say more about street tests as a format for the introduction of technology to society.

7.2 Street Testing as a Mode of Innovation: Towards the Beta-Testing Society?

On-the-road testing of automotive technology resembles some other experimental formats for innovation in public or social environments, like the urban laboratory (Evans and Karvonen 2014), living labs (König and Evans 2013), and field experiments (Kelly 2012). Deployed in different fields of research and innovation, from IT to sustainability and medical research, each of these approaches involve experimentation in everyday settings, away from scientific laboratories. In the words of Evans and Karvonen (2014: 415), to experiment in social environments "is to create a space apart from the norm, and by bounding space, urban laboratories [..] inscribe a privileged space of innovation." The particular urban test that Evans and Karvonen describe here is a city corridor, a section of the Oxford Road in Manchester, which was designated a zone of low-carbon innovation by the municipal government around 2009, and was turned into a site for testing a variety of technologies, from air quality sensors embedded in pavements, to new road lay-outs, designed to favor pedestrians and cyclists over cars. This points to another distinctive feature of urban labs – or street tests: they do not only offer a site for technology testing, for trying out new or still unstable devices. Tests have a double function, presenting both a practical arrangement for trying out technical systems and an organizational space for configuring new forms of governance (See on this point also Leonardi 2010). Thus, Evans and Karvonen characterize street tests as "a strategy for local governments" to pursue both economic and societal goals (sustainability, economic growth) "in partnership with public and private property owners" (Evans and Karvonen 2014: 413). As noted above, Laurent and Tironi (2015) also highlight the capacity of street tests to assemble heterogeneous actors, and argue that they exemplify a negotiated form of governance based in partnership between government and industry.[3] Other scholars have emphasized that urban tests do not only operate in organizational and technical dimensions, but also in epistemological and mundane ones. Calvillo et al. (2016) propose that urban experiments materialize an imagined future in which technology operates upon "the intimate details" of everyday life – from doing the laundry to the air we breathe (see also Wentland 2016). Street experiments, then, create an 'in-between' space in which different types of actors come to interact – from companies to local government and mundane figures like "pedestrians" – and they operate in multiple registers, of innovation, politics, intimacy, imagination, and so on. By carving out bounded environments in society where existing or 'normal' relations and obligations may cease to apply, street tests allow for experimentation on the level of (a) technology, through the implementation still

[3] Importantly, tests have long been ascribed this relational capacity, irrespective of whether they take place in laboratories or in the field. In the case of automotive testing, Leonardi (2010) has described tests as a relational device for managing relations of accountability between government, industry and society. However, street tests bring social actors into play in a different way from laboratory tests, a point I elaborate below.

unproven or unstable propositions, (b) politics, by suspending established rules of public accountability; (c) society, by attempting shifts in existing ways of doing things. As such, however, it is clear that real-world testing does not just enable participation with innovation, it may also undo forms of engagement that make up or made up 'life in the street.'

A distinctive feature of field tests is their widespread uptake across a range of different areas of research, development, governance and activism – from social experiments in community cohesion to the piloting of new technical infrastructures, like responsive traffic light systems. Automotive technologies – cars – are by no means the only innovations to be trialed by means of street testing. But over the last few years computationally-enhanced cars have featured prominently in media reports of on-the-road experimentation with technology. Most readers have probably heard of the fatal crash that occurred in Florida in the spring of 2016 and involved a Tesla car in self-driving mode, which mistook an on-coming truck for the sky and killed its driver "while he was watching Harry Potter" (Levin and Woolf 2016). There have been other crashes, fatal and non-fatal, and many other more minor "kerfuffles" involving computationally-enhanced cars reported in the news in recent years. In July 2015, the Daily Mail asked the rhetorical question: "Is your car safe?", announcing that "Hackers had taken control of a Jeep Cherokee and crashed it into a ditch by gaining access through the entertainment system." (Bates 2015) The article went on to report that The Cherokee "hackers" made their intervention "while sitting on a sofa" (although the same article also refers to these actors as "security experts"). One reason street tests of computerized cars have been at the center of public attention in recent years is no doubt the way in which different technological systems – and imaginaries, and industries, – intersect in the area of automotive innovation today. The development of what in the UK are called CAV's – connected and autonomous vehicles – explicitly crosses domains: it brings together the more traditional automotive industry (car companies) with newer players from the digital economy (the tech sector). Furthermore, these technologies are ascribed the capacity to address a broad range of issues from environmental health (reducing emissions) to economic regeneration (bringing new jobs to previously de-industrialized regions) (on this point, see also Marres 2017). If it follows from the preceding discussion that ***a distinguishing feature of field laboratories and street tests is that they enable the curation of an 'in-between space', it should be added that cars today, too, present an eminently suitable vehicle for the pursuit of this strategic aim.***

There is a further reason why street trials of intelligent automotive technologies present an especially intriguing case for those concerned with the changing relations between technology, science, and society. As noted, automotive technology testing is or was supposedly firmly anchored in the laboratory: does the testing of technology in the street signal that the gap between laboratory and society is being bridged? I here would like to argue that today's "real-world tests" of driverless cars in everyday environments like public roads **cannot** be understood in terms of a straightforward re-location of experimentation from the laboratory to the street. Street tests do *not* simply involve the un-doing of Leonardi's (2010) historical narrative, which proposed that automotive testing had moved away from testing in 'the open road' to the closed spaces of the lab and 'virtual simulation' over the course of the twentieth Century. While car testing is today being explicitly located in everyday environments like public roads, many of *these tests do not really involve a move away from the lab, and this insofar as they bring "the laboratory' with them, into the streets*. Take this most well-known of cases from the area of connected and autonomous driving: Tesla's Autopilot, the driver-assist feature produced in the form of a software plugin directly downloadable into this fashionable brand of electric cars, which was released in 2014. It remains the most popular of driverless street experiments today, with an estimated number of 100,000 vehicles participating by November 2016. (Lambert 2016) Importantly, this technology does not just facilitate automated driving, it equally involves location-based *data capture in the field:* as a reporter put it tellingly, Tesla uses "its fleet of vehicles owned by its customers to collect data" and "uses the data even when the Autopilot is not active in order to feed its machine learning system." (Lambert 2016) It may not be an exaggeration to say that the driver-assist application Autopilot is designed as much to establish new types of connections between data-capture in the field and analysis in data centres managed by big tech companies, as it is to facilitate new forms of (automated) driving. Insofar as current street tests of intelligent automotive technology entail the re-validation of the 'open road' as a test environment, it is then of an 'open road' that is now intimately connected to, if not exactly the laboratory, to centers for data aggregation, crunching and algorithmic learning. *Street tests enable an expansion and intensification of interaction between the laboratory and the street, and* ***not*** *the displacement of innovation from the one to the other.*

Given this, it should not surprise us that various social, ethical and political problems have been pointed out with street testing as a way of configuring relations between innovation and society. It has been argued that urban laboratories are "simply 'dropped into' urban areas rather than integrated with their local contexts [..] with little regard for social issues" (Evans and Karvonen 2014 (citing Hodson and Marvin 2009): 415–416).[4] From this perspective, street testing is marked by a lack of engagement with societal contexts and concerns, and by a lack of accountability towards the populations enlisted in tests (see also Kelly 2012; and Lindtner 2015 on ethical problems with introducing relatively untested technologies into social environments). Street tests, then, may be valuable arrangements from the standpoint of innovation and governance, but it is far from self-evident they serve social and societal needs effectively. However, a very different way of understanding the relation between technology testing in societal settings and social issues is opened up by recent work by Steve Jackson et al. (2014). In their account, technology testing in

[4] Evans and Karvonen point out further problems: "A further issue regarding the corridor that has not been addressed is the unevenness of laboratorization; in short, the experimental capacities of cities are not distributed evenly [..]This is particularly evident in the Oxford Road corridor, where adjacent low-income communities are being framed as beneficiaries of the infrastructure upgrades but not considered as participants in the experimental process. [..] "As such, the partnership and the laboratory tend to reinforce the divide between the knowledge community and the surrounding neighborhoods rather than integrate these." (Hodson and Marvin 2009: 425–426)

everyday settings does not at all entail a disregard for social issues: to the contrary, it provides a way of rendering such issues amendable to an experimental approach. Jackson et al. argue that, in the tech industry, it has become customary to release experimental products and services to users at an early stage in their development, as companies increasingly rely on user trials and field tests to identify *not only technical problems with the applications in question, but also ethical, social, political and legal problems with their functioning in society.* Such a **beta-testing approach to the introduction of new technology to society**, in their account, is "intensely relaxed" about releasing relatively un-tested, un-stable devices into everyday environments, and the societal disruption this may cause (see on this point also Neff and Stark 2004). Indeed, it actively welcomes controversy, expressions of concern and outrage, insofar as this may be an effective way of cultivating an audience for a new product or service (see on this point also Geiger et al. 2014). Jackson and colleagues offer the example of a location-aware smart phone app called Girls Near Me: its release raised concern and indeed outrage online about the privacy and gender implications of the app. And this controversy translated into prompt changes in this application's design and functionality, offering the company an opportunity to demonstrate its willingness and capacity to "listen and learn." While the beta-testing of technology in society then provides opportunities to practice, and put on display, responsive governance, it also indicates a laissez-faire attitude towards harmful societal implications of technology: the suggestion is that such consequences must occur before they are addressable (Marres 2017).

It is certainly *not* self-evident that street tests of intelligent automotive technology, as a general category, fit with Jackson et al's description of the beta-testing approach. It *does* appear to apply to the release of Tesla's Autopilot, in which case, too, an evidently unstable technology was being released into society, onto public roads. Here too, this was framed as an experimental opportunity to facilitate learning and product improvement (Stilgoe 2017). Since its release, Autopilot has become the subject of a barrage of videos of accidents and near-misses involving Tesla cars in self-driving mode on YouTube,[5] which in turn have generated an avalanche of online commentary and news media reports, something which the company did not seem too interested to discourage. Following another, non-fatal crash involving a Tesla car in Pennsylvania in the summer of 2016, the company CEO Elon Musk repeatedly took to Twitter with provocative statements to the effect that "accidents happen".[6] However, Jackson et al's account is not straightforwardly transfer-able to

[5] This is not a new genre, On YouTube, the video mentioned in footnote 2 has around 700 K views and shows two guys trying to get the collision avoidance software on their Volvo cars to kick in, and failing, which apparently led Volvo more recently to cancel advertising campaigns publicizing said collision avoidance features in its higher-end models Volvo S60 Pedestrian Detection System Test Failure, August 2010, https://www.youtube.com/watch?v=w2pwxv8rFkU

[6] "The Tesla CEO responded to a second non-fatal crash involving the company's Model X vehicle in a Tweet explaining the driver's 'autopilot' feature was not engaged at the time of the crash. The company has faced controversy after a fatal crash in Florida in late May." Twitter moment, July 14, 2016 https://twitter.com/i/moments/753703538094505984?lang=en

other automotive street tests, such as for example those undertaken with government support in the UK, where the word "relaxed" does *not* adequately capture the significant caution with which these tests are being approached by industry and government – more about which below. The point then is: we need to empirically examine the different ways in which street tests of intelligent automotive technology configure relations between society, government and innovation in different contexts, and their adaptability to different normative objectives in this regard. Such tests seem to have both the capacity to facilitate societal engagement with new technology and to threaten or undermine existing arrangements of the evaluation of innovation from a societal perspective. Whether and how this ambivalence is exploited, exposed, ignored, rendered tractable, and/or addressed in street tests of intelligent automotive technology is what I will examine in the remainder of this chapter.[7]

7.3 Emissions Testing: Intelligent Cars as Problems-of-Accountability on Wheels

In the previous section I suggested that in the cause of Tesla's Autopilot, street testing was used to side-step the type of regulatory processes of evaluation that one would expect new technologies to undergo prior to their introduction to society. In this section, I would like to demonstrate that street tests not only generate problems of accountability, they may also be deployed to expose and address these very problems. I will do so by examining a by-now classic street test of a computationally enhanced vehicle. The on-the-road emissions tests that resulted in the VW emission rigging scandal otherwise known as Dieselgate were originally undertaken by the Council for Clean Transportation of the University of West Virginia (Shiermeier 2015), but an interesting variation on this trial has since then undertaken by two computing experts or hackers in Germany to highlight the role of software in the scandal. Two self-described "curious individuals", Felix Domke and Daniel Lange, a computer programmer and a former IT strategist of a large Bavarian car company, presented their findings of this trial in a public presentation to the Chaos Computer Club in Hamburg in December 2015. The object of their experiment was to demonstrate the workings of the "defeat device" – the piece of "cheating code" that was revealed in the summer of 2014 to be running on the *engine control*

[7]The ambivalence of technology – that it may serve both good and bad, legitimate and illegitimate purposes – is a well-established theme in the sociology of technology (Woolgar and Cooper 1999). In societal testing of technology, this theme emerges anew, and, one could say, with a vengeance, as it here becomes caught up with issues of legal and indeed criminal liability for societal and environmental harm, as in the case of the emissions tests I discuss in the next section. I do not address this directly in this paper, but have argued elsewhere that there is no reason to suspend the notion of ambivalence in relation to criminal technology: while harm is rightly punish-able, this does not suspend the question of how it is to be defined and demonstrated. Marres, N. Evil artefacts have ambivalence too, 4S Conference, Barcelona, August 2017. See for a discussion Kelty 2017.

boards – ECUs – of an estimated 11 million of VW cars. This software is able to detect when the car in question is being driven under so-called 'test conditions' – by monitoring things like speed and the position of the steering wheel– in which case the ECU adjusts the car's performance, dramatically reducing its emissions of CO2 and cancerous NOX. When these artificial test conditions no longer apply, – that is, when the car is driving in the street and not in a lab – emissions radically go up. It was to expose this behaviour that Domke and Lange undertook their street trial: First, they purchased this particular ECU on Ebay. To demonstrate its workings, they then used a method called 'real-time logging': hooking the ECU up to his own car, Felix ran the firmware while driving around, first in his own neighbourhood, then on a so-called dyno in a garage. Monitoring the cars emissions over time, the defeat device can be seen to kick into action: when test conditions apply (slow, straight driving), the display shows reduced emissions. When moving out of this mode, emissions significantly increase.

It should first of all be noted that, in presenting their findings, Domke and Lange granted exceptional capacities to the "technological device," the Engine Control Unit and the firmware running on it, as the protagonist of the VW Dieselgate controversy. Their trial was designed to put the computational element in VW Diesel cars to an engineering test, and their test did not explicitly define the car as a social phenomenon (practice, infrastructure, system). Nevertheless, Domke and Lange's street test also has a number of designed features which enables it to inform societal evaluations of technology. Their test clearly was configured as a field demonstration, as an "experiment in society" – meaning in this case, that it relied on technologies and settings that are mundane, part of the fabric of everyday life, and generally accessible in society. As I said, they used Domke's own car; they purchased their ECU from the non-specialist platform Ebay; they ran the test driving in Domke's neighbourhood and then in a neighbourhood Garage; their "live logging" method made it possible to show how these cars go about defying regulations and damaging health as they go – "on the go," – by virtue of their "intelligent" ability to detect test conditions. Their street trial, one could say, was designed as a way of keeping it real, of demonstrating the car's inner workings in the "real world," beyond the closed environment of the laboratory. As such, Domke and Lange managed to bring to life what to some might still seem a theoretical point: *it is in our streets, in the social environment, that VW diesel cars and defeat devices are wreaking havoc.* And this circumstance can be demonstrated in those very streets, using mundane technology. Domke and Lange, furthermore, also made a more abstract political point, about the problems of accountability that arise when software comes to mediate the relations between "the roads" – the infrastructural environments in society – and the laboratory – the more or less virtual tests sites and monitoring centres on which research and governance rely.[8] More specifically, they showed how the computerization of the car makes it possible to inscribe the test conditions (the laboratory) into

[8] See CAVs the recent UK government consultation *Driverless vehicle testing facilities: call for evidence* issued by the Department for Business, Energy & Industrial Strategy and Innovate UK in April/May 2016.

automotive systems – the car "knew" when it was undergoing a test, and thereby was able to game it – and this seriously undermines the ability of government to regulate automotive behaviour in society by way of laboratory-based emissions testing. Cars may "change their behaviour without telling us."

More generally speaking, the introduction of intelligent automotive technology is intimately connected with efforts to computerize wider infrastructural arrangements in society – like automotive systems. These efforts are likely to give rise to new forms of interactivity between settings of everyday life and laboratories, or centres of management and control, and these forms of interactivity pose a challenge to laboratory-based forms of 'empirical governance.' Domke and Lange's street trial highlights some of these constraints placed on regulatory regimes in computationally intensive societies: the car's performance in the lab, may be strictly regulated, but its behaviour on the street is a different matter, and seems beyond the control of empirical forms of governance that are anchored in laboratory-based test regimes. And this problem accountability does not just pertain to the relation between government and industry, it equally implicates consumers and the wider public. However, there is also a sociological point to be made here, which received less attention in Domke and Lange's demonstration. In presenting their findings, these IT experts granted special capacities to technology, to the Engine Control Unit and the firmware running on it, but it is clear that the scandal cannot be solely attributed to these technologies. As Domke and Lange also pointed out, many experts and government insiders were aware of the existence of defeat devices in diesel cars by 2011 and the type of test results they presented had been known *for several years* by those familiar with the automotive industry (Lippert 2016). It was only after sustained attention from journalists and other actors acting in the name of the public – including scientists and experts undertaking eminently reportable on-the-road emissions tests – that these "technical" results gained the capacity to cause a scandal. The public exposure of the ECUs dodgy behaviour required extensive labour of interpretation by teams of scientists, policy experts, governmental agencies and NGOs (see Marres 2016). The 'known' fact of emission rigging could only be proven to be unacceptable on a societal level. The passage of Dieselgate through everyday conversation, just as the passage of the toxic NOX through our lungs and bodies, were critical to achieving this scandal. Accordingly, we may define *the street trial as an interface between innovation, industry, government, society and the public – a site that allows for mutual engagement between sectors, but where, at the same time, crises of accountability can become detectable.*

7.4 Self-Driving Cars: Changing Relations Between the Lab and the Street, and the Public

Recent street tests involving driverless cars, or as they are known in the UK, Connected and Autonomous Vehicles (CAVs), could be said to take these questions of accountability to the next level. Whereas in the VW emissions scandal the limited reach of regulatory frameworks features as a matter of concern, and indeed, scandal, in the case of CAVs the limits of existing regulatory regimes have been publically affirmed as given. While the UK government has adopted a code of practice for the real-world testing of CAVs (Department for Transport and Centre for Connected and Autonomous Vehicles 2015), many regulatory aspects remain under-explored. Alongside its code of practice, the UK government has endorsed what it called in its 2016 evidence check "a non-regulatory approach" to connected and autonomous vehicles, (Science & Technology Select Committee 2016) an approach that celebrates the uniqueness of the UK as the only country in the world in which it is legal to take your hands off the steering wheel while driving on public roads, meaning that there is no need to go through a formal approval process to test driverless cars. There are currently several street trials underway in UK cities, and these trials range from widely publicized corporate street tests undertaken with direct government support, as in the case of the driverless pods currently being tested on selected public roads in Milton Keynes and in London's industrial North Greenwich quarter, (Holdsworth 2016; Olewitz 2016) to more small-scale tests in London with so-called "non-descript mules" – "an industry term for a car that has the technical components of a future model, [but] doesn't necessarily have the production exterior components," and are thus likely to go unnoticed. (Jaynes 2016,) Driverless cars are also set to arrive in the city of Coventry, where I currently work, although at the time of writing it is still uncertain in what form. (Mullen 2016).

However, it is already clear that in this non-regulatory context, the introduction of driverless cars affects relations not just between the laboratory and the street, but also between the laboratory and the street and the public. These relations are multi-faceted. On the one hand, the first reports are in of public protests against driverless cars in the UK, as cyclists criticize municipal governments for trialling "driverless motorcars [..] on pavements while cyclists were regularly vilified for pedalling on them." (Loeb 2017) On the other hand, street trials currently happening in Greenwich and Bristol come with extensive provisions for public participation. As the homepage of the current Gateway Trial in Greenwich loudly summons: "GET INVOLVED". Its "September Update" reports: "In May we opened the door for members of public to register take part in the Gateway project. Since then over 5,000 people have registered for a chance to take part which really demonstrates the enthusiasm and interest in automated vehicles." Several of these participation initiatives have a distinctively experimental flavour as in the case of the locative opinion mapping exercise undertaken by digital agency Common Place in Greenwich, which invites Greenwich residents to report their views on driverless cars in a location-specific way, by annotating a Greenwich map, and which recorded various, rather open-ended resident comments such as "this area can get foggy at times" (see Fig. 7.2).[9] In line with this, it is also clear that participation exercises in Greenwich are undertaken with a research-led framework. The stated aim is to "gain insight into people's attitudes towards the use of automated vehicles and their operation in cities."

[9] With thanks to Jimmy Tidey for pointing this out to me.

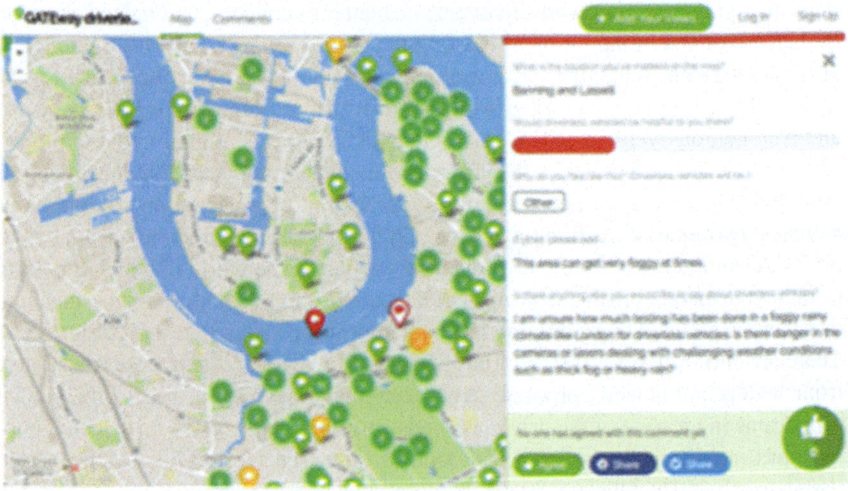

Fig. 7.2 Common Place, Opinion Mapping, April 2016, https://gateway.commonplace.is/comments

(Gateway 2016) As such, these initiatives are in line with Javier Lezaun and Linda Soneryd (2007) account of public participation exercises as forms of knowledge-production: governmental initiatives to foster public engagement with innovation, they note, often double as social research. The production of data and expertise demonstrating "levels of societal acceptance" of new technology is as important an object as the facilitation of public engagement itself. What is striking about the UK context, is how relaxed the organizers appear to be about publically endorsing this instrumental approach to public participation in street trials. If there are any tensions between the roles of citizen, tech enthusiast and research subject in street trials, they are not considered to pose any trouble.

Arguably, a rather paternalistic logic of participation appears to be at work in some of the recent UK trials, as when these trials are explicitly framed as instruments for "increasing public acceptance." As a representative of the Transport Catapult that organised a recent demonstration of driverless pods in Milton Keynes explained the rationale behind public street trials: "A lot of it is about gaining the trust of the public. If people can see that these vehicles are capable of driving themselves, they can gain trust and make sure that all the correct safety measures are in place to allow them to drive themselves, *and then that is exactly what we should be doing* (italics mine)." (Davies 2016) These are rather managed events. Reports of this LUTZ pathfinder trial in Milton Keynes were under embargo until the day *after* this street test took place: it appears that publics are only allowed to engage qua publics after the fact. It suggests that street trials in the UK are certainly *not* in all respects conducted in the beta-testing spirit identified by Jackson and colleagues: it is *not* the case that social, ethical and legal problems are allowed to freely emerge, 'in the wild.' Rather, participation initiatives appear to be designed to achieve particular pre-determined operations upon public perceptions of intelligent automotive technology.

This raises a further question: to what extent do these street tests of driverless cars satisfy the definition of an experiment, where their social and public dimensions are concerned? Javier Lezaun, Manuel Tironi and myself (2016) have written elsewhere about 'experiments in participation' which we define as "the deployment of settings, devices and things to curate processes and moments of participation in which, under at least partly controlled conditions, taken-for-granted ways of doing are unsettled, and which elicit expressions of public affairs that would otherwise remain under-articulated or exist only *in potentia* (Lezaun et al. 2016)." Considering driverless street trials from this perspective, it can seem that fairly little is put the test in these events, at least not where the social aspects of driverless cars – their capacity to operate in society – are concerned. In view of the framing of these trials as instruments for increasing public acceptance of this technology, it can seem that very little can happen here, in the sense of the curation of a situation in which a proposition can be challenged, or new perspectives can emerge. There can seem to be little attempt at problem articulation. However, such an analysis should be considered incomplete or even lazy, insofar as it only considers the public media framings of these trials. I want to probe further to see what experiments in participation, exactly, *can be* enabled through street tests of driverless cars.

7.5 Eliciting Social Aspects of Driverless Cars in the West Midlands

To explore this, my colleagues and I have been developing street tests of our own, with the specific purpose of eliciting social aspects of driverless cars.[10] In this research, we used a range of creative methods – including digital methods of issue mapping (Rogers and Marres 2001) and practice-based approaches to design research (Kimbell 2009) – in order to analyse and enable conversations about driverless cars. During a post-graduate workshop at Warwick in September 2016, we ran an initial test, in which we mapped issues raised by driverless cars in social media in a location-aware way, focusing in the West Midlands, the region where I work and where driverless cars have been said to be arriving soon. (Gilbert 2017) This exercise focused on a specific platform, namely Twitter, where much publicity around driverless cars occurs. We asked: Can we use social media analysis, not to conclusively represent the debate, but as a heuristic for identifying issues raised by driverless cars in our region? We delineated a particular Twitter data set with the aid of the Twitter Toolset for the Capture and Analysis (T-CAT), consisting of all tweets

[10] With thanks to Nerea Calvillo, Rebecca Cain, Ana Gross, Lucy Kimbell, Alessandro Brunetti, James Tripp and Arun Ulahannan. For more info see the workshop report "Surfacing Social Aspects of Driverless Cars with Creative Methods", University of Warwick, April 2017 http://www2.warwick.ac.uk/fac/cross_fac/cim/events/driverlesscarswithcreativemethods/

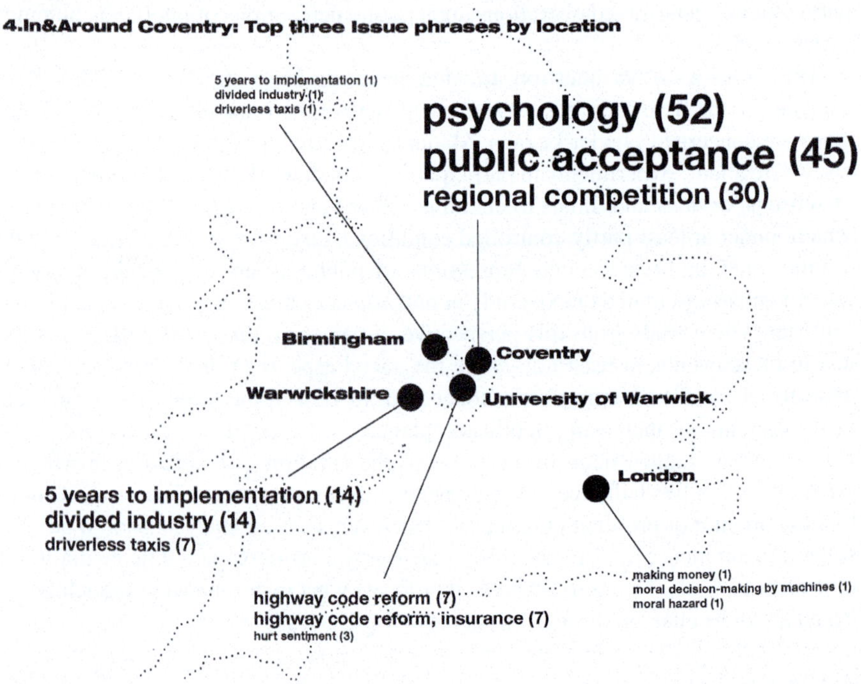

Fig. 7.3 Issues raised by street trials of driverless cars in the West Midlands, December 2016, Design: A. Brunetti

containing the term driverless – and similar terms like self-driving, autonomous vehicle, CAV – during June and September 2016.[11] For the purposes of this exercise, we included only the tweets traceable to the region, namely tweets sent from accounts that listed Coventry, Warwick and other locations in this region. We then asked: What issues do driverless cars raise across locations in this region? To answer the question, we manually extracted issue terms from the data using a loose interpretative framework, which we then visualised using different criteria such as uniqueness and frequency (see for example Fig. 7.3). In doing so, our aim was *not* to produce a conclusive representation of the issues, but rather, to produce issue maps that could subsequently be deployed as "devices of elicitation" (Lezaun and Soneryd 2007): displaying concerns offers a way to elicit views during informal encounters in the street and other social settings, such as the Warwick Campus café and Coventry city centre, inviting actors to informally account for issue formations.

The issue visualisations we produced in this way, I think, do demonstrate a capacity of UK street trials to elicit social aspects of driverless cars. The issue terms that made it onto our maps display a broad range of concerns – from "mobile living

[11] We used T-CAT, the Twitter Toolset for the Capture and Analysis of Twitter data, developed by Erik Borra, Bernhard Rieder and others (Borra and Rieder 2014).

room for shopping" to "how does it feel?", "lethal robots," "new types of testing needed" and "divided industries" – especially the conjunction of such specific heterogeneous terms suggests more is going on than either the promotion or rejection of known proposition, i.e. issue formation. However, for these expressions to qualify as issue *articulations* more is required than their visualisation in a map. To gauge the salience of these formations, further operations are required. To this end, we made an attempt during a second workshop to curate an environment for the exploration of issue-scapes: together with Rebecca Cain and others of the Warwick Manufacturing Group, we organised a creative participation exercise, inside the aforementioned driver-in-the-loop simulator, in which we invited participants to annotate the simulator, using issue terms featured in our social media maps. Participants received instructions as to how to produce an issue scape by annotating the simulator, using sticky notes to mark up objects present in the simulator – like the car itself, or the stretch of Coventry road projected on the 360 screen surrounding them (Fig. 7.4). There were also cardboard figures available for annotation, representing human actors and cardboard boxes representing non-humans (machines; technologies; institutions; etc.). While participants were clearly fascinated by the simulator itself, they also made significant efforts to locate some of the issues mentioned on social media in the driverless simulator environment – in doing so generating further issue articulations. For example, 'the elderly' were introduced in the setting in the form of an 'old lady' cardboard figure, which was settled into the back seat of the car, with a note on the window noting 'a dashboard that says 'old lady on board': stigma'.

Fig. 7.4 Annotating the Driver-in-the-loop simulator, Warwick Manufacturing Group, University of Warwick, December 2016. (Photo by L. Oliviera)

The design of these social tests of driverless cars is currently on-going, but it is already clear that these exercises are decidedly interpretative in nature: the objective is not to document concerns as they 'naturally' exist out there, in society, but rather, to determine how existing experimental settings – like the social media platform Twitter, or the driver-in-the-loop simulated at the University of Warwick – can be configured to enable the collective articulation of social issues in an interpretative register. In both cases, the researchers addressed and actively involved others – passers-by in the street, in the first exercise, invited workshop participants from government, academia and the non-governmental sector, in the second – to explicate and elaborate issues associated with driverless cars. As such, they implement part of the definition of an experiment in participation mentioned above, namely to generate a form of engagement that did not already exist prior to the research. However, even as both these exercises aimed to move beyond a descriptive or 'representational' approach to researching public perceptions of technology, the aim is certainly not to replace this with a narrow interventionist, or actionist understanding of participation. Participation can*not* be adequately understood as a phenomenon that can be straightforwardly 'created' or 'designed' (Marres et al. 2018). Instead the point is to generate interpretations of an exploratory kind, or better put, to explicate issues raised by driverless cars that are latent in the setting. This is why a generative test environment is all important to the success of the exercise, and also why the term 'experiment in interpretation' might be more appropriate in this instance than "experiment in participation". In a context in which participation is decidedly ambivalent from a normative perspective – being so eminently deployable to multiple, even contradictory, ends – this seems to be an especially important question: has the test produced a situation in which new qualifications of the social phenomenon in question can be produced? Does it enable experiments in interpretation (Marres 2016)?

7.6 Conclusion

Research on the social capacities of driverless cars is in its early stages, and it is certainly too early to draw any conclusions about the type of issue formations that driverless street trials are eliciting or could elicit in the UK context. But some more speculative conclusions about the role of street trials as devices for the societal introduction of intelligent automotive technology can be drawn. Firstly, and most importantly, there can be little doubt that street trials do not just serve to test to technical aspects of intelligent automotive technologies, they also enable the evaluation of their functioning in society. More specifically, street tests of these innovations involve the configuration of "environments of participation" in society. I have argued that the operations upon participation that these tests enable are decidedly ambivalent: while intelligent technology tests in social environments like public roads offer opportunities for, and indeed necessarily include, public and societal engagement with the evaluation of new technology, they also enable the side-stepping existing arrangements for the public accountability of new technology. Under these conditions, it becomes important to determine whether and to what extend street tests have the demonstrable capacity to facilitate issue articulation. I have proposed that driverless street tests are and can be configured to enable the elicitation of social, political and ethical aspects of new technology that are not already apparent, but that they are *not* on the whole already configured this. The explication of social and societal dimension of intelligent technology requires the deliberate adaptation of test environments in society to enable participatory forms of evaluating innovation and societal change.

Ann Kelly (2012) has proposed a very useful term to characterize the forms of participation enabled through field experiments: in her study of experimental huts used in malaria research in the Gambia, she proposes that fields experiments involve the curation of a "semi-field", a "controlled yet un-contained setting" [..] [which] "fuses a generic scientific space and a specific inhabited space." (Kelly 2012: 6). And: these experiments happen in the field but aren't 'of the field' (Kelly, pers. Communication). Similarly, the street tests of intelligent cars that I have discussed in this chapter involve the construction of a location between laboratory and society that is artificial enough for the functioning of technology to be controllable and manageable, while "natural enough" for social engagement with technology to become possible in these settings (see also Derksen and Beaulieu 2011). It is this 'semi-naturalism' of field experiments that will be crucial to recognize in further investigations – and tests – of the evaluative capacities of street tests, of their ability to serve as instruments for the evaluation of not just the technical but also the social, cultural and political aspects of the introduction of intelligent technology to society. While urban tests, social experiments and field laboratories feature prominently in societal programmes for the introduction of new technology to society, these forms of engagements are generally understood as *complementary* to regulatory approval processes. Rightly so. Nevertheless, the role of technology testing in society has radically expanded over the last years, assuming a prominent role in the public communication of innovation, and as part of strategies for promoting 'societal acceptance' of technology. Tests, experiments and social laboratories increasingly function as de facto conditions for public engagement with innovation, and as such, it becomes crucial to move beyond external assessments of their legitimacy. We must also step 'inside the bubble' to undertake internal tests of their generative capacities: are street trials able to facilitate the evaluation of innovation from a soci etal point of view?

If we consider the ways in which street testing configures infrastructural environments in society – or are deployed to that end – it can seem plausible, from a sociological or philosophical perspective, to interpret these tests *as threats to democracy, and/or the culture of public accountability*: the experimental introduction of new automotive technologies into society then presents itself as a vector of the commercialization, privatization and managerialization of public space (for a discussion of this point, see also Dickel, this volume). To adopt such a critical frame of interpretation is to think along with Hannah Arendt (1998 (1958)), for whom the rise of the technological society amounts to the dis-organization of the public realm, understood as a space for political action. From this perspective, street trials signal the emptying out of the spaces of public action, and political accountability, and the invasion of "administration." However, in this chapter I have tried to show that this "threat" does not adequately sum up driverless street trials, as they are configurable to serve other purposes. Street experiments can be criticized for many reasons, but it *cannot* be said that they inherently lack of capacity for problem articulation. If we consider the range of issues that street trials elicit, from stigmatization to securitization, and it is clear the insights generated in these experimental settings have freshness and relevance for multiple audiences. One indication of this is the range of entities that were invoked in the experiment in interpretation we undertook in the driver-in-the-loop simulator: road kill, the visually impaired, Brexit, industrial regeneration, new trends in road rage, and so on. It is not enough, then, to determine whether street tests do or do not satisfy requirements of participation and accountability from an external perspective. The question is also whether the methodology and design of street tests of intelligent technology are configure-able so as to facilitate experiments in participation: the challenge is not only to 'interpret' the role of technology testing in society but to specify requirements for the adjustment of this role to the task of societal evaluation. This indicated yet another way in which street trials are decidely double-edged: on the one hand, disparate entities are brought into relation through these exercises, in ways that contribute to the elucidation of the societal challenges involved. On the other hand, this process of cataloguing the entities and concerns involved also highlights the extend of their misalignment, the immense challenge of bringing them together into a shared frame that could accord them a degree of commensurability.

References

Arendt, H. 1998. The Public and the Private Realm, Chapter 2. In *The Human Condition*. Chicago: University of Chicago Press.

Bates, D. 2015. *Is YOUR Car Safe?* Hackers Take Control of a Jeep Cherokee and Crash it Into a Ditch by Gaining Access Through the Entertainment System. http://www.dailymail.co.uk/news/article-3169724/Hackers-control-Jeep-Cherokee-crash-ditch-gaining-access-entertainment-amid-concerns-cars-vulnerable.html.

Borra, E., and B. Rieder. 2014. Programmed Method: Developing a Toolset for Capturing and Analyzing Tweets. *Aslib Journal of Information Management* 66 (3): 262–278.

Calvillo, N., O. Halpern, J. LeCavalier, and W. Pietsch. 2016. Test-Bed as Urban Epistemology. In *Smart Urbanism: Utopian Vision or Smart Dawn?* ed. S. Marvin, A. Luque-Ayala, and C. McFarlane, 146–168. London/New York: Routledge.

Davies, R. 2016. *Self-Driving Car Tested for First Time in UK in Milton Keynes*. https://www.theguardian.com/technology/2016/oct/11/self-driving-car-first-uk-test-milton-keynes-driverless-lutz-pathfinder.

Dennis, K., and J. Urry. 2009. *After the Car*. Cambridge: Polity.

Department for Transport and Centre for Connected and Autonomous Vehicles. 2015. *Guidance Testing Automated Vehicle Technologies in Public*. https://www.gov.uk/government/publications/automated-vehicle-technologies-testing-code-of-practice.

Derksen, M., and A. Beaulieu. 2011. Social Technology. In *The SAGE Handbook of the Philosophy of Social Science*, ed. I. Jarvie and J. Zamora-Bonilla, 703–720. London: SAGE Publications.

Evans, J., and A. Karvonen. 2014. 'Give Me a Laboratory and I Will Lower Your Carbon Footprint!'—Urban Laboratories and the Governance of Low-Carbon Futures. *International Journal of Urban and Regional Research* 38 (2): 413–430.

Gateway. 2016. *Project Update: September 2016*. https://gateway-project.org.uk/project-update-september-2016/.

Geiger, S., D. Harrison, H. Kjellberg, and A. Mallard. 2014. *Concerned Markets: Economic Ordering for Multiple Values*. Cheltenham: Edward Elgar Publishing.

Gilbert, S. 2017. *Driverless cars coming to the streets of Coventry in 2017*. http://www.coventry-telegraph.net/news/video-driverless-cars-coming-streets-10817021.

Hodson, M., and S. Marvin. 2009. Cities Mediating Technological Transitions: Understanding Visions, Intermediation and Consequences. *Technology Analysis and Strategic Management* 21 (4): 515–534.

Holdsworth, R. 2016. *Driverless Pods Are Coming To Greenwich*. http://londonist.com/2016/01/driverless-pods-are-coming-to-greenwich.

Jackson, S.J., T. Gillespie, and S. Payette. 2014. The Policy Knot: Re-Integrating Policy, Practice and Design in CSCW Studies of Social Computing in CSCW '14. In *Proceedings of the 17th ACM Conference on Computer Supported Cooperative Work & Social Computing*, 588–602. New York: ACM.

Jaynes, N. 2016. *Faraday Future Gets Green Light to Test Driverless Cars in California, Report Says*. http://mashable.com/2016/06/22/faraday-future-driverless-car-license-california/?utm_medium=twitter&utm_source=twitterfeed#jJkkU.bZZiqR.

Kelly, A.H. 2012. The Experimental Hut: Hosting Vectors. *Journal of the Royal Anthropological Institute* 18 (s1): 145–160.

Kelty, C. 2017. *Evil Infrastructures*. Theorizing the Contemporary, Cultural Anthropology Website. https://culanth.org/fieldsights/1117-evil-infrastructures.

Kimbell, L. 2009. Design Practices in Design Thinking. *European Academy of Management* 5: 1–24.

König, A., and J. Evans. 2013. Introduction: Experimenting for Sustainable Development? Living Laboratories, Social Learning and the Role of the University. In *Regenerative Sustainable Development of Universities and Cities*, 1–23. Northampton: Edward Elgar.

Lambert, F. 2016. *Tesla Has Now 1.3 Billion Miles of Autopilot Data Going into Its New Self-Driving Program*. https://electrek.co/2016/11/13/tesla-autopilot-billion-miles-data-self-driving-program/.

Laurent, B., and M. Tironi. 2015. A Field Test and Its Displacements. Accounting for an Experimental Mode of Industrial Innovation. *CoDesign* 11 (3–4): 208–221.

Leonardi, P.M. 2010. From Road to Lab to Math: The Co-Evolution of Technological, Regulatory, and Organizational Innovations for Automotive Crash Testing. *Social Studies of Science* 40 (2): 243–274.

Levin, S., and N. Woolf. 2016. *Tesla Driver Killed While Using Autopilot was Watching Harry Potter, Witness Says*. https://www.theguardian.com/technology/2016/jul/01/tesla-driver-killed-autopilot-self-driving-car-harry-potter.

Lezaun, J., and L. Soneryd. 2007. Consulting Citizens: Technologies of Elicitation and the Mobility of Publics. *Public Understanding of Science* 16 (3): 279–297.

Lezaun, J., N. Marres, and M. Tironi. 2016. Experiments in Participation. In *The Handbook of Science and Technology Studies*, ed. C. Miller, L. Smith-Doerr, U. Felt, and R. Fouche, 4th ed., 195–221. Cambridge: MIT Press.

Lindtner, S. 2015. Hacking with Chinese Characteristics: The Promises of the Maker Movement Against China's Manufacturing Culture. *Science, Technology & Human Values*: 1–26.

Lippert, I. 2016. Umwelt – "Version 2b": Das Programmieren ökologischer Fehlentscheidungen und Grundlagen für eine neue Umweltpolitik. *Leviathan* 44 (3): 399–427.

Loeb, J. 2017. *Pedestrians Rage at Autonomous Pods and Delivery Bots on Pavements*. https://eandt.theiet.org/content/articles/2017/07/pedestrians-rage-at-autonomous-pods-and-delivery-bots-on-pavements/.

Marres, N. 2016. Experiments in Interpretation. In *Innovative Methods in Media and Communication Research*, ed. S. Kubitschko and A. Kaun, 75–79. New York: Springer.

———. 2017. *Digital Sociology: The Re-Invention of Social Research*. Cambridge: Polity.

Marres, N., R. Cain, A. Gross, L. Kimbell, and A. Ulahannan. 2017. *Surfacing Social Aspects of Driverless Cars with Creative Methods*, University of Warwick, Workshop Report, April 2017. https://warwick.ac.uk/fac/cross_fac/cim/events/driverlesscarswithcreativemethods/.

Marres, N., M. Guggenheim, and A. Wilkie. 2018. Introduction: From Performance to Inventing the Social. In *Inventing the Social*, ed. N. Marres, M. Guggenheim, and A. Wilkie. Manchester: Mattering Press.

Mullen, E. 2016. *Driverless Cars are Coming to the Streets of Coventry in 2017*. http://www.coventrytelegraph.net/news/coventry-news/driverless-cars-coming-streets-coventry-10814799.

Neff, G., and D. Stark. 2004. Permanently Beta. In *Society Online: The Internet in Context*, ed. P.N. Howard and S. Jones, 173–188. Thousand Oaks/London/New Dehli: Sage.

Olewitz, C. 2016. *Forget Driverless Cars, London is Going to Test Automated Pods as a Public Transportation Option*. https://www.digitaltrends.com/cool-tech/london-testing-driverless-public-transportation-pods/.

Rogers, R., and N. Marres. 2001. Landscaping climate change: A mapping technique for understanding science and technology debates. *World Wide Web, Public Understanding of Science* (9): 141–163.

Shiermeier, Q. 2015. *The Science Behind the Volkswagen Emissions Scandal*. http://www.nature.com/news/the-science-behind-the-volkswagen-emissions-scandal-1.18426.

Stilgoe, J. 2017. Machine Learning, Social Learning and the Governance of Self-Driving Cars. *Social Studies of Science* 48 (1): 25–56.

Wentland, A. 2016. Imagining and Enacting the Future of the German Energy Transition: Electric Vehicles as Grid Infrastructure. *Innovation: The European Journal of Social Science Research* 29 (3): 285–302.

Woolgar, S., and G. Cooper. 1999. Do Artefacts Have Ambivalence: Moses' Bridges, Winner's Bridges and Other Urban Legends in S&TS. *Social Studies of Science* 29 (3): 433–449.

Noortje Marres is Associate Professor and Research Director in the Centre for Interdisciplinary Methodologies at the University of Warwick (UK). She was trained in the sociology and philosophy of science and technology at the University of Amsterdam and the École des Mines in Paris. She is the Author of two books, *Material Participation: Technology, the Environment and Everyday Publics* (2012/2015) and *Digital Sociology: The Reinvention of Social Research* (2017). The edited collection *Inventing the Social* (with Michael Guggenheim and Alex Wilkie) came out with Mattering Press (2018).

Chapter 8
Biofacts, Bioprospecting, Biobanking: A Reality Check of Seed Banks

Nicole C. Karafyllis

8.1 Seed Banks in TechnoScienceSocieties[1]

8.1.1 Outline

Agriculture depends on, produces and mediates hybrid forms of nature and technology, in short: technonature. The organisms concerned grow neither by themselves nor for themselves. They *are grown* by means of technology and are thus termed "biofacts" (Karafyllis 2003). Human interfering in the somehow autonomous growth process is the main criterion for demarcating a biofact (see Sect. 8.2), and historically it starts with agriculture. As agriculture is the prototype of culture as such, it may be a fruitful candidate for investigating TechnoScienceSocieties. Rather than dwelling on the valuable insights on industrial agriculture from rural sociology (e.g. Hoban et al. 1992), the following focuses on the material objects of technological change, i.e. high tech plants, and the premodelling of their beginnings (seeds) by seed banks. At the outset, imagine thousands of jars filled with wheat, barley and corn seeds. They line up on shelves in cooling chambers and are long-term stored at −18 °C. In addition, we might think of tiny bits of potato tissue, which is cryoconserved in liquid nitrogen tanks. That one cannot conserve potato cultivars type-identical by storing their seeds, which generates problems for long-term storage in conventional cold chambers, is one of the paradoxes of modern agriculture, which are outlined in the following.

[1] I wish to thank two anonymous reviewers for their valuable comments. This chapter presents results from the cooperative research project „The language of biofacts: semantics and materiality of high-tech industrialized plants", project A (2015–2017 funded by the German BMBF, No. 01UO1501B). See www.biofakte.de (accessed on January 31, 2018).

N. C. Karafyllis (✉)
Technische Universität Braunschweig, Institute of Philosophy, Braunschweig, Germany
e-mail: n.karafyllis@tu-braunschweig.de

S. Maasen et al. (eds.), *TechnoScienceSociety*, Sociology of the Sciences Yearbook 30, https://doi.org/10.1007/978-3-030-43965-1_8

A historical sketch, running through the whole article, highlights that large-scale seed banks as agricultural institutions profited from the "gene rush" since the 1960s and particularly from the Convention on Biological Diversity (CBD) in 1992 (see Sect. 8.3). During those decades, seed banks became both ecologically enframed and high tech, while at the same time, they began to face several kinds of restrictions and resistance, not last by the complex nature of the very different plants, which were meant to be banked. Other technical problems addressed in this chapter are the digitalization of collections and expenses for cooling technologies. Moreover, in order to emphasize the relation between specific politics of time and the different narrations of how an organism becomes a biofact (i.e. by planting and breeding), I will comment on how the concept of biofact relates to the future-oriented idea of bioprospecting, i.e. the targeting, collecting and commercializing of new products and their biological predecessors (a process in which seed banks are bottlenecks). Considering politics of time, I critically emphasize seed banks as instances of both making and annihilating presence for the sake of the future (Sect. 8.4).

Using seed banks as an example of institutions, which premodel and preselect objects for future growth processes, embedded in a theory of biofacts, my aim is to open a new strain for STS researchers, enabling them to analyse agriculture as a crucial or even existential mode of technology. In this mode, the triangular relation of nature, technology and society continuously rearranges and generates both new materialities and semantics. Based on growth processes while being high tech, modern agriculture allows one to narrate very different stories about TechnoScienceSocieties' pasts and futures. Certain terms – for example, here: "gene bank", "biodiversity" and "bioprospecting" – and technologies (e.g. radiation technology and genetic engineering) are involved, resembling specific politics of time and space, to be unveiled. This will be done from a philosopher's point of view, thus contributing also to a philosophy of biotechnology. In Sect. 8.5, I will do a reality check of seed banks and demarcate some of their urgent problems (also based on expert interviews). These insights will question the idea of seed banks as biocybernetic models for an industrialized agriculture that newly labels itself as sustainable and concerned with biodiversity issues. Section 8.6 concludes with a short summary and reconsiders the relation of STS and agriculture.

8.1.2 *What Is a Seed Bank?*

By definition, seed banks operate in both history- and future-oriented ways. "A seed bank is an organization or facility used to store and preserve plant seeds, in particular varieties that are rare, have fallen out of commercial use, and/or may have unique desirable genetic characteristics." (Cohen 2014: 1642). This definition shows the ambivalent position between nature conservation activities and genetic research for innovative breeding options, in which modern seed banks are situated. By ex-situ conservation, they should prevent gene erosion while forming an infrastructural alliance of banking and breeding, including genetic engineering and synthetic biology.

Since the 1960s, seed banks are also called *gene banks*, broadened to *plant germplasm systems* and *pBRCs* (plant Biological Resource Centers), as seed banks increasingly rely on bioinformatics (on the different labels, see Karafyllis 2018b). Furthermore, from the 1990s onwards, seed bankers have technically been able to long-term store plant reproductive units known as "recalcitrant seeds" (i.e. germplasm other than "orthodox seeds"[2]) by cryoconserved tissue, and tissue from bulbs and tubers (e.g. potato). However, the common grape vine (*Vitis vinifera*) resists cryoconservation and has to be 'stored' as a whole plant (i.e. in a vineyard, as a collection termed "field gene bank"), like most fruit-bearing trees and shrubs. Their blossoms are pollinated by insects, which means that the genetic constitution of each seed within one fruit (e.g. an apple) is different from each other. Therefore, apple trees and alike are propagated by cuttings (i.e. by clones). This also applies to potato, which do not grow "true to seed" and are thus propagated by "seed potatoes" (again: by clones). From a modern breeding perspective, insect pollination is a problem, as is plant sexuality as such. It undermines the genetic identity or 'purity' of plant generations, which also matters for seed banking. Besides, important agricultural crops like rapeseed and potato can use different pollination systems (wind and insects). In consequence, a seed bank is an institution that specializes on certain plant types and the related agri- or horticultural practices. The physiological variety of plant reproductive units is one of the severe technical challenges for standardizing ex-situ conservation and for centralizing the many seed banks, while the totipotency of the plant cell nourishes hopes for 'banking them all' in the future.

When in the following, plant names like "wheat", "maize", etc. are mentioned, they do not mean a distinct cultivated sort rather than a liminal identity of a plant species (the gene pool), which is including all of its varieties, be they 'wild' mutations or primitively bred varieties, i.e. so-called *landraces*. Particularly the latter are of interest for future breeding options. Landraces are already adapted to farming techniques (though low-tech ones) and have withstood the rigors of biotic and abiotic stresses. They also reflect socio-cultural eating preferences.

In general, seed collections belong to agriculture as intrinsically as the highway belongs to the automobile. In the twentieth century, seed banks have become less and less farmer-led or community based. Nowadays, they are governmental institutions related to both bioscience programs and – somehow on the contrary – programs to protect cultural heritage. With regard to biotechnology and genetics, seed banks are the bottlenecks for transforming living plants into a controllable stock of *instruments*. Against this aura of fixation, however, we might understand seed banks also as cooperative agents and creators of an *infrastructure* (see Barry, this volume) that turns seeds into mobile objects. These infrastructures, emerging from various transnational networks of seed banks and bankers, allow for the exchange of material objects and ideas over space and time. In practice, the materiality of the concerned objects causes severe challenges for technological standardization in the

[2]This is a *terminus technicus* in seed banking, opposing "unorthodox" or "recalcitrant seeds", i.e. seeds that do not survive drying and freezing procedures needed for ex-situ conservation, e.g. the fatty seeds of the cocoa tree.

bank (see Sect. 8.6). The banked biofacts partially resist the idea of complete control. This poses the question for STS, if the two main characteristics of infrastructures – facilitating the flow (of people, information, goods, etc.) and catalysing standards – are synergistic or rather antagonistic.

Since their very beginning, improving and guaranteeing food security, safeguarding biodiversity, and building research inventories are the three main strategic aims of modern seed banks. The major technical means to reach these ends are different cold-storage units for keeping germplasm in a prolonged dormant state, which can last up to 30 years on average. After plant-specific time spans and by means of highly standardized and controlled procedures, the seeds are allowed to grow up as plants and generate new seeds – a kind of rejuvenation. Then storage for the next decades proceeds again, accompanied by research on the stored objects. Consequently, banked seeds are *agricultural* objects, *epistemic* objects, *technical* objects and *cultural-heritage* objects all at the same time. In terms of intellectual property and access rights, seeds have also become *legal* objects in the last decades (Merson 2000; Schubert et al. 2011). By the 1980s, public resistance against the banking of plant genetic resources reached a first peak (Juma 1989; see also Plucknett et al. 1987; Fowler and Mooney 1990).

In fact, the resources of interest are – seen in the light of genetics – not *new* but rather "underutilized" and "neglected", as Western agroscience puts it (Hammer 2003). However, often old varieties are still in use by local communities in distant rural regions, many of them in less developed countries. In industrialized countries, amateur horticulturalists and home gardeners have preserved an impressive amount of rare plant varieties. Other important actors are farmers who resisted agricultural policies imposed on them. As in the case of farmer-led seed banks, these rare varieties have been preserved as *plants* and neither as *genes* nor as *genetic resources*. Conservation strategies do not necessarily imply external programs, but are a matter of place, agency and people. Seeds can hence be understood as symbols of both resistance and hope (Nazarea et al. 2013) for the 'cultivation' of the world.

8.1.3 History and Institutions of Modern Seed Banking: From the Start in Russia into the West

Historically, the implementation of governmental seed banks relates to the imperialist politics of nation states. In the first third of the twentieth century, Russian, US-American, German, Swedish, British and other scientists began systematically collecting – mainly in Latin America, Africa and Asia – *intra-species diversity* relevant for agriculture, i.e. the genetic varieties of plant species like potato, wheat and maize (agrobiodiversity). At that time, the first national seed banks were founded. The mother of all modern, science-driven seed banks is the *Bureau of Applied Botany and Plant Genetics* founded at the Botanical Garden of St. Petersburg in late nineteenth century Russia. From the 1920s on it was known worldwide for its famous geneticist and plant collector *Nikolaj I. Vavilov* (1873–1943), later under its

Soviet name *Vavilov Research Institute of Plant Industry* (VIR). In his groundbreaking article "The Law of Homologous Series in Variation" (1922), Vavilov argued that in order to understand variation regularities behind the polymorphism of plant species, one has to collect and investigate plant varieties and races of *cultivated* plant species *together* with its *wild* relatives. This mixture of the cultivated and the wild (the gene pool) has remained the collecting policy of modern seed banks ever since. Vavilov's concept of a plant's genetic *center of origin*, which is simultaneously its center of diversity, demarcated the hot spots for collecting, and still does. To conclude, the very beginning of modern seed banks originated in science-driven politics of time that generated liminal spaces: the sites of collecting and the sites of storage. The main actors synchronized the past of domesticated plants with future breeding options by means of territorial searches for diversity at 'present'.

Furthermore, Vavilov already understood the territorial increase of bred cultivars, i.e. the *artificial*, as a problem for understanding the full range of the *natural* and its potentials: "Artificial hybridization threatens considerably to enlarge the external diversity of forms." (Vavilov 1922: 52). His typology of similarities between different polymorphisms allowed one to connect present with future knowledge by means of bioprospecting: "But it is already evident that the similarity in series of polymorphism [...] is so regular that it becomes possible to forecast, on this basis, the existence of forms and of varieties [...], not yet discovered. Some such unknown forms might be obtained by artificial hybridization of corresponding varieties [...]." (ibid.: 72). For the same reason, the artificial can – inside the laboratory – complete the natural manifoldness of forms, even if the "natural", "wild" or "primitively" bred form has not been collected yet. Therefore, modern seed banks are research inventories for fundamental and applied research purposes. They help to generate patterns of genetic potentiality, for which the missing links of assumed variations can be induced, i.e. provoked, by different kinds of technology.

This idea gave rise to radiation technologies for induced mutations carried out in seed banks before and after WWII. Provocation moved closer to bioprospecting. X-irradiation of dry seeds was the first standard method, which later expanded to nuclear irradiation. Radiation technologies were used, for example, from the 1930s onwards by the German geneticist *Hans Stubbe* (1902–1989), director of the seed bank of the German Reich, which continued with Stubbe as national seed bank of the GDR; by the plant breeding scientist and collector *Rudolf Freisleben* (1906–1943) at Halle University/Germany; by the plant breeding scientist, collector and initiator of the West-German gene bank, *Hermann Kuckuck* (1903–1992) at Hannover University/FRG, and by the technological path-setters, the Swedish plant geneticists *Hermann Nilsson-Ehle* (1873–1949) and *Åke Gustafsson* (1908–1988) in Svalöf/Sweden. They used cultivars from the Swedish Seed Association (Lundqvist 2014). Both Freisleben and Kuckuck visited the Swedish radiobiologists for training. The Swedish mutant collection of barley is still kept at the *Nordic Genetic Resource Center* (NordGen) in Alnarp/Sweden; the German mutant collection of the ornamental plant *Antirrhinum majus* (snapdragon; Stubbe's epistemic object inherited from Erwin Baur) is likewise still available at the German national gene bank for crops at the IPK.

The option "technology or nature?" for creating genetic diversity *ex-situ* materializes on the personal level. Directors of modern gene banks usually have to mediate between two scientific actor-groups: the "mutationists" believing in laboratory experiments and technical apparatus, and the "collectors" emphasizing the natural sources of genetic diversity. For example, when geneticist Hans Stubbe conceptualized the seed bank of the German Reich in 1943, he stressed that it would function as a unifying place for ordering and specifying the many plant seeds having already been collected at the Hindukush and in Tibet, while it would also be a place for fundamental genetic and innovative breeding research.[3] Forty years later, and after gathering many more objects for the GDR, the retrospect of his "world assortment" (German: *Weltsortiment*) still cherished the same aims (Stubbe 1982). Stubbe used this term already in the early 1940s. Until 1980, the GDR institution refused to adopt the Western term "gene bank" (Müntz and Wobus 2013: 203). Instead, the IPK used the Vavilov inspired idea of *one world* of plant genetic resources, which could be collected, stored and traded as an assortment. Though obsolete, "world assortment" is the most accurate term for describing the concept of a modern seed bank.

At the turn of the millennium, more than six million accessions (samples of plant varieties) were stored in seed banks worldwide, one third of them cereals (Scarascia-Mugnozza and Perrino 2002: 1). From the 1750 institutional crop collections existing worldwide (FAO 2010), the high-tech ones are in focus here, i.e. seed banks with sophisticated cooling and cryoconservation systems and institutionalized collecting practices. Among them are big transnational institutions supported by the CGIAR (Consultative Group on International Agricultural Research; founded 1971), which were set up with the strategic aims of improving food security, nutrition and health in the 'Third World' since the 1940s. The CGIAR members consist of country governments, institutions like the *World Bank* and the FAO (Food and Agricultural Organization of the UN), and foundations (e.g. Rockefeller Foundation, Ford Foundation). The CGIAR originally supported four centres still operating: the *International Wheat and Maize Improvement Center* (CIMMYT) in Mexico, purported to have the mission of helping starving Latin America; the *International Rice Research Institute* (IRRI) in the Philippines; the *International Center for Tropical Agriculture* (CIAT) in Colombia; and the *International Institute of Tropical Agriculture* (IITA) in Nigeria. Governments of the NATO-countries had to face the fact that after 1945 the then two biggest seed banks of the world, the *Vavilov Institute* and the *Zentralinstitut für Genetik und Kulturpflanzenforschung* (the later IPK) in Gatersleben/East Germany, were located behind the Iron Curtain in the domain of the Warsaw Pact.

During the 1960s and in the light of the negative impacts of industrialized agriculture, collecting plants for the sake of protecting biodiversity came under renewed discussed in terms of "plant genetic resources" (PGR) in the Western hemisphere. An ecologically and scientifically motivated movement emerged, which was easily

[3] See in Germany's National Archive *Bundesarchiv*, R 4901/15195: 18f, letter of intent (typoskript) by Hans Stubbe to the Reich's Ministry for Science and Education in Berlin, 26 July 1943, referring to the Kaiser-Wilhelm-Institut für Kulturpflanzenforschung (i.e. Germany's first seed bank).

soaked up by key industrial players in plant breeding and agrobiotechnology (see Pistorius 1997). The methods of genetic engineering were already beckoning on the horizon. One of the major figures of establishing modern seed banks in the West is the British botanist and potato-collector "Jack" (John) Gregory Hawkes (1915–2007), an instigator of the *plant genetic resources conservation movement* of the 1960s. Following Vavilov whom he had personally met, Hawkes envisioned a global network of seed banks that would be able to conserve all plant genetic varieties available. From the very beginning, this movement got support from experts consulting the FAO such as Hawkes, the plant collector Erna Bennett and the cereals expert Jack Harlan (Jackson 2013). For the West-German side, the professors of plant-breeding Hermann Kuckuck and Dieter Bommer (1923–2010) have been dominant actors at the FAO, both dreaming of a *world-network* (German: "Weltnetzwerk") of plant genetic resources including global documentation strategies (Bommer 1972). Their international activities resulted in implementing the West-German seed bank *Braunschweig Genetic Resources Collection* (BGRC) in 1970 (Hondelmann 1974; see also Karafyllis and Lammers 2017). At that time, the BGRC was one of the only three national gene banks in Western Europe, allying the Germplasm Institute (GI) in Bari/Italy and the seed bank of Kew Gardens in London/UK (newly termed: Kew Millennium Seed Bank).

In the meantime, CGIAR has embraced institutions particularly in Africa and India and has expanded its networking activities to genetic resources relevant for forestry and fishery. While European governments and the Italy-based Bioversity International (formerly known as IPGRI) – a global non-profit research for development organization – are part of the CGIAR, the seed banking of and in Europe historically also followed paths which differ from the CGIAR model (Saraiva 2013). Dominant reasons are WWII and its aftermaths, and the coexistence of two political systems on European territory during the Cold War. However, Europe was and still is heavily interested in the genetic resources of the less developed countries, supported, among others, by the FAO and the World Bank since the 1960s. At that time, the *European Association for Plant Breeding Research* (EUCARPIA) was founded. Its members backed the Worldbank's and FAO's idea of a world-network of bio-banked PGRs, in which also European countries should be central hubs. On the one hand, European seed banks emerged as national "introduction centres" from which local breeders should profit, on the other hand, actors of those centres should supervise and assist transnational "exploration centers" located at the crops' centers of origin (i.e. in the less developed countries). For example, the West German BGRC hosted vast collections of cereals from Ethiopia, based both on collecting expeditions and stored "safety duplicates" from the Ethiopian seed bank, which was set up by development aid from the German government in the 1970s. In 1980, the ECPGR (European Cooperative Program for Plant Genetic Resources) became operative, aiming at mobilizing funds for conserving PGR. In particular with regard to fund-raising and technical developments (incl. IT), European seed banking shows solo attempts within the world-network idea, partially opposing strategies of the USA.

Against the aforementioned international framing, based on ideas of donation and exchange, it is important to know that most seed banks have remained *national* entities, financed by a country's government and following national interests with regard to economy and research politics. For the following, I will continue to use examples from Germany. Many of them are associated with the genebank at the *Leibniz Institute of Plant Genetics and Crop Plant Research* (IPK) in Gatersleben in central Germany (formerly belonging to the GDR). It is one of the ten major seed banks in the world, Germany's national seed bank and currently Europe's biggest bank of live plant material. Considering its predecessor, the *Kaiser-Wilhelm-Institut für Wildsippenforschung* (simultaneously also named: *für Kulturpflanzenforschung*), which was founded 1943 in Vienna (from which the seeds were evacuated to the area of Gatersleben in early 1945), the IPK has been a red thread through the history of three political systems: Nazi Germany, East Germany/GDR, unified Germany/FRG (Müntz and Wobus 2013). After Germany's reunification, the IPK swallowed the smaller West German seed bank BGRC once located in Braunschweig. In 2003, the approximately 50.000 West German accessions were transported 90 kilometers to the East and fused with the roughly 100.000 'socialist' seed samples (Karafyllis and Lammers 2017). Soon after, the IPK, like many other large-scale seed banks, entered the stage of capitalism on verge of bioeconomy. During the last decade, it became the host of the Biotech Campus Gatersleben.[4] There, companies such as Bayer CropScience and Orgentis Chemicals attempt biotech-innovations in close neighborhood to the stored seeds, which are seen as inventory of genes or genetic resources.

Interestingly, the former and the new German IPK sets a high value on the term *Kulturpflanze*, literally translated as "culture plant", which in English is simply referred to as "crop plant". The term "Kulturpflanze" highlights the alliance between culture and agriculture, and is used by almost all plant agricultural institutions in Germany; in Austria and Switzerland, too. The questions for STS scholars engaged in the field of agriculture are pending: How do the materialities and semantics of "culture" and "technology" interrelate on both the societal and the scientific level? Moreover, do we have the adequate conceptual framework for analyzing these interactions? This question becomes particularly pointed when we ask how the spheres of the artificial and the natural interact.

8.2 What Denotes "Biofact" and Why? On the Difficult Relation of STS and Biotechnology in the Light of Modern Agriculture

The term biofact, a neologism comprised of "bios" and "artefact" (Karafyllis 2003, 2006), refers to a being that is both natural and artificial. The term denotes a two-level approach: empirical and conceptual. On the *empirical* level, a biofact is a growing self that is not growing *by itself*. It is brought into existence by purposive human action. While the result is ontologically still a *being* (alive), it is, at the same

[4] See IPK's portrait at http://www.ipk-gatersleben.de/en/institute/portrait/ (accessed on February 29, 2016).

time, a *fact* (from Latin: *factum*, German: Tatsache). The actions involved in making biofacts are, however, all but obvious. In most cases, they are disguised and melt into planting, transplanting and stimulating growth processes – processes that are usually interpreted as natural.

On the *conceptual* level, "biofact" denotes a hermeneutic strategy. Social scientists would maybe term it "deconstructive". The rhetorics, processes and procedures of generating and proliferating organisms (or their parts) which have been technically set and altered should be unveiled and made debatable. As a hermeneutical concept, it does not impose the question "What is a biofact?" but rather: "How is this entity (which is supposed to be a biofact) made?" Therefore, the technical biographies of biofacts are under investigation, following a thorough analysis of when and how the growth potentials (= the mediality) of a living being have been turned into technical means for agriculture, engineering and design (just to name a few). Biofacticity covers a spectrum of technically altered entities in a certain time-span in which complex forms of technonature were developed and continue to develop. A seed in a cold-storage system, kept semi-living at −18 °C for decades, is a biofact because the *latency* of growth, usually actualized seasonally, is prolonged technically. In seed banks, biofacts really live only in a state of *interpassivity* – when the seeds are rejuvenated by plant growth, before they are stored again. In the concept of biofact, the idea of nature is reserved to immediacy or unmediatedness, i.e. when growth is the medium for the living being itself and remains its mode of continuity without the interference of human action (see Table 8.1).

At each step of 'creating' a biofact, the interrelation of natural processes and technical procedures (German: *Verfahren*) are analysed, resulting in complex higher-order forms of technonatures. Among the first steps of generating technonature are actions like sowing, planting, transplanting, grafting or injecting. Not astonishingly, this approach commonly reveals that a linear narrative of technogenesis and innovation does not make sense; there are byways, detours and setbacks to consider. Some of them result in new knowledge orders that are relevant for biotech practices (see Table 8.2). For taking specific technological pathways, material signposts are important, such as the gathering of collected seeds in the form of seed banks. They are livestock – a technical term used in animal production. In Heideggerian terms, a seed bank gathers the growth potentials as (German) *Bestand* (English: stock), literally a standing reserve. A Bestand is not an assembly of things; it consist of dispositions that might actually become means.

In sum, "biofact" can be applied to both levels of speech: object language and meta-language. The concept should help to criticize the various forms of biological *essentialism*, moreover in its unquestioned alliance with "nature". In the last decade,

Table 8.1 Growth as medium and mean

	Nature	Technonature	Technology
Entity	Living being	Biofact	Artefact
Mean	–	Growth	Action
Medium	Growth	Action	Action

Table 8.2 The politics of time and changing semantics of Seed Banks/Gene Banks

Level	History oriented	Future oriented	Mediating terms	Instance of making presence
Object of interest	Seed	Gene	Reproductive unit/germplasm	Farm/Garden/ Natural Habitat/ Seed Bank
Scarcity	Conservation	Preservation	Gene erosion, biodiversity	Ex-situ/in-situ conservation
Utility	"Kulturpflanze" (culture plant)	Plant genetic resource (PGR)	Agrobiodiversity, bioeconomy	Demand
Metaphysics	Heritage of Mankind	Options for the future	World	Narratives of agriculture, domestication and the first settlers; overpopulation of the Earth; world-hunger; climate change; global commons
Technology	Expanding seed collections by storage and low-tech breeding	Expanding and prolonging gene collections by high tech methods: radiation, cold-storage and cryopreservation, genetic engineering, "new plant breeding technologies" (incl. CRISPR), synthetic biology	**Biofact**	Gene bank
Science	Completion of seed collections with regard to the existing world ("world assortments")	**Bioprospecting** of biofacts for a new and/or future world	**Gene pool** as pool of breeding options for the future	Models of universal breeding, prototyping and synthetic biology
Bioprospecting	**Morphospecies/ Phenotype**	**Ideotype** (= ideal model plant type)	**Genotype**	Survey collecting, exchange of seeds by "world-networks"
Infrastructure	Network of seed banks with material exchange	Network of seed banks with data exchange	Data bank and information systems ("Big Data")	In silico conservation

the concept of biofact has been used for investigating themes as different as the idea of *naturalness* (Birnbacher 2014), the phenomena of doping in sports, the creative use of biomaterial for the works of art (bioart) and, of course, the design of transgenic organisms and objects of synthetic biology.

The term "biofact" was coined at the beginning of the new millennium,[5] when the public debates on green genetic engineering had first reached a peak. Then it had become more than obvious that referring to a whatsoever "nature" of transgenic plants always implies two vulnerable premises: a nostalgic view of agriculture in which the very idea of experience with life forms is embedded, and a reliable ontology of 'what there is' concerning nature. On the contrary, a term was needed which can denote both what there *was* and what there might *be* in the future in a long-term perspective of breeding and cultivating.[6] To some extent, "biofact" already encompasses the idea of prospecting the future of (and by means of) biological entities and technological pathways, though the single meaning of "fact" refers to what *is* actually done at present in the laboratory and on the field. However, what is designed at present is able to reproduce and progress in form of living beings. Philosophically speaking, "biofact" plays with the semantics of immanence and transcendence that occur in the ideas of "life" and "technics", and in its scientific disciplines biology and technology, respectively.

Additionally, "biofact" was born from a terminological bias in academia. In the publications of peers engaged in STS, the newly designed life forms were all too easily named as epistemic *things*, technical *instruments* or just formable *matter*, although these terms occurred in the mode of critique. When looking at language, a number of science researchers actually reproduce the limited understanding of life which they want to criticize. This terminological downgrade turned out to be a 'dead end' regarding the aim to investigate biotechnology not only for their reductionist understanding of life, but also for their aims and scopes which usually relate to societal values, risk policies and legal transformations. In science research, there were no other than technomorphic terms to describe what's going on between laboratory, industry and the farmers' fields.

At the same time, transgenic organisms were publically criticized for not being "natural" anymore. This has led to the recurring question: When have these organisms – domesticated plants and animals – ever been natural since the Stone Age? Or do we indeed face a continuity of "10.000 years of manipulation" with genetic engineering as an almost natural outcome, as proponents of genetic engineering regularly have stated over the last 40 years? Astonishingly, against the present rhetoric of mobility, globalization and migration, present biotech discourses conceptualize their object of concern with the abovementioned archaic enframing. This is true for both opponents and proponents of genetic engineering. Obviously, biofacts should

---

[5] More than 50 years after the term was coined within the epistemic context of preparatory studies by Bruno M. Klein in the Natural History Museum of Vienna (defining dead structures, e.g. shells, from living or once living entities) and was long forgotten, it reoccurred again with a different semantics first in Karafyllis (2001), ch. 4.

[6] Thereby also contrasting "onto-politics" (see Tyfield, Chap. 13, this volume)

still serve their original functions: cultivating the land, domesticating and segregating life forms, and finally making territories. The seed as biofact in seed banks plays a crucial role in this context, as the history of science has shown (Flitner 1995; Heim 2008).

Undeniably, problems of ontological terminology are involved. Neither the semantics of "thing"[7] nor of "object" covers the crucial idea of biofact which relates to historicity and thus to the entangled histories of both nature (evolution) and technology (technogenesis). So far, there is no better option. "Hybrid" and "chimera" just semantically disguise the ontological problem that a mixture always is a mixture of *something*, which needs to be predefined. Having these problems in mind, I decided to focus on technology rather than ontology when conceptualizing "biofact". It is a neutral term that can express a wide spectrum of in-betweenness, ranging from natural living entities to technical artefacts. To reformulate the central ideas of "biofact": the mediality of *growth* is managed as a technical means to certain ends. Against the "makeability"-idea, *that growth is what you can make grow*, the biofact-concept shows that growth processes are never – and never have been – able to represent the ideal of a complete and controllable stock of instruments. Working with and relying on biofacts is necessarily risky.

In a philosophy of science perspective, "biofact" relates to an *instrumentalist* interpretation of the life sciences (understood in a broad sense), including the debate if there is a reality 'behind' experience or not. This debate has prominently been opened by instrumentalist and pragmatist John Dewey and is still alive, for example, in the present works of Richard Sennett on the spirit of craftsmanship. Different than still in the 1970s, it is today's academic consensus that technology and its means are *not* neutral. In the context of STS, instrumentalism is particularly debated with a view to understanding and criticizing the different concepts of *technological determinism*. But even if a range of soft and hard explications of technological determinism exist and are tested on various, mostly modern innovations, there is still no serious doubt that *breeding techniques* of the Holocene have determined social change.[8] As it seems, there never was any real alternative to breeding and agriculture, at least not 'for us'. In narrating the development of civilization, one faces the alliance of *historical determinism* and *technological singularity* which shapes the perspective of the history of ideas in which agricultural biofacts usually become discourse objects. While this grand anthropological narrative is unquestioned, one can easily state that biofacts have helped to transform humans from hunters and gatherers into settlers and farmers. Over the course of time, the agricultural life form has led to the division of labour, and the concept of ownership and altered kinship structures (in total also described as sociocultural evolution).

[7] In German, "thing" has two terms: "Ding" and "Sache", similar to the ancient Greek word *pragma*.

[8] How long this social change took is a matter of scientific debate, criticizing the 1930s term „Neolithic Revolution". "Revolution" signifies a misleading rapidity of social change, which probably covered around 5000 years.

This narrative, anchoring modern biotechnology in agricultural practices of the archaic past, is tacitly expanded into the future by the present actors of bioeconomy.

Different from machines and urban planning, breeding and agriculture are hardly ever understood as "technology" by twentieth century thinkers, unveiling many of them as intellectual victims of the modernity paradigm. Their focus is still on artefacts and on the related actions of *making* and *using*. These categories can only partially be applied for understanding the technogenesis of the short-lived goods of agriculture, which are *consumed* rather than used. Many academics involved have become victims of the modernity paradigm that goes along with machinery and industrialisation. Alternatively, when thinkers as different as José Ortega y Gasset, Langdon Winner, Hans Blumenberg and Don Ihde have proclaimed the phenomenological insight that technology makes *worlds* (including worlds of imagination and language), they mostly forget to mention that in this process of world-making the former worlds are not merely transformed (in a diachronic perspective), but are partially still *there* and synchronically existing – and not only in international forms of a 'Third World'. The rural worlds of agriculture in highly industrialised and post-industrialised countries is but one example for worlds thought to be outdated, though existing parallel to the urban worlds of computerization and virtual reality ("multiple modernities"). Glocalisation and the coexistence of worlds is the reason why agriculture actually profits from technological innovations and is all but low tech. To the majority of citizens, however, agriculture and the rural land appear as heterotopy.

A related problem which shows on the level of public awareness is that the vast majority of people in highly industrialized countries live in cities and rarely have contact with farmers any more. The long duration of conventional breeding procedures and the patience of farmers and foresters are underestimated. As a result, genetic engineering is not understood by the public as a strategy of *accelerate breeding*, but instead as a strategy of invading the "natural" sphere. Ontologies of species and kinds dominate the public discourse. Hence, the problem of identity and its alienations is in focus – from "foreign DNA" in genetics discourses to "foreign species" in nature conservation discourses.

Against this background, modern seed or gene banks can be regarded as instances of *liberalization* in two respects: firstly, when highlighting the common heritage of a nation's or region's plants, cultural identity is stabilized – e.g. by means of stressing the eating habits over the last centuries –, even if the crops concerned did not originate in this region (e.g., the potato eaten in Europe). Secondly, the character of PGR as *cosmopolitan* commons is emphasized (both by NGOs and actors of agro-industry), i.e. that agrobiodiversity is a *common heritage of mankind*. However, from there, it is just a small step to state that banking the *total* of agrobiodiversity is a must for mankind's future. In this perspective, the difference between the wild and the cultivated, and between present and future generations, is dissolved both spatially and temporally (see Sect. 8.4).

8.3 The Gene Rush in Light of Risk Policies

8.3.1 Gene Erosion and the Convention on Biological Diversity

Throughout the twentieth century, genetic information that derives from the wild, uncivilized or at least rural regions of the world, is more and more understood as (a) a *scarce* resource and (b) a *non-renewable* resource. At the same time, seeds in banks tended to be described with technical terms, such as: "resource", "information", "documentation", "safety copy" – which Pistorius (1997) sees as one crucial failure of the FAO and World Bank politics in the 1960s. Due to the technical enframing of the living objects, the practical problems in controlling and standardizing life forms suddenly hit the new emerging gene banks of the 1970s (see Sect. 8.6).

In a way, the stored seeds seem to exist like artefacts in a museum – despite the fact that they still host the potential to live (germability) if managed properly. It comes as no surprise that seed banks employ rhetorics of *museums*, for example when terming the person who cares for the storage and reproduction of a species' variety "curator". However, present seed bankers feel almost insulted when referring to their institution as "museum" instead of "research inventory" or "gene bank".[9] While the above-mentioned technical terms employed by the FAO – foremost: "plant genetic resource" (PGR) – evoke the aura of control, they are accompanied by several risk policies that stress the urgent *need* for control. To quote some dominant questions of protagonists of modern seed banking: Will the few highbred crop varieties on which global food security is based survive the stresses of climate change? Shouldn't we better take action right now and utilize the gene pool, i.e. collect and store the rare landraces that are already adapted to the various stressors as *options for the future*? Shouldn't "we" (actors from the Western, highly-industrialized countries) take care of the vanishing plant diversity in the less-developed world, particularly in light of the technical and scientific inferiority *there*? – Since the 1960s, the PGR-discourse's demarcation line for separating the "natural" from the "cultivated" does not refer anymore to the plant 'species' itself, but to the *technical standard* for preserving it.

Another term of risk policies applies. The ongoing process of extinguishing the materialized forms of genetic variability is termed *gene erosion*. The term was coined at the 1967 FAO conference (Pistorius 1997: 9). Gene erosion highlights the slippery slope towards species extinction. Reasons for gene erosion are, among others, unsustainable land use and cropping patterns (e.g. monocultures) and the vast introduction of high-tech plant varieties including their heavy machine-based and agrochemically supported cultivating techniques. Hence, globally dominant agricultural practices and key corporations and institutions are part of the problem, and it is debatable if their technological pathways can be part of a solution. With regard

[9] Personal interviews (Dec. 2015–Feb. 2016) with staff working at the IPK Gatersleben (Germany) and at the former West German BGRC (1970–2006).

to the big technology-fashioned 'World Seed Bank' which opened 2008 in the polar region of Norway (see below), the *Svalbard Global Seed Vault* (SGSV), historian of science Tiago Saraiva (2013: 185) remarks: "The facility at Svalbard was designed to resist major disasters, which is paradoxical because the crop diversity it harbours is endangered by the very standardization of agriculture at a global scale that breeding based on seed banks has facilitated." As the author shows, the breeding politics and technologies of early Soviet Russia and Nazi Germany in particular set the pace for diminishing crop diversity in twentieth century Europe and beyond. Both passed early Seed Acts[10] to prevent farmers from growing their traditionally bred plants for the sake of implementing the sophisticated cultivars that were developed by means of scientifically staffed seed banks, including statistics-based genetic research on mutations, preselection and hereditary laws (see Sect. 8.1.3).

In consequence, gene erosion is anything but a problem to be negotiated around the geographical axis North/South or the developed versus the developing countries, as the political discourses on biopiracy and intellectual property rights insinuate. Loss of crop diversity is a prominent problem in Europe. The predominant solution suggested is, however, essentially the same as in the period between the two World Wars: genetically 'enriching' national crop diversity by means of utilizing the gene pool. This implies collecting genetic varieties from the rural and distant parts of the world, which harbour the genetic origins of most cash crops. Indeed, this strategy generates an axis of political geography and encompasses areas struck by wars and terrorism: Central America (particularly Mexico and the Caribbean), the Andes, the Middle East from the Levante up to the Caucasus, the Himalaya region and the mountainous areas of central East Africa, including Kenya and Ethiopia.

Since 1992, the collecting and storing of seeds is *ecologically* enframed with the UN Rio Earth Summit's *Convention on Biological Diversity* (CBD; UN 1992), which states that global biodiversity has to be protected. Accordingly, further gene erosion should be prevented. Primarily, the CBD continues to trigger critical arguments of intellectual property rights and secure trade regulations: "This convention challenged the assumption that the earth's biological and genetic resources are part of the 'global commons' by giving property rights over these resources to the nation-states." (Merson 2000) Also under debate is, if CBD and the whole idea of protecting biodiversity is serving neo-colonial interests. Secondly, questions on *how* to conserve biodiversity arose – by in situ- or ex-situ conservation. They are continuously discussed by NGOs and industry, farmers and home gardeners, actors of botanical gardens and zoos, ethnologists and also 'gene bankers' (see Nazarea et al. 2013). A third bunch of arguments deals with the terminological question, what "biodiversity" denotes for different contextual purposes: inter-species diversity (ecology-driven term), intra-species diversity (genetics-driven term), crop diversity (agriculture-driven term) or *agrobiodiversity* (term fusing approaches from agriculture, genetics, ecophysiology and technology). Depending on the semantics of *biodiversity*, 'gene hunters' target very different kinds of living objects.

[10] For Nazi Germany, the controlling institution *Reichssortenamt* was opened 1934 (today: Bundessortenamt, located in Hannover).

Of particular relevance is Article 9 of the CBD, which explicitly stresses ex-situ conservation methods. They should complement in-situ measures in order to sustain natural habitats. The approach and its clear-cut dichotomy of in-situ/ex-situ is science-driven and tacitly ignores the farmers, i.e. the conservation of biodiversity *on-farm*. Hence, strategies for supporting organic agriculture were, at that time, politically underestimated. As the plant scientist Karl Hammer (2003) argues, the CBD overburdened the then existing seed banks. They neither had the spatial and material capacities to conserve all forms of biodiversity requested for protection by the CBD, nor had they adjusted their traditional narrative of preserving *cultural heritage* (or even "common heritage of Mankind", see Hammer 2003, 4) to the new conceptual changes in plant genetic research. Pace Hammer: the "paradigmatic shift" in seed banking did not only result from developments in "molecular biology" in the 1990s, as most long-term seed banks always have incorporated high-level plant genetics research; but rather from what is known in STS as "mode 2 science". This denotes forms of science that are project-based and head for short-term utilization of knowledge, interdisciplinarity and external funding. This innovation-driven way of doing science conflicted with the *modes of continuity* in which seed banks were used to operate and which they once shared with breeders and farmers.

8.3.2 IT Systems: Material Versus Digital Seeds

Agreeing with Hammer, another drastic change in seed banking "came into being through the swift development of electronic data documentation, management and dispersion. This led in less than two decades [...] to highly developed, internationally oriented information systems." (ibid.) The first computerized gene bank information system (Seidewitz 1979) was set up by the horticultural scientist *Lothar Seidewitz* (born 1935), working at the West German BGRC since its foundation 1970. His thesaurus for cereals' descriptors became operational in 1974. Still based on a punchcard system, the seed banks of Western Europe adopted it soon, aiming at establishing the FAO-envisioned "world-network" of PGR at least within the small world of the European Community. Referring to technical infrastructure: bottlenecks, still in the 1980s, were the different telecommunication systems of the various countries. This lack of European standard complicated data transmission.[11] Another bottleneck concerning hardware was to bring the German SIEMENS-standard in alliance with the IBM-standard (Karafyllis and Lammers 2017; Karafyllis 2018a).

Referring to personal infrastructure: although prohibited, Lothar Seidewitz, Walter Hondelmann (1923–2015) and others from the West-German gene bank

[11] Personal interview with Lothar Seidewitz, Feb. 2016.

have been in close contact[12] with their East German colleagues of the IPK throughout the 1970s and 1980s (e.g. with Helmut Gäde, born 1929; and with Christian O. Lehmann, 1926–1992).[13] They exchanged not only seeds, but also descriptor lists, passport data and technical details about Western and Eastern computers for establishing a world-network without political borders.

Today, each major seed bank transforms the characteristics of the collected seed and germplasm samples ("accessions"), termed "passport data" and usually catalogued in inventory lists, into computerized data and codes. A selection of these data – the ones that are regarded relevant at present – are transferred into meta-data stored in international data banks; a standardization process which runs anything but smoothly and has its own problems on the practical level. Several parallel data banks exist, specializing on different plant features. One recent outcome is the global data bank and web portal GENESYS which is online since 2011, serving actors in agriculture by also including geo-data. GENESYS emerged due to a cooperation of Bioversity International, CGIAR and the Crop Diversity Trust.[14] It generates meta-data from the data collections of CGIAR, the European catalogue EURISCO (see below) and the US-American NPGS (National Plant Germplasm System). "Genesys features Universally Unique Identifiers (UUID) for accession records and makes records accessible through Permanent URLs using PURL.org service. The archived records remain resolvable and accessing the PURLs of archived records will return the archived data."[15] This statement on protecting the historicity of codes became necessary, when the different datasets of Russia (VIRDB)[16] and EURISCO were melted into one in fall 2015, thereby replacing the Russian mixed code-system (letters VIR + numbers) with the numbers-only code used in EURISCO.

The ECPGR initiated EURISCO,[17] which is a web-based search catalogue providing information on ex-situ plant collections in Europe. It was hosted by Bioversity International in Italy from 2003 to 2014. Since 15th April 2014, it is hosted at and maintained by the IPK in Germany. EURISCO retrieves its data from National Inventories (NIs) from member countries, aiming at a regular update. Moreover, and to differentiate itself from the new GENESYS, it now collects *phenotypic* data.

---

[12] Personal interviews with Lothar Seidewitz, Christopher Otto, Lothar Frese (Feb. 2016); all formerly employed at the West German gene bank/BGRC located at the Bundesforschungsanstalt für Landwirtschaft Braunschweig-Völkenrode (FAL).

[13] Lehmann was 1969–1990 head of the IPK's cross-sectoral research unit "Kulturpflanzenweltsortiment" (transl.: world assortment of cultivated plants); Müntz and Wobus 2013: 203. On Hondelmann, leading scientist at the West German gene bank in Braunschweig, see Karafyllis and Lammers 2017.

[14] I thank Suzana Alpsancar for notifying me about the GENESYS data bank.

[15] Post "Accession records of Vavilov Institute (RUS001) match EURISCO" (22 Oct. 2015) on the GENESYS website. See https://www.genesys-pgr.org/de/content/news/43/accession-records-of-vavilov-institute-rus001-match-eur (accessed on March 5, 2016).

[16] The Vavilov Research Institute Database (VIRDB) can be found at http://91.151.189.38/virdb/ (accessed on March 5, 2015).

[17] A predecessor is AEGIS (A European Genebank Integrated System).

Currently, the catalogue consists of passport data of more than 1.5 million accessions. Interestingly, the highest number by far (43.43% of all accessions) results from the NI of Great Britain. Germany is second rank with only 9.47% of documented accessions total.[18] The plant with the highest number of accessions in Europe is not a crop but a weed: *Arabidopsis thaliana*, the model plant for genetics. To conclude, seeds have become digital objects, which serve also – and probably foremost – scientific purposes. A direct intersection with agriculture might provide the new concept of "digital farming" (incorporating geo- and space data), in which key companies in agribusiness engage at present.[19]

8.3.3 The SGSV and the Idea of Safety Duplicates

In close relation to this virtualization process is the implementation of the first transnational seed bank Svalbard Global Seed Vault (SGSV), located on an arctic archipelago on the territory of Norway.[20] After a long planning process initiated by the FAO, the SGSV opened its doors in the ice in 2008. In 2014, it reached a capital stock of 400 million US Dollars, mostly deriving from the Bill Gates Foundation, global players of the breeding industry and Western governments (incl. Germany). The SGSV regards itself as a "final back-up" in case of wars and nuclear catastrophes, and with regard to the many seed banks that still operate in a low tech way or with unstable electricity conditions for cold-storage, assumingly located only in the poor countries (see Sect. 8.6).

In fact, the SGSV is expecting "donations" of duplicates of seed samples, which have already been collected in national seed banks. Consequently, the SGSV names its seed suppliers "donors". It functions as a meta-seedbank. A high number of bilateral agreements with various big seed banks exist (e.g. with the German IPK, the Russian Vavilov Institute and the Mexican CIMMYT). In terms of international acknowledgement and funding opportunities, these agreements make the donations almost compulsory. However, when looking at the SGSV's web portal for the number of donated accessions,[21] it is astonishing that biodiversity-rich countries like Russia and Brazil have donated comparably few yet, while Mexico, Yemen and Afghanistan show high donation rates. The advanced infrastructure for seed transport and storage provided by the SGSV is obviously not regarded as useful by everyone, depending on political power and imperial identities. By the staff of donating seed banks, it is commonly stressed that the SGSV is not a real seed or

[18] http://eurisco.ipk-gatersleben.de/apex/f?p=103:30 (accessed on March 4, 2016).

[19] See e.g. BAYER's concept at http://www.digitalfarming.bayer.com/ (accessed 16 Sept. 2016).

[20] The following is based on information presented on the SGSV's websites at the Crop Diversity Trust (www.croptrust.org, accessed on October 1, 2015).

[21] For the Seed Portal, see http://www.nordgen.org/sgsv/ (accessed on March 4, 2016).

gene bank, but a "storage".[22] A euphemistic term for the SGSV used by seed bankers is "passive gene bank" – referring to (a) the fact that the SGSV, with regard to the seeds' germability, does not have *active stock* from on-farm or greenhouse-based reproduction, which leads (b) to a strictly limited concept of access and benefit-sharing.

The idea of depositing safety copies in another country is all but new. Noteworthy are political differences, though. While the seed banks of the former Warsaw Pact states seem to have operated *without* extraterritorial safety duplicates and backup-systems (e.g., the former IPK in the GDR never had safety copies from the Vavilov Institute in Russia,[23] and vice versa),[24] the West German seed bank BGRC contributed to this FAO-driven risk policy for both Ethiopia and Costa Rica during the 1970s and 80s. Among seed bankers, the practice of safety-duplicates is known as *black-boxing*, i.e. the material seed 'copies' are transported in a closed storage-device to a seed bank in another country with the obligation not to open it, or even use its content. Referring to the BGRC, at least one case is known, where the director of the West German seed bank (Manfred Dambroth; 1931–1994) did not adhere to this principle. He tried to propagate Ethiopian wheat on Germany's territory. The result was the end of the German-Ethiopian genebanking cooperation,[25] once set up by development aid from West Germany.[26] The Norway-located SGSV operates with the blackboxing-concept. Incoming accessions of seeds in small aluminium-coated packages are deposited without controlling the content.

However, one might ask how risk relates to geopolitics. While on the SGSV's website the North Pole region is marketed as a peaceful and detached site that can guarantee the security of seeds "forever" – suggested by the aura of Norway, the country which awards the Nobel Prize for Peace –, the geopolitical interest in the region (e.g. by Russia) and the search for oil and gas would suggest otherwise. History shows that in many wars seed collections were objects of robbery and dislocation, similar to art collections (see e.g. Heim 2008). Most likely this is also true for Baghdad's century-old collection of date palm seeds which vanished during the second Iraq war (2003) and is officially labelled as destroyed by the SGSV website to highlight a tragic example for already lost seeds.

[22] Personal interviews with Andreas Graner (director of the IPK), Lothar Frese (former director of the BGRC), and others (2015/16).

[23] The Vavilov Institute has a network of research stations throughout Russia, keeping safety duplicates. During the Cold War, the VIR had also appendices in Kazakhstan and Azerbaijan (if they kept safety duplicates, is not known by author).

[24] Personal interview with Andreas Börner, head of the gene bank unit, IPK Gatersleben, 18 Dec. 2015; and with Lothar Frese, former director of the West-German gene bank BGRC, Jan. 2016.

[25] Personal interviews with staff of the former BGRC, Jan/Feb. 2016.

[26] See *Bundesarchiv Koblenz*, register B 213, folder no. 33192 (*Errichtung einer internationalen Genbank*, 1973–1979; series title: *Technische Hilfe für Äthiopien*).

8.4 Results: Politics of Time and Semantics of Biofacts in Seed Banks

All biobanks including seed banks are instances of *turning the past into a presence* that actually does not exist (similar to museums and archives). At the same time, the future seems to be already 'here' to grasp. This happens by referring to the (assumingly) still existing growth potentials of origins, addressing "the wild" rather than "the cultivated". However, in the concept of agrobiodiversity and the related narratives of breeding and domestication (see Sect. 8.2), the wild and the cultivated are close allies. Objects of interests are hence crops *and* crop wild relatives (CWRs, e.g. grasses as genetic predecessors of bred cereals), as well as farms *and* natural habitats. Cultivated plants can thus be regarded as both cultural heritage of mankind ("Kulturpflanze") and options for the future ("plant genetic resources"). Seed banks incorporate biofact histories and transform them into narratives of future options for "the world" (e.g. backed by world population projections). They synchronize the past with the future by means of (a) the biographies of the collected, semi-living objects, (b) the technological options ahead, and (c) the metaphysical world-concept. Table 8.2 summarizes the different conceptual levels and mediating terms, on and by which pasts and futures of living organisms are transformed into a *present stock of instruments*. For example, in the light of climate change, a seed of a drought-resistant landrace semantically turns into a genetic reservoir for genetic engineering and synthetic biology. It is debatable, if seeds or genes are stored – or if, in light of digitalization strategies, maybe data will be the objects of long-term storage. However, the "making of" presence involves different locations and translocations of life forms (newly termed "gene pool"): for example, the sites of bioprospecting and 'hunting', the sites of gathering, the sites of comparing and collecting, the sites of identifying, cataloguing and data processing, and the site where the collection itself is located and might become a research inventory. This spatial approach is enframed with certain politics of time that show on the semantic level.

Here, I can comment on just one more of the many semantic shifts presented above. Whereas the United Nations' CBD 1992 still suggested the *con*servation of genetic resources, the current political rhetoric in agricultural bioeconomy stress their *pre*servation. The latter emphasizes the character of (breeding) potential and option for the future. Moreover, the term "preservation" is in tacit alliance with the ethical precautionary principle and the "ethos of openness" (see Schneider, this volume), which has been applied in IT-contexts and here also means: open into the wild. Wilderness is thus becoming a sphere of preservation. As Table 8.2 shows, the taking over of wilderness as a preservation zone is also encouraged by technological ambitions, as molecular targeting and genetic engineering aim at controlling the complete *gene pool* of a species. It is against this worldview, a total world of all possible breeding options, that the idea of bioprospecting makes sense.

The ideas and rhetorics of modern seed banking, with its high time during the 1960s and 1970s, show remarkable input from systems theory. It developed out of biological systems theory or *biocybernetics*, and is deeply intertwined with the development of genetic engineering. As engineer and mathematician Mihajlo

D. Mesarović stated: „Systems theory is the theory of formal (mathematical) models of real life (or conceptual) systems“(Mesarović 1968: 60). Formal structures of the model are set invariant. They build a relation, which can be mathematized and used for „bio-engineering“. On the one hand, this biocybernetic model results in „bio-instrumentation“, on the other hand, it enables biology to search for universal laws, „which govern the behavior and evolution of living matter in a way analogous to the relation of the physical laws and non living matter.“(ibid.). In sum, the dead is the paradigm for modeling and designing the living.

8.5 Reality Check: How the Biocybernetic Model of Seed Banks Fails

Contrasting the seed banks' aforementioned biocybernetic enframing of control, regulation and makeability, the following lists some of the real problems still occurring:

- Referring to infrastructures of "the wild", one of the major problems of bioprospecting and collecting is *naturalisation* of cultivars. Many plant species and varieties have escaped, sometimes centuries ago, from places of ex-situ conservation; for example, botanical gardens. Hence, a bioprospected wild relative of a crop species can turn out to having been a cultivar. This results in problems for genetics and breeding research. Moreover, naturalisation is under critical scrutiny by nature conservationists who fight *bio-invasions* (van der Weijden et al. 2007) and try to protect a dominion of culturalized nature – a reminiscence of the garden – by means of identity- and health-politics.
- Referring to the language of biofacts when identifying and naming the collected plant, different taxonomical methods for cultivated plants existed worldwide up to the twentieth century. On the contrary, the universal botanical taxonomy uses the *wild* plant as reference. Consequently, names of crops in old collections show high variability. Furthermore, the once aggregated passport data of the varieties (referring also to specific collection sites) might not include all information relevant for present agricultural use. Even today, there is no universal standard for the act of plant collecting (e.g. some seed banks include collecting soil samples from the original growth location in order to facilitate reproduction later, others regard this dispensable).
- This problem shows on the infrastructural level of IT, when different data collections are fused and transformed into meta-data. Related are also the terminological problems of *synonyms* and *homonyms* – when the same name (homonym) is given at certain times and places for different plant varieties, or, reversely, the same variety is listed under synonyms.
- Therefore, seed banks need cross-reference systems usually consisting of old-fashioned forms like books (library) and herbarium collections of type specimens. Whereas the old and major seed banks of the world (e.g. Kew in the UK, IPK in Germany) encompass rich libraries and herbaria, the smaller and

newer ones often lack these reference systems. For example, the former West German BGRC never received funding for a herbarium from the German Ministry of Agriculture.[27]

- In the light of the blackboxing-system practiced at the SGSV in Norway, you never know which kind of plant variety you actually deposit as safety duplicate and preserve in cold-storage. One has to rely on the taxonomic expertise of the donating institution – though many seed banks do not host taxonomical experts. New methods operating with molecular markers (Börner 2006) are costly.
- A major problem is reliable cold-storage technology, *also* in the industrialized countries. Whereas the FAO generated its gene bank hype already in the 1960s, it took until 1994 to implement a gene bank standard for technology (FAO and IPGRI 1994). It has to remain an open question if all parties involved actually practice these standards (see below). Furthermore, when looking at the storage units, there exist at least three parallel material objects for conserving orthodox seeds: tinned seeds in *metal* cans (e.g. used by China and West Germany), seeds in preserving *glasses* (similar to jam jars, used by the IPK), and seeds in vacuum-sealed *aluminium*-packets (used by the SGSV).

Thinking of the common narrative that the Third World's seed banks should better deposit safety duplicates outside their countries (because of unreliable cold-storage technology and poor electricity supply at home), the following is noteworthy. Two national institutions of the so-called First World, the BGRC in West Germany and the Germplasm Institute (GI) in Bari/Italy, faced continuous technical challenges quite recently. Interviews with the technical staff of the BGRC revealed that due to lack of funding, the total of seeds have never been cooled at the obligatory −18 °C, but by only −10 °C, in order to save money (this was practiced during the almost four decades of BGRC's existence).[28] The replacement of the outdated cold-storage systems of the 1970s never happened. This need became urgent during the process of reunification of Germany, when the fusion with the East German IPK seemed possible – and the Ministry of Agriculture postponed the decision for modernization perpetually, though leaking water led to the tins' corrosion. Furthermore, the head institution where the seed bank was located (the national research institute for agriculture FAL in Braunschweig-Völkenrode) did not have a *motorized* technical maintenance service unit on the huge campus until the mid 1980s. There was also no specialized refrigeration engineer or technician available. Consequently, when the alarm-system of the seed bank signalled a dangerous increase in temperature, nothing used to happen for quite some time. The technicians arrived by bicycle, and if they had forgotten a tool at their service station, the time for biking back and forth put the seeds at considerable risk.[29]

[27] Personal interview with Lothar Seidewitz (Feb. 2016) who remembered that the first director of the BGRC, *Dieter R. F. Bommer* (1923–2010), had applied for funds for a herbarium in 1970.

[28] Personal interview with Otto Baars-Hibbe, technical assistant of the former BGRC, Braunschweig, Germany (Feb. 2016).

[29] Personal interview with Christopher Otto, former head of the director's office at the FAL, Braunschweig, Germany (27 Feb 2016).

Even more dramatic is the presentation given by Pietro Perrino at the *Sustainable World Conference (14–15 July 2005 in Westminster).* Perrino was long-term director of the Italian *Germplasm Institute* (GI) at Bari, which was, as the BGRC, opened in 1970 as a national gene bank and resulted from FAO's "world-network" politics. 2005, in the light of a new merge of the GI with another research institution specialized in plant genetic engineering (IGV), Perrini highlights how the lack of maintenance of cold-storage facilities might have followed strategic interests:

> Since November 2002, the management of the IGV [Istituto di Genetica Vegetale] has created a lot of trouble for the ex GI. The most serious concerns the cooling system for seed storage in the gene bank, which did not function for a few months and therefore the temperature of cold storage rooms went up for a considerable number of days. The Magistrate of Bari has already made an intervention with the result that I was nominated judicial custodian of the gene bank. Only then was the cooling system repaired. […] I, as Director of the ex-GI, now Research Manager, together with a few remaining colleagues are fighting to defend the gene bank and the stored plant genetic resources from the Director of the IGV and his lobby, who want to have full control of the germplasm in order to use it as a pretext for getting large research funds, as they are not interested in biodiversity but are fully involved in GMOs or even worse, in the opinion of some of us (including outsider supporters), the lobby, linked to multinationals, wants to destroy the germplasm and prevent farmers from using them. This last hypothesis is not so strange if one considers that the Director of the IGV did not respond positively to the request of repairing the cooling systems of seed storage rooms when they were not functioning. (Perrino 2005)

The above-mentioned problems all relate to the historicity of biofacts, which partially resist synchronization. Moreover, the stakeholders seem to have different judgements on the different areas and duties of seed banking, and if they deserve high tech or low tech. Obviously, also on the political level, integrating the different modes of in-situ, ex-situ and on-farm preservation of PGR is a difficult task. Recently, European stakeholders in the conservation of PGR (Frese et al. 2014) released a discussion paper on the need of a *European Plant Germplasm System* in order to make the different political resorts responsible for nature conservation, seed banking, farmers' rights, botanical gardens, IT-systems et al. finally act in concert. The national germplasm system of the USA functions as prototype for the European hopes. Different stakeholders including NGOs took part in the discussion process. The gene bank stakeholders' result is (Frese et al. 2014: 7):

- "Most gene banks are not independent with respect to their program, funds or staff. Crop experts are thus forced to prioritise research work at the expense of conservation work.
- Ex situ management of PGR of many EU gene banks is not based on defined, written quality management standards. Limited funding exists for basic tasks of maintaining and regenerating PGR.
- Gene banks are seldom leading the development of national in situ management and on-farm management strategies. Many different governmental authorities are responsible and national strategies for in situ/on-farm management are often missing.
- Insufficient support of germplasm conservation activities at all governmental levels."

8.6 Summary

When analysing seed banks, alienating dichotomies occur: genealogies versus network models, robbery versus donation of seeds, the indigenous versus the cultivated world. It remains to be debated in STS, if agriculture can still be regarded as culture or should better be understood as technology. In other words, seed banks highlight a form of culture that derives its name from an agricultural metaphor but is deployed in the service of a capitalist ideology. Therefore, seed banks and their related biodiversity politics have become instances of transient standardization (see Delgado and Rodríguez-Giralt 2014). At the same time, seed banks remain *heterotopias* – imaginary places where one cannot actually live, but which stabilize the continuity of low-beneficial places in the life world. The time politics to which seed banks contribute, praising the benefits for the future, might trigger a general dehistorization of agricultural knowledge and practice. As a side effect, this is putting the concept of culture – including the "culture plant" (German: *Kulturpflanze*) – at risk.

It was shown, however, that the presence of nature, whether we call it "wild" or "original", is not indispensable for aiming at ex-situ conservation. This chapter has emphasized that growth of biofacts is a means for reproduction serving as both a tool ready-at-hand and a tool presence-to-hand.[30] In order to become a tool, growth potentials have to be transformed into a stock of instruments. Instances of stock building are the acre, the garden and the seed bank. Only the latter relates to the both material and science-driven politics of infrastructure for agriculture on a global scale. The created livestock is not a mere assembly of biofacts. Instead, the stock, because of its inherent instrumentality and potentiality, generates the idea of completing and diversifying itself. The final goal is having a world-assortment. Having said that, the seed bank seems to be in a theoretical opposition to some of the general principles of modern agriculture, as agriculture's economic success story is based on the concentration of only a few genetic varieties and their broad dispersal. This paradoxical structure is but one explanation for seed banks' poor funding, and for the lack of political support on the governmental level (at least in most cases).

Finally yet importantly, seed banks allocate resources. Together with related science policies and corporate investors, and somehow unconscious of their own massive impact, seed banks make material politics of infrastructure for, on the one hand, actors in biotechnology and bioinformatics and, on the other hand, actors in agricultural politics that separate seeds from lands. Hence a question omitted by all parties concerned is: Who owns the land on which the preserved plant genetic resources will grow in order to actually be options *in* the future, not for the future?

[30] These terms denote the Heideggerian „Vorhandensein"and „Zuhandensein".

References

Birnbacher, D. 2014. *Naturalness*. Lanham: University Press of America.

Bommer, D. 1972. Entwicklung und Bedeutung von Genbanken in der Welt und in der Bundesrepublik Deutschland. *Saatgutwirtschaft (SAFA)* 24 (15/16): 391–394.

Börner, A. 2006. Preservation of Plant Genetic Resources in the Biotechnology Era. *Biotechnology Journal* 12 (1): 1393–1404.

Cohen, S. 2014. Seed Banking, Seed Saving, and Cultivating Local Varieties. In *Encyclopedia of Food and Agricultural Ethics*, ed. P.B. Thompson and D.M. Kaplan, 1642–1648. New York: Springer.

Delgado, A., and I. Rodríguez-Giralt. 2014. Creole Interferences: A Conflict Over Biodiversity and Ownership in the South of Brazil. In *Beyond Imported Magic. Essays on Science, Technology, and Society in Latin America*, ed. E. Medina, I. da Costa Marquez, and Ch. Holmes, 331–348. Cambridge, MA/London: MIT Press.

FAO. 2010. *The Second Report on the State of the World's Plant Genetic Resources for Food and Agriculture*. Rome: FAO.

FAO, and IPGRI. 1994. *Gene Bank Standards*. Rome: FAO. ftp://ftp.fao.org/docrep/fao/meeting/015/aj680e.pdf. Accessed on 14 Feb 2016.

Flitner, M. 1995. *Sammler, Räuber und Gelehrte: Die politischen Interessen an pflanzengenetischen Ressourcen 1895–1995*. Campus: Frankfurt am Main.

Fowler, C., and P. Mooney. 1990. *Shattering. Food, Politics and the Loss of Genetic Diversity*. Tucson: University of Arizona Press.

Frese, L., A. Palmé, L. Bülow, and C. Kik. 2014. Towards a European Plant Germplasm System. *The Third Way*. http://www.nordgen.org/ngdoc/plants/Samarbeten_och_natverk/PGR_secure_workshop2013/TowardsEuropeanPlantGermplasmSystem_policyPaper.pdf. Accessed on 5 Mar 2016.

Hammer, K. 2003. A Paradigm Shift in the Discipline of Plant Genetic Resources. *Genetic Resources and Crop Evolution* 50: 3–10.

Heim, S. 2008. *Plant Breeding and Agrarian Research in Kaiser-Wilhelm-Institutes 1933–1945. Calories, Caoutchouc, Careers*. Boston: Springer.

Hoban, T., E. Woodrum, and R. Czaja. 1992. Public Opposition to Genetic Engineering. *Rural Sociology* 57 (4): 476–493.

Hondelmann, W. 1974. Das Weltnetz der Genbanken und die Deutsche Genbank in Braunschweig-Völkenrode. *S.A Umschau* 74 (19): 605–609.

Jackson, M.T. 2013. Hawkes, John Gregory. In *Oxford Dictionary of National Biography 2005–2008*, ed. L. Goldman, 499–500. Oxford: Oxford University Press.

Juma, C. 1989. *The Gene Hunters: Biotechnology and the Scramble for Seeds*. Princeton: Princeton University Press.

Karafyllis, N.C. 2001. *Biologisch, natürlich, nachhaltig. Philosophische Aspekte des Naturzugangs im 21. Jahrhundert*. Tübingen/Basel: Francke.

———., ed. 2003. *Biofakte – Versuch über den Menschen zwischen Artefakt und Lebewesen*. Paderborn: Mentis.

———. 2006. Biofakte – Grundlagen, Probleme, Perspektiven. *Erwägen Wissen Ethik* 17 (4): 547–558.

———. 2018a. Vom Biofakt zum Cyberfakt. Die Samenbank als digitales ‚Weltnetzwerk pflanzengenetischer Ressourcen'. In *Mit Biofakten leben. Sprache und Materialität von Pflanzen und Lebensmitteln*, eds. B. Gill, and K. Zachmann, 87–128. Baden-Baden: Nomos.

———., ed. 2018b. *Theorien der Lebendsammlung. Pflanzen, Mikroben und Tiere als Biofakte in Genbanken*. Freiburg: Alber.

Karafyllis, N.C., and U. Lammers. 2017. Big Data in kleinen Dosen: Die westdeutsche Genbank für Kulturpflanzen ‚Braunschweig Genetic Resources Collection' (1970–2006) und ihre Biofakte. *Technikgeschichte,* 84(2): 163–200.

Lundqvist, U. 2014. Scandinavian Mutation Research in Barley – A Historical Review. *Hereditas* 151 (6): 123–131.
Merson, J. 2000. Bio-prospecting or Bio-piracy: Intellectual Property Rights and Biodiversity in a Colonial and Postcolonial Context. *Osiris* 15: 282–296.
Mesarović, M.D. 1968. Systems Theory and Biology – View of a Theoretician. In *Systems Theory and Biology*, ed. M.D. Mesarović, 59–87. Berlin: Springer.
Müntz, K., and U. Wobus. 2013. *Das Institut Gatersleben und seine Geschichte*. Berlin: Springer.
Nazarea, V.D., R.E. Rhoades, and J.E. Andrews-Swann, eds. 2013. *Seeds of Resistance, Seeds of Hope. Place and Agency in the Conservation of Biodiversity*. Tucson: University of Arizona Press.
Perrino, P. (2005), 'Italy's Genebank at Risk. Institute of Science in Society', ISIS-Report 03/08/05. http://www.i-sis.org.uk/ItalysGenebankAtRisk.php. Accessed on 12 July 2017.
Pistorius, R. 1997. Scientists, Plants and Politics. In *A History of the Plant Genetic Resources Movement*. Rome: IPGRI.
Plucknett, D.L., N.J.H. Smith, J.T. Williams, and N. Murthi Anishetty. 1987. *Gene Banks and the World's Food*. Princeton: Princeton University Press.
Saraiva, T. 2013. Breeding Europe. Crop Diversity, Gene Banks, and Commoners. In *Cosmopolitan Commons. Sharing Resources and Risks Across Borders*, ed. N. Disco and E. Kranakis, 185–212. Cambridge, MA: MIT Press.
Scarascia-Mugnozza, G.T., and P. Perrino. 2002. The History of Ex Situ Conservation and Use of Plant Genetic Resources. In *Managing Plant Genetic Diversity*, ed. J.M. Engels, V. Ramanatha Rao, A.H.D. Brown, and M.T. Jackson, 1–23. New York: CABI.
Schubert, J., S. Böschen, and B. Gill. 2011. Having or Doing Intellectual Property Rights? Transgenic Seed on the Edge Between Refeudalisation and Napsterisation. *European Journal of Sociology* 52 (1): 1–17.
Seidewitz, L. 1979. Data Base Concepts for a Rational Handling of Information in Relation to Plant Genetically Valuable Material. In *Technical Aspects of Information Management and Means of Communication in Plant Genetic Resources Work for a Future Utilization of Genetic Material in Plant Breeding*, 31–71, Proceedings of a Colloquium held in Braunschweig on 7th and 8th November 1978. Institut für Pflanzenbau und Saatgutforschung, FAL Braunschweig-Völkenrode.
Stubbe, H. 1982. *Geschichte des Instituts für Kulturpflanzenforschung Gatersleben der Deutschen Akademie der Wissenschaften zu Berlin (1943–1968)*. Berlin: Akademie.
UN. 1992. *Convention on Biological Diversity*. Rio de Janeiro: UN. https://treaties.un.org/doc/Treaties/1992/06/19920605%2008-44%20PM/Ch_XXVII_08p.pdf. Accessed on 1 Feb 2018.
van der Weijden, W., R. Leewis, and P. Bol. 2007. *Biological Globalisation. Bio-invasions and Their Impacts on Nature, the Economy and Public Health*. Utrecht: KNNV.
Vavilov, N.I. 1922. The Law of Homologous Series in Variation. *Journal of Genetics* 12 (1): 47–89.

Nicole C. Karafyllis is a Trained Philosopher and Biologist and Full Professor of Philosophy at Technische Universität Braunschweig in Germany. Her areas of specialization are philosophy of science and technology, phenomenology, environmental ethics, philosophy of agriculture, and intercultural philosophy. Based on her theory of biofacts, she recently introduced a "Theory of Living Collections," focusing on gene banks and biorepositories (published in 2018 by Verlag Karl Alber, Herder, Germany).

Chapter 9
Music, Recommender Systems and the Techno-Politics of Platforms, Data, and Algorithms

Jan-H. Passoth

9.1 Introduction

It is early in the morning, the sun has just started rising, and the smell of coffee fills the air. "Hey Google," I shout into the empty kitchen, "play my summer flashback playlist on Spotify" and David Bowie starts singing "Ziggy Stardust." The songs that mix with the sounds of the city come from a dynamic playlist, curated by me and my algorithmic companions. It gets a bit boring from time to time, because playing a song from that list makes it even more likely to be played later. I have, however, developed a habit of hand-picking different titles so that Spotify's algorithms have more diverse data to crunch. Sometimes I also need to skip a title – "Hey Google, next song," but most of the songs capture my mood and taste quite okay – not really great, but okay.

Music is an interesting case for studying the contours of how our contemporary post-industrial societies have become techno-logical: We like to think of musical taste, preferences and the habits and idiosyncrasies of the "music lover" (Hennion 2001) as a last resort of a culture untouched by technoscience – although decades of research have mapped its techno-scientific intermediation (Hennion and Meadel 1986; Hennion 1989). But digital technologies have amplified that exponentially: sound and voice modification, computer aided acoustic design of theaters, streaming systems, recommender systems – the production, distribution and consumption of music has been transformed bit by bit by the design, implementation and use of software. Such transformations are deeply controversial, and the shape and contours of contemporary music production, distribution and consumption are the fragile outcome of complex technological and political struggles. Music is therefore an interesting, but not a unique case in terms of its techno-logical rebirth.

J.-H. Passoth (✉)
Munich Center for Technology in Society, Technical University Munich, Munich, Germany
e-mail: jan.passoth@tum.de

S. Maasen et al. (eds.), *TechnoScienceSociety*, Sociology of the Sciences Yearbook 30, https://doi.org/10.1007/978-3-030-43965-1_9

In this chapter, I argue for an engaged approach to the techno-politics[1] of platforms, data and algorithms, drawing on three limited, but hopefully nevertheless interesting empirical vignettes on music recommender systems. The three vignettes are neither about individual taste, nor about its colonialization by a metaphorical machinery of the cultural industry, but about how the logistics of music consumption have been transformed into data science and data engineering problems. Based on these three vignettes, I will highlight what is at stake with the current upgrade of our TechnoScienceSocieties infrastructural nerves when we reinstall them digitally: we are neither making them more rigid nor more participatory, neither more foreseeable nor more personalized, we are simply making them computational. Following a suggestion by Ted Striphas in his commentary on Galloway's description of "algorithmic culture" (Galloway 2006; Striphas 2015) and Callon and Muniesa's work on calculation (Callon and Muniesa 2005), I will argue that by making our common world computational, we reorder it by taking it apart into little pieces and make it reassemble-able into different, contextually specific, and adoptable ways. Unpacking computing and mapping how exactly we are increasingly making our world computational will, I suggest, offer insights into how to deal with different forms of (material, if we follow Barry 2013) techno-politics of platforms, data and algorithms that undergird contemporary techno-societies.

But that, I will argue, forces us to be more serious about the term "politics" in "politics of platforms, data, and algorithms." An engaged and political approach to the ontological politics of digital platforms, data collection and handling as well as algorithms and procedures of processing them is more than needed. Focusing on computation will help us understand the (re-)arrangements that are designed, produced and implemented in computer science departments, software engineering companies and ICT service start-ups. It can also help us trace the formats, the models, and the blueprints that are used, rewritten or evaded. But does it help us to turn this inquiry into an experimental, collaborative, interventionist endeavor? By making use of Latour's and Rancière's "disturbing" understandings of politics (Latour 2003; Rancière 2006) as well as of Stenger's and Michael's reading of Deleuze's technoscientific political and conceptual persona of the idiot (Stengers 2005; Michael 2012), I will conclude this chapter with a proposal for working on these other worlds inside that array of projects, processes and practices of reassembling the infrastructural nerves of our TechnoScienceSocieties.

[1] I use the term 'techno-politics' in the way that Timothy Mitchell has used it in his book on experts and Egypt to describe a "transformation at the material level as much as the level of representations" (Mitchel 2002). Mitchell does not really define his use of the term, but by opposing it to James C. Scott's analysis of the uses and misuses of science and technology by modern nation states, he points out that his use of the term is to focus on the "social and political practice that produce simultaneously the powers of science and the powers of modern states" (Footnote 77: 43). I use the term in this way, although I use the term "politics" in a much broader sense that treats state-oriented politics as only one techno-political arrangement amongst others (see Passoth and Rowland 2010, 2015).

9.2 On Music, Machine Learning and Recommender Systems

But let's start with some music. Contemporary music is at the same time standardized and atmospheric, normalized and affective, mass produced and fitting the needs of the connoisseur. There is Adorno's characterization of contemporary music as industrialized and commodified, perversely manipulative and seductive (Adorno 2001), and the common idea that music rests on some arcane knowledge, is deeply cultural and, despite of everything that happened in popular culture and electronic music, there is something about (real) music that will forever stay somehow untouched by science and technology. This is disputable, of course, as Hennion's work (Hennion and Meadel 1986; Hennion 1989, 2001; see for a summary also Looseley 2006) on taste, radio programming and music production as well as studies on the manufacturing of audiences in audience research (Ang 1991; Bermejo 2009; Bourdon and Méadel 2011, 2014) have demonstrated: There is a deep and intense connection between contemporary music and philosophy, formal logic, math and various aspects of technoscience regarding the construction and design of standardized instruments and formal notation schemes that produce the score or, as with contemporary experimental music, alternatives to it. The misconception of music as being untouched by technoscience makes it an especially interesting case for studying the transformation of its creation, logistics and consumption by technoscientific means. In what follows I will focus on such a transformation: the attempt to turn the selection of music, the evaluation of listening habits and the categorization of listeners into problems of data science and data engineering.

Music distribution and its algorithmic rebirth might seem like a harmless case, but the nuts and bolts of its digital infrastructure are disturbingly similar to those in other, more serious cases: audio pattern recognition systems developed and implemented here are also in public security systems based on recording and automatically analyzing sound. Machine learning algorithms tweaked here to identify user behavior are transplanted into information retrieval systems and content selection systems for libraries and scientific search engines. Some of the programs to automatically extract and generate meta data from connection logs implemented in the case of music distribution were originally developed for the automatic monitoring of tweets, comments and Facebook posts and used by various agencies, in advertising and intelligence alike. Recommender systems, an application of machine learning technologies like naïve Bayes classifiers, decision tree classifiers or support vector machine are about to be enwoven into many other fields of social practice such as financial decision making or romantic matchmaking. They are praised for their ability to find multi-dimensional patterns in transactional and user data and for their empowering work as personal assistants. They are also blamed for their habit of producing confirmation bias bubbles (or "filter bubbles," to use Eli Pariser's (2012) term) and to assist not individual users, but shady interests and patronizing forms of governing behavior through smart "nudging" (Thaler and Sunstein 2009; McLaughlin 2016) activities. Music distribution, as strange as that might sound, is a banner case for the reconfiguration of a common practice — may it be listening to music, trading, policing, health care or keeping your house warm — through the implementation of data collection and processing software.

Whether this reconfiguration of practice into a data science problem of gathering and formatting data, crafting and optimizing algorithms and building and maintaining infrastructures of recurring computing has to be considered as an "expression of power, not of someone having power over someone else, but of the software making choices and connections in complex and unpredictable ways" (Beer 2009: 997) is, despite the public debate on the power of and through algorithms, an open, empirical question. Fortunately for us, music distribution is, even by those that work in this field, considered important but harmless – which makes field access much easier than in the case of security systems or algorithmic trading. When we approached startups that implement music recommender systems as their core product or use them in the background as tools for editorial work in 2009, there was hardly any public or political debate on the social and political role of data analytics or the power and potential biases of algorithmic systems – a situation that has dramatically changed since then. From 2010 to 2012 we studied a sample of companies based on a collection of publicly available documents – from technical blueprints to media coverage – and interviewed around 30 of their spokespersons, employees and users on tech fairs, music festivals as well as in their offices and homes. We used this material to select cases for an in-depth study, and we managed to convince two of them to open up for 2–3 months of fieldwork and participant observation each. The following three vignettes are based on this fieldwork that allowed us to study the everyday work with and for the recommender systems in the offices of product managers, artists and repertoires (A&R) managers and, most importantly, software engineers.

9.2.1 First Vignette: From File Sharing to Streaming

In 2010, online music distribution was in a process of rapid and experimental change. Before content streaming became a mass market, music was distributed as files – ripped, mixed and burned by users for users and shared using hard drives, CD Rom, Zip drives, FTP (file transfer protocol) servers or legal, quasi-legal and straightforward illegal file sharing tools such as Napster, Limewire or Kazaa. But in 2010, Spotify, the first successful subscription-based music streaming service had just been launched in Sweden and in the UK, and a move to the US market was announced for 2011. Youtube had just started to stream not only user generated content, but cricket matches of the Indian Premier League. The first attempt to relaunch Napster, the former champion of file sharing, as a legal, ad-based music streaming service had just failed and Apple's music application iTunes had just dropped the CD from its app icon.

2010 was thus an interesting year in online music distribution. But in 2009, those interesting developments were still to come. Music publishers and record companies blamed file sharing for killing music, MTV had just switched from airing music videos to reality shows, and the Pop-Komm, Germany's top industry meeting for anything commercially related to music had been canceled. Last.FM, the reason for us to look at music distribution as a banner case for the then still young data-analytics industry, had just been sold to CBS. But startup funding flourished and when we started mapping the actors, music recommender startups were all over the place. Many of them were just websites without any actual service, while others were merely technical descriptions of recommender engines without any practical application, but in 2009, that worked quite well to get a first round of venture capital. Rumors emerged from Sweden and the UK that a new service called Spotify had just found the holy grail of online music distribution, but only very few had actually seen it.

Just a few years ago the music discovery service Last.FM had very successfully created a new practice. While its streaming service required a computer and a web browser and could only play a limited number of songs, its plug-in called "Audioscrobbler" was everywhere. The tool logged the common ID-3 metadata (song title, artist's name, album name, genre, album track number, release date) of music played on desktop and portable music players and uploaded it to a common database used for creating recommender-based playlists in those players or on the Last.FM website. In a time when streaming was not yet a mass market, music recommendations were a distinct service and it was used as an add-on for various forms of listening to music. Users we interviewed to prepare for fieldwork told us that they used the Audioscrobbler extensively although they did not listen to music online: the ability to automatically curate playlists was the only way to cope with the masses of ripped and legally or illegally acquired files sitting on their hard disks. It was hard enough to catch up with storing, sorting and re-naming those files manually or semi-automatically using publicly available database systems like Music Brainz or the CDDB and self-written or self-copied scripts. Music recommender systems, like those scripts and databases, were part of the regular tool box of the serious music connoisseur using a computer – and Last.FM had the data.

In 2009, however, this started to turn into a problem. Last.FM had been bought by CBS and its original founders had left the company. Rumors emerged that the Recording Industry Association of America had asked CBS to help track down illegal file sharing services and users by handing over the Last.FM data. The rumors were never really clarified, but their existence alone points to an important detail. In 2010, new music distribution services tried very hard to not become associated with one of the old industry players. This was especially true for data analytics and recommender systems: using data to create playlists and manage huge music collections was considered a blessing, but only when such services were not in the hands of record companies and publishers. They were tools for those interested in an open and distributed infrastructure for music that was free of costs and digital rights management systems. And even those who subscribed to more commercial and less emancipatory interests positioned their software as at least aligned to such a free and open web idea. One of the companies we studied in detail – a Berlin based startup whose core product was a clean green website with a single input field and a

"Go!" button that played related songs (and commercials) – only had two further links on its website: one for imprint and legal details and one linking to a self-written description of the Fraunhofer Institute project that created the technology running its recommender system. And although the company offered a subscription service that allowed listening to music without ads, only very few users bought that service. The FAQ section for those paid subscriptions contained questions on payment, cancelation policies and the like, but the question most read was: "Does XX use the items I listened to recommend new music?" – and the answer was a straightforward: "No. Unlike other services, XX does not use user data to recommend music."

9.2.2 Techno-Politics of Platforms (and Infrastructures)

Both in terms of digital technologies and approaches to study them it is helpful to remember that a decade of work can change a lot. In 2009, Google, Facebook and Apple were still busy establishing their digital ecosystems, Netflix was still mostly shipping DVDs by mail and Youtube was still running on Adobe's Flash Player. Recommender systems were still basically tools to compile lists of items for users to choose from, and experiments with auto-play and non-linear content programming were still in very early stages of development. When we started our fieldwork in 2009, the empirical and conceptual link between Science & Technology Studies and Media Studies, Communication Studies and historically or philosophically oriented works on New Media was not as established as it is today. The 2008 *Handbook of Science & Technology Studies* contained a chapter by Leah Lievrouw and Pablo Boczkowski that called for a "bridging of STS and Communication Studies" (Boczkowski and Lievrouw 2008) and Judy Wajcman and Paul K. Jones could still describe the outcome of their ARC linkage grant on mobile phone use as a "conceptual mine field" (Wajcman and Jones 2012).

This has changed tremendously during the last decade: The now established connection between STS and Media Studies has produced a huge range of case studies of nearly everything between wearables (Crawford et al. 2015) and killer drones (Asaro 2013), algorithmic journalism (Anderson 2013), wikipedia bots (Geiger 2014) and twitter (re)tweets (Marres and Gerlitz 2014). STS and Media Studies created a common space to argue about the scripts and values inscribed into very specific digital technologies by design (Steen 2015), and they have seen the rise and fall of a "code as fetish" (Chun 2008) and a similar foregrounding and then fading away of software studies (Fuller 2008) and platform studies[2]. Moreover, after a few years of blaming algorithms and the rise of massive data handling for security concerns (Kitchin 2014), filter bubbles (Pariser 2012) and the financial crisis (Lenglet 2011), they have just started to develop a new subfield of "critical algorithm studies"

[2] https://mitpress.mit.edu/books/series/platform-studies

around a curated reading list by Tarleton Gillespie and Nick Seaver of already nearly 150 titles.[3]

That this now established field at the intersection of STS and Media Studies seems to be extraordinarily volatile and inconsistent in the choice of its preferred focus of analysis – infrastructure, code, software, algorithms, data, platforms – is not (only) an effect of the various conceptual minefields, but of the shifts, turns and upgrades of digital technologies. What they are, how they work, who is using them and how – these are questions that are always only partially and provisionally settled in the controversies that drive and shape their development and use. The algorithmic rebirth of music distribution started as experiments with data-driven and software-based solutions for the problems that its earlier reconfiguration as a distributed infrastructure of filesharing systems, portable file storage systems and mobile mp3 players had created: incomplete meta data, parallel manual, semi-automatic and automatic systems for sorting and classifying content and incompatible standards for audio compression.

The ubiquity of music streaming platforms today makes it easy to forget that less than a decade ago both being a platform and not being part of the old infrastructure of music distribution were highly political issues. Many of early empirical case studies and interventions into a public and political discourse on the role of software, data, and algorithms draw on early works by Susan Leigh Star, Geoffrey Bowker or Paul Edwards on classification systems (Bowker and Star 1996, 1999), Computer Supported Collaborative Work (Bowker et al. 1999; Bowker and Star 2001; Edwards et al. 2011) and knowledge infrastructures (Bowker et al. 2007; Ribes and Lee 2010). This is not a coincidence, as the issues at stake around 2010 were exactly raised concerning questions of standards and tools, collaboration and sharing and the competition between old industries and new approaches.

What Plantin et al. later called the "'platformization' of infrastructures and an 'infrastructuralization' of platforms" (2018: 6) – the simultaneous splintering of a common, centralized and standardized infrastructure leaving us with an ecology of platforms like Youtube, Netflix or Spotify combined with the rise of these platforms replacing any alternative infrastructure – was not fully developed. Critical voices in Media Studies and STS have focused predominantly on the second part of that development – the infrastructuralization of platforms – by pointing out that the use of the term 'platform' for huge services like Youtube is deeply political as it hides commercial, regulatory or content-related interest under a cover of neutrality (Gillespie 2010). But it is also the more technical use of the term that Montfort and Bogost offer in their work on the Atari VCS which points to the politics interwoven into the platformization of infrastructures. A "platforms' essential characteristic is the programmability which permits users to go beyond the original designers' project" (Plantin et al. 2018: 5). This opens up a first level of techno-politics of platforms by connecting "the fundamentals of digital media work to the cultures in which that work was done and in which coding, forms, interfaces, and eventual use

[3] https://socialmediacollective.org/reading-lists/critical-algorithm-studies/

are layered upon them" (Montfort and Bogost 2009: 147). In 2009, data analytics, recommender systems and shared databases like the CDDB were part of a flexible infrastructure for file sharers, music lovers and MP3 hoarders. Turning them into elements of a platform meant a lot of work and a controversial issue: which pieces should one assemble, how to make sure the various parts hold together, how to draw boundaries between one's own platformization project and competing initiatives?

9.2.3 Second Vignette: Too Much and Not Enough Data

People listen to music on their home computer, their mobile phone, in their car, using web radio boxes and the like. For traditional FM radio that did not make a huge difference, as a radio in a car is basically very similar to a radio in a kitchen. But for streaming services this does make a huge difference, questions of connectivity even left aside. Some of these devices have detailed interfaces, some are just streaming services, some, like the car, the tv or the home stereo have additional input devices – a remote control, a microphone, switches on a steering wheel – making user feedback – like/ban, next/back, etc. – different each time. While it is easy to ban a song during the first 20 s (after which the song is tracked as "played" by an unspoken law in recommender systems construction first installed by Last.FM) when you are in your car, hands on the wheel, it is difficult when the stream is running on your kitchen radio while you are in the bathroom. For music distribution services this creates a paradoxical problem: on the one hand, they seem to know their users better than any traditional radio has before, and knowing them has turned into a question of what activity counts as listening in the first place: Actively skipping songs in the car, not allowing songs to0 be played more than 5 s each? A continuous stream of an hour of not skipping or liking any song on a kitchen radio while taking a bath and not noticing it switching from Tom Jones to Black Sabbath? FM radio did not know its listeners any better, but based on the system of twice-a-year CAPI telephone surveys asking "do you listen to …" and "do you recognize this jingle?", it for sure knew it had a dispersed, anonymous and standardized audience.

With the exception of the still existing download farms of the warez and dark web there is hardly any online music distribution service today that does not in one way or the other rely on data analytics. Even seemingly traditional "web radios" rely – in the background – on the same systems of data collection, processing and of selecting and combining playlists for the otherwise moderated programs. We were incredibly surprised when after a month of fieldwork at another company that we studied in detail – a Hamburg based web radio service that positioned itself as the rebirth of art house radio with shows hosted by famous underground musicians and legendary radio hosts – we accidentally participated in one of the weekly data analytics meetings. Data, we learned, was used extensively in the background to generate playlists later used by showrunners and hosts and to evaluate the resulting traffic patterns for the purpose of moving shows to optimal time slots. The issue at stake was not if data analytics should be used but which algorithms and which methods of data handling would be appropriate for organizing and positioning the new service.

Having all that data – usage data, personal profiles, content meta data – seems like a data scientist's heaven, a gold mine for generating all kinds of reports, predictions, speculations. But in practice, that opportunity was less easy to take. No matter how much data was collected, it was only very rarely the right kind of data for the task at hand. Having data opens up a range of speculations and even a few wilder dreams about how one could "do more with numbers" (Vollmer 2007). This was less evident inside the companies we studied, but a repeating issue in meetings with outsiders such as music production and distribution companies, media planners and marketing agencies. Customers buying ad-space – visually presented on the website or played as sound snippets inserted into the music stream – and affiliate links – paid partnerships where a click on a link leads to a small payment when it leads to a sale on the partner's side – expect click-counts and conversion-rates – the percentage of users turned into paying regulars –, but streaming services do not have clickable interfaces. This caused trouble, manual work and controversies: Can the problem be solved by routinely reloading the player and thereby producing an event similar to page impression? If so: how often should that reload happen? Is the reload even necessary? Would it be legitimate to just let three streaming minutes count as a click to get a rough estimation? If not: Would 7 min be enough? How to create and deliver the numbers that existing and potential partners expected was a constantly controversial issue.

9.2.4 Doing Data and Dealing with Legacy Systems

Data, in sharp contrast to the etymology of the term, is not given at all. A decade of Critical Data Studies (see for example Boyd and Crawford 2012; Iliadis and Russo 2016) has provided case-based evidence and rigid arguments for the now widely shared understanding that data is always made – created, handled, arranged and transformed – in practice and that the very idea of "raw data" is, to use Geoffrey Bowker's often cited phrase "both an oxymoron and a bad idea" (Bowker 2005). But in that exact practice, this "bad idea" is (still) common and as Kitchin and Lauriault have argued, it has been a key element of the politically and economically driven hype on Big Data. Framed this way, data is "benign, neutral, objective, and nonideological in essence" and it is "'collected', 'entered', 'compiled', 'stored', 'processed', and 'mined'" (Kitchin and Lauriault 2018) – it is framed as a resource, a commodity and, technically, something given.

Various scholars in Critical Data Studies – a field that is both inspired by classic STS arguments about the performative character of scientific models, measurement standards and devices (for example Black 1962; Harré 1976; Hacking 1983) and fueled by contemporary STS research on infrastructures, standards and ICT (Lievrouw and Livingstone 2006; for example Bowker et al. 2007; Ribes and Lee

2010) – have argued that such framing of data is only possible on the very foundation of complex and large-scale socio-technical arrangements, embedded in institutional and technical legacies and a structured and often biased ecosystem of actors, historical trajectories and cultural patterns. Any attempt to handle data – to make, collect, arrange, employ and use something as an integer, boolean, string, array, set, graph or tree – requires dealing with complex data assemblages and 5/17/2019 3:20:00 PM "the technological, political, social and economic apparatuses and elements that frame their nature, operation, and work" (Kitchin and Lauriault 2018). Shaping and utilizing these data assemblages requires a second level of techno-politics of data dealing with the issues and struggles involved in the practices and processes of their rearrangement.

During fieldwork we noticed some of these rearrangements: on a level of services (web-based, mobile, streaming infrastructure) we noticed that what counts as a listener, a member of an audience differs both between services and in comparison to the imagined and addressed audience of traditional wave-based radio; on the level of songs and playlists we noticed that what counts as listening can differ between different contexts and that rather than just collecting usage data and statistically identifying patterns, data is explicitly and artificially produced for high valued contexts; on the level of results we noticed that, to use a phrase very loosely adopted from Bittner and Garfinkel (Garfinkel 1967: 186ff), there are good organizational effects of (dealing with) bad data analytics. Audiences, top artists, music experts – they all still look similar to their pre-algorithmic siblings, but in fact they are not. These are small, but drastic rearrangements of data, analytics and the technical and institutional legacies of music production, distribution and consumption: audiences turn into friends, users and by-passers; different modes of listening are extrapolated from server-logs and by data-clean up; new music curator jobs are created for dealing with the results of a black boxed recommender system.

9.2.5 *Third Vignette: Recommender Systems and the Art of Curation*

During the last decade, more and more services for online music distribution have turned from being large, legal or illegal collections of music files for users to browse and search into more or less automated systems for selecting, recommending, and curating content based on usage patterns (when, where and under which conditions do users listen to which kind of music) and content characteristics (loudness, tempo, instrumentation, etc.). Early services like Amazon's 'Users that bought this book also liked…' or Pandora's use of the Music Genome Project are good examples for both the strengths and weaknesses of the two main strategies for automatically recommending music. Collaborative filtering systems that recommend content based on usage patterns work very well when they create lists of items that users can actively choose from. If enough listeners of heavy metal music are also parents and therefore also listen to nursery rhymes, then recommender systems based on usage patterns will present mixed lists of Metallica, Slayer and Teletubbies. Content-based filtering systems based on metadata on content characteristics, on the other hand, recommend items that sound alike and produce separate playlists for heavy metal and nursery rhymes, but they require consistent classification systems as they cannot recommend items without data.

Recommender systems have various user groups: smart-phone users with headphones (they definitely listen), smart-tv users (but how many listeners share a tv?), users across devices (but what if someone uses two at the same time, liking a song on one but banning it on the other device?). Modeling all these differences improves the quality of usage data – for example, by identifying hours of listening while the speakers are muted – but also segments the data set so that the accuracy of the latent factor models decreases. Even with a huge user base the problems persist – but the music distribution services we studied were small and had a limited, even not very loyal user base. In 2010, relying on distinct patterns in usage data was only an option for a platform like Last.FM. But even a company like Netflix that had huge amounts of DVD order records at their disposal struggled with existing systems based mainly on "Straightforward statistical linear models with a lot of data conditioning."[4] To solve that problem, they created the now famous "Netflix price" – 1,000,000 $ for beating Netflix' own Cinematch algorithm, a linear regression based collaborative filter, by 10% – in 2006. It was won exactly one time in 2009 by a team that blended a vast collection of available individual predictors and was canceled in 2010 due to a privacy dispute.

But as mentioned before: our companies were small and had neither the user data nor the money to operate like Netflix. One of the reasons why the FAQ of the Berlin based service could so openly state that no user data was used for recommendations was that there simply was no way to use the limited data and the limited financial and computational resources in a meaningful way. Instead, an algorithmic musical expert was hired: The algorithm was supposed to be a true music genius that classified and selected related songs only by using automatically generated meta data on instrumentation, speed and sound. Sound data was available: offering a streaming service in 2009 meant hiring a team of students and low-wage workers to convert CDs into MP3 files – a delicate issue in terms of licensing music, but a very common practice that Spotify implemented as well.[5] The algorithm was used to analyze the files and to calculate similarities based on a range of dimensions mostly unknown to those who worked with it. Its documentation and specifications were abstract and with the exception of very few people in general management and in IT nobody really knew the parameters used to configure it in the first place. Using it was like a guessing game: when it calculated a 0.72 similarity between Iron Maiden and Spice

[4] https://web.archive.org/web/20070821194257/http://www.netflixprize.com/faq

[5] https://techcrunch.com/2018/01/02/spotify-faces-1-6-billion-lawsuit-from-music-publisher-wixen/

Girls or Tom Jones and Knorrkator, those converting the files often started to look for 'hidden' similarities to guess what was going on in that opaque black box.

But here is a fun fact about big data and a sad truth about popular music: to cluster well and distinctively, data needs to be vast; to judge similarities, on the other hand, the data needs to be distinguishable in the first place – and sadly, with respect to contemporary music, that is rarely the case: songs have a pretty average length, they have drums and base and strings, a voice track or two, overall, they are pretty similar to pattern recognition systems. In fact: based on the limited data available – the songs were converted into files by hand – Tom Jones, a Swedish hard rock band, and a German pop rock artist all can have similarity rates between 0.69 and 0.71 – enough for a human listener, but not enough for the algorithm. And while it seems easy to change that by tweaking the algorithm itself: that can be hard to do if its code is proprietary or the software engineers that coded it work in an academic institution not prepared for service provision. Without the necessary engineering skills and the relevant institutional resources there was only one manageable workaround: changing the results by hand to make sure artists like Tom Jones and Knorrkator are at least 0.4 points apart. For this, a former limitation turned into an opportunity: As every song that could be streamed by the service had to be manually added to the system, connections and relations between songs could be added, too. This turned the students converting CDs into background curators, music scouts that scan their personal surroundings and all kinds of expert knowledge on making playlists. Playlists could be based on music magazines, top lists, and the expert knowledge of Turkish barbers on Istanbul's greatest hits. Once those lists were compiled, data for training the recommender models was modified so that they produce results similar to those playlists – a repetitive and boring task as the similarities are constantly recalculated when new content is added, producing playlists connecting Iron Maiden and the Spice Girls again.

9.2.6 *Algorithmic Culture and the Practice of Calculation*

"Algorithmic culture" is a term coined by Alexander Galloway as a subtitle of his 2006 book on gaming (Galloway 2006) which he, although not using the term in the book at all, positions in the lines of Deleuze's societies of control (Deleuze 1992): learning to win the (video) game, Galloway argues, is a process of getting used to think in terms of the game, think like the game. And in a deep reading of Sid Meyer's Civilization, a game that is based on a simulation of social and cultural history, he points out that even in its complex versions, "the complexity, variation, and rich diversity of human life are erased" (Galloway 2006) This mantra of "Algorithmic Culture" as the outcome of a "silent revolution" (Bunz 2013) or the rise of the "secret algorithms that control money and information" in a "black box society" (Pasquale 2015) has since then increasingly dominated public debates. Following Galloway, Striphas sums this mantra up quite nicely when he concludes that "algorithms are becoming decisive, and why companies like Amazon, Google and

Facebook are fast becoming, despite their populist rhetoric, the new apostles of culture" (Striphas 2015). Or, in the words of Kushner, "the cultural configuration where humanity's round pegs are forced into the square holes of computer subroutines" (Kushner 2013). The brute force attempts to solve the problem of video recommendations using huge amounts of money and huge amounts of data look like a perfect example of this contemporary tendency of making culture computational. Liking and watching movies on Netflix has nothing to do with the passions and tastes of the cineaste, but instead, they are reduced to the prediction of potential five-star ratings for movies not yet watched.

Gillespie, on the other hand, indicates a decisive entry point for an STS approach to algorithmic culture, an approach that "must not conceive of algorithms as abstract, technical achievements, but must unpack the warm human and institutional choices that lie behind these cold mechanisms" (Gillespie 2013). If "Algorithmic Culture is characterized by 'the sorting, classifying, and hierarchizing of people, places, objects, and ideas using computational processes'" (Striphas 2015), then "computational processes" is what we need to understand. Computing is, like counting and calculating, not at all a mere technical term. "Com-putare" in Latin means putting together, and counting is derived from "comit-" which already in Latin had the double meaning of "something that goes together with..." and the companion. Calculating refers to the calculus, the pebble used for counting and hints at the very procedure that makes up computing: taking things apart until all you have is little pieces, little white stones, and putting it back together in an ordered, but reorderable way. "Computing" was turned into a term for the conditions and the logic of turning a problem into one that can be solved step-by-step in the aftermath of Turing's "Theory of Computation" (1936), on the one hand, and into a name for those who calculate – human or non-human – after the exhibition of de Colmar's (1785–1870) Arithmometer at the 1855 *Exposition Universelle* in Paris. This is what happens when implicit concepts of musical similarity and tacit knowledge about listening habits and assumed preferences are translated into a framework of closeness and similarity parameters by hand just to be able to 'teach'[6] the recommender system what a proper Turkish pop playlist looks like.

Following Callon and Muniesa's analysis of the calculative collective devices of contemporary financial markets, such an understanding of processes of calculation can only be an empirical understanding, but it starts by unpacking the very concept of calculation itself: Calculation can be understood as a three-step process in which whatever should be calculated first needs to be detached from all its relevant contexts, then "subjected to manipulation and transformation" before finally a result, something that follows from being "taken 'into account'" (Callon and Muniesa 2005), has to be re-embedded into a context. In all three steps calculation is deeply techno-political: How should we take apart? What should stay connected? What should be left out? What needs to be considered? What are the procedures to

[6] In the strict terms of machine learning concepts, such a practice would not count as 'teaching' as this would require the careful collection of training data and (counter) examples to 'learn' the models in the first place.

consider? Are there more than one? Less than many? And of course: What are the effects of dealing with the results of such a transformation instead of with what has been taken apart in the first place? How to calculate, how to build the algorithms that shape our 'algorithmic culture' is a third level of techno-politics of algorithms. It is far subtler and it works in the background, but it is as decisive as the other two.

9.3 Political Ontologies and Idiotic Interventions

These three levels of techno-politics of platforms, data, and algorithms are not intrinsic politics of digital technologies, inscribed into them by the big players — Google, Facebook, Apple — of our digital industries and now shaping and determining the outlines and details of our contemporary algorithmic culture. They are also not a shortcut for arguing that there needs to be a political and public debate or that the shape of our algorithmic culture needs to be regulated. The politics of platforms, data and algorithms are first and foremost ontological politics in which the "conditions of possibility are not given" (Mol 1998). These politics refer to a set of practices to rearrange not only how we understand, focus on or regulate audiences, listeners, users, producers, experts, organizations, they re-arrange what they are. Without the machinery of audience research, there is no audience, just an amorphous mass of potentials. Without the apparatus of user tracking and profiling, it is nothing but the audience, or, as Ian Ang has argued so fittingly, it is an audience that needs to be desperately searched (Ang 1991). The three levels of politics are political because they are practical metaphysics, they are attempts to explore the controversies over agencies that "populate the world with new drives and, as ceaselessly, contest the existence of others" (Latour 2005). And of course, as there are more than one, less than many attempts of platform-building, data handling, and algorithm design, we are faced with "many practical metaphysics, many different practical ontologies" (Latour 1999).

But there is something very tricky about these kinds of politics: it is about "transforming the several into one, initially through a process of representation - (…) and subsequently through a process of retransformation of the one into several (…)" (Latour 2003). It is, following Rancière (2006), the struggle and controversy about installing an arena for arranging our collectives. There is no way to shortcut the circle without getting rid of politics itself. And here Galloway's Deleuzian algorithmic culture and the outline of the three levels of techno-politics presented above share both a fear and a hope: Installing patterns of calculation into our ecology of practices strengthens what Latour has called a "double click," a shortcut to the specific politics of platforms, data and algorithms that is even riskier than "the absurd dream of a substitution of skew talk by straight talk" for bringing the "particular form of repetition and agitation of politics" (Latour 2003) to a standstill. That is the shared fear.

Here is the hope: Galloway argues, in a good post-structural sense, for a form of critique that focusses on and embraces subversion, irritation, counter-gaming and on an empowerment of the users in the game of the serious games he studied. But there is more that can be done and it is us – as researchers, as practitioners, as infrastructural engineers – that are responsible for it. There are at least two potential ways to keep the political circle turning, to continue opening up the process of turning one into many, many into one, of helping the two chambers of our parliament of things continue their work. One follows Jensen and Moritas' point that by highlighting the infrastructural character of the current transition towards TechnoScienceSocieties (in STS research), our inquiries can serve as an entry point for considering the core questions for a non-humanist politics: "what does reality consist of? Which other creatures are we living with? And how can we live differently?" (Jensen and Morita 2015) Another option is, to follow Marres' call for an "experimental approach to political ontology" and to "purposefully design politics and morality into material objects, devices and settings" (Marres 2013). How can that be done?

When arguing for ontological politics in the first place, Mol noticed that "talking about reality as multiple depends on another set of metaphors. Not those of perspective and construction, but rather those of intervention and performance" (Mol 1998). Let us go one step further: for talking we need metaphors, for dealing with it as multiple we might need attitudes, attitudes of intervention and performance. We might need an attitude that does not only map controversies, but produces them in the first place by resisting "the consensual way in which the situation is presented and in which emergencies mobilize thought and action" (Stengers 2005) and to engage in what Fraser called "inventive problem making" (Fraser 2010). In the case of technoscience, however, this is not an easy task. If we follow Stengers' argument about Gallileo's dictum that "one man will win against a thousand rhetoricians, whatever their gift of persuasion or the authority of their references, if this one man has the facts on his side" (Stengers 2000), we need Deleuze's favorite "conceptual persona" (Deleuze and Guattari 1994), the idiot. Only the idiot is brave and ignorant enough to say that "there is something more important (that) cannot be taken into account" (Stengers 2005), "the idiot can neither reply nor discuss the issue … (the idiot) does not know…the idiot demands that we slow down, that we do not consider ourselves as authorized to believe we possess the meaning of what we know" (Stengers 2005). Can an experimental approach to political ontology include what Michael called "proactive idiocy," an approach that "is sensitive to and indeed actively seeks out, that which is empirically and practically nonsensical" (Michael 2012)? If so, we need to design our projects so that we can make the claim of the idiot that there is "something more important" exactly where calculation is done, where a practice — of music consumption, health care, industrial work – is taken apart, manipulated and turned into a calculated result to work with. If the techno-politics of platforms, data and algorithms are inevitable, they also open up spaces for critique, intervention and (re-)design. John Law and Vicky Singleton's (fictional?) dialogue in one of their fish pond texts is instructive in this regard: "(…) good research is passionate because it is disturbing" (Law and Singleton 2013). Such a project would be an experiment with collaboration that opens up spaces of reflexivity in which the idiot can raise his concern. Software engineering is a practice that is by design open for such forms of collaboration.

References

Adorno, T.W. 2001. *The Culture Industry: Selected Essays on Mass Culture*. New York: Routledge.
Anderson, C.W. 2013. Towards a Sociology of Computational and Algorithmic Journalism. *New Media & Society* 15: 1005–1021.
Ang, I. 1991. *Desperately Seeking the Audience*. London/New York: Routledge.
Asaro, P.M. 2013. The Labor of Surveillance and Bureaucratized Killing: New Subjectivities of Military Drone Operators. *Social Semiotics* 23: 196–224.
Barry, A. 2013. *Material Politics: Disputes Along the Pipeline*. Oxford: Wiley-Blackwell.
Beer, David. 2009. Power through the algorithm? Participatory web cultures and the technological unconscious. *New Media & Society* 11 (6): 985–1002.
Bermejo, F. 2009. Audience Manufacture in Historical Perspective: From Broadcasting to Google. *New Media & Society* 11: 133–154.
Black, M. 1962. *Models and Metaphors*. Ithaca: Cornell University Press.
Boczkowski, P., and L.A. Lievrouw. 2008. Bridging STS and Communication Studies. Scholarship on Media and Information Technology. In *The Handbook of Science and Technology Studies*, ed. E.J. Hackett, O. Amsterdamska, M. Lynch, and J. Wajcman, 951–977. Cambridge: MIT Press.
Bourdon, J., and C. Méadel. 2011. Inside Television Audience Measurement: Deconstructing the Ratings Machine. *Media, Culture & Society* 33: 791–800.
———. 2014. *Television Audiences Across the World*. London: Palgrave Macmillan.
Bowker, G.C. 2005. *Memory Practices in the Sciences*. Cambridge: MIT Press.
Bowker, G.C., and S.L. Star. 1996. How Things (actor-net) Work: Classification, Magic and the Ubiquity of Standards. *Philosophia* 25 (3/4): 195–220.
———. 1999. Sorting Things Out. In *Classification and Its Consequences*. Cambridge: MIT Press.
———. 2001. Social Theoretical Issues in the Design of Collaboratories: Customized Software for Community Support Versus Large-Scale Infrastructure. In *Coordination Theory and Collaboration Technology*, ed. G.M. Olson, T.W. Malone, and J.B. Smith, 713–738. Mahwah/New York: Lawrence Erlbaum Associates Inc.
Bowker, G.C., S.L. Star, W. Turner, and L. Gasser. 1999. Social Science, Technical Systems, and Cooperative Work: Beyond the Great Divide. *IEEE Transactions on Professional Communication* 42: 132–134.
Bowker, G.C., K.S. Baker, F. Millerand, and D. Ribes. 2007. Towards Information Infrastructure Studies: Ways of Knowing in a Networked Environment. In *International Handbook of Internet Research*, ed. J. Hunsinger, M. Allen, and L. Klastrup, 97–118. Dordrecht: Springer.
Boyd, D., and K. Crawford. 2012. Critical Questions for Big Data. *Information, Communication & Society* 15: 662–679.
Bunz, M. 2013. *The Silent Revolution*. London: Palgrave Macmillan.
Callon, M., and F. Muniesa. 2005. Economic Markets as Calculative Collective Devices. *Organization Studies* 26: 1229–1250.
Chun, W.H.K. 2008. On "Sourcery," or Code as Fetish. *Configurations* 16: 299–324.
Crawford, K., J. Lingel, and T. Karppi. 2015. Our Metrics, Ourselves: A Hundred Years of Self-Tracking from the Weight Scale to the Wrist Wearable Device. *European Journal of Cultural Studies* 18: 479–496.
Deleuze, G. 1992. Postscript on the Societies of Control. *October* 59: 3–7.
Deleuze, G., and F. Guattari. 1994. *What is Philosophy?* London: Verso.
Edwards, P.N., M.S. Mayernik, A.L. Batcheller, G.C. Bowker, and C.L. Borgman. 2011. Science Friction: Data, Metadata, and Collaboration. *Social Studies of Science* 41: 667–690.
Fraser, M. 2010. Facts, Ethics and Event. In *Deleuzian Intersections: Science, Technology and Anthropology*, ed. C.B. Jensen and K. Roedje, 52–82. Oxford/New York: Berghahn Books.
Fuller, M. 2008. *Software Studies: A Lexicon*. Cambridge: MIT Press.
Galloway, A.R. 2006. *Gaming: Essays on Algorithmic Culture*. Minneapolism: University of Minnesota Press.

Garfinkel, Harold. 1967. *Studies in ethnomethodology*. Englewood Cliffs, N.J: Longman Higher Education.
Geiger, R.S. 2014. Bots, Bespoke, Code and the Materiality of Software Platforms. *Information, Communication & Society* 17: 342–356.
Gillespie, Tarleton. 2010. The politics of 'platforms'. *New Media & Society* 12 (3): 347–364.
Gillespie, T. 2013. The Relevance of Algorithms. In *Media Technologies*, ed. G. Tarleton, P. Boczkowski, and K. Foot, 167–194. Cambridge: MIT Press.
Hacking, I. 1983. *Representing and Intervening: Introductory Topics in the Philosophy of Natural Science*. Cambridge/London: Cambridge University Press.
Harré, R. 1976. The Constructive Role of Models. In *The Use of Models in the Social Sciences*, ed. L. Collins, 16–43. London: Travistock.
Hennion, A. 1989. An Intermediary between Production and Consumption: The Producer of Popular Music. *Science, Technology & Human Values* 14: 400–424.
———. 2001. Music Lovers Taste as Performance. *Theory Culture Society* 18: 1–22.
Hennion, A., and C. Meadel. 1986. Programming Music: Radio as Mediator. *Media, Culture & Society* 8: 281–303.
Iliadis, A., and F. Russo. 2016. Critical Data Studies: An Introduction. *Big Data & Society* 3 (2): 1–7.
Jensen, C.B., and A. Morita. 2015. Infrastructures as Ontological Experiments. *Engaging Science, Technology, and Society* 1: 81–87.
Kitchin, R. 2014. Big Data, New Epistemologies and Paradigm Shifts. *Big Data & Society* 1: 1–7.
Kitchin, R., and T. Lauriault. 2018. Towards Critical Data Studies: Charting and Unpacking Data Assemblages and Their Work. In *Thinking Big Data in Geography: New Regimes, New Research*, ed. J. Thatcher, A. Shears, and J. Eckert, 3–20. Lincoln: University of Nebraska Press.
Kushner, S. 2013. The Freelance Translation Machine: Algorithmic Culture and the Invisible Industry. *New Media Society* 15: 1241–1258.
Latour, B. 1999. *Pandora's Hope: Essays on the Reality of Science Studies*. Boston: Harvard University Press.
———. 2003. What If We Talked Politics a Little? *Contemporary Political Theory* 2: 143–164.
———. 2005. *Reassembling the Social: An Introduction to Actor-Network Theory*. Oxford: Oxford University Press.
Law, J., and V. Singleton. 2013. ANT and Politics: Working in and on the World. *Qualitative Sociology* 36: 485–502.
Lenglet, M. 2011. Conflicting Codes and Codings: How Algorithmic Trading is Reshaping Financial Regulation. *Theory, Culture & Society* 28: 44–66.
Lievrouw, L.A., and S. Livingstone. 2006. How to Infrastructure. In *Handbook of New Media: Social Shaping and Social Consequences of ICTs*. Thousand Oaks: Sage.
Looseley, D. 2006. Antoine Hennion and the Sociology of Music. *International Journal of Cultural Policy* 12: 341–354.
Marres, N. 2013. Why Political Ontology Must Be Experimentalized: On Eco-Show Homes as Devices of Participation. *Social Studies of Science* 43: 417–443.
Marres, N., and C. Gerlitz. 2014. Interface Methods: Renegotiating Relations Between Digital Research, STS and Sociology. *The Sociological Review* 64 (1): 21–46.
McLaughlin, K. 2016. *Empowerment: A Critique*. New York: Routledge.
Michael, M. 2012. "What Are We Busy Doing?": Engaging the Idiot. *Science, Technology, & Human Values* 37: 528–554.
Mitchel, T. 2002. *Rule of Experts: Egypt, Techno-Politics, Modernity*. Berkeley: University of California Press.
Mol, A. 1998. Ontological Politics. A Word and Some Questions. *The Sociological Review* 46: 74–89.
Montfort, Nick, and Ian Bogost. 2009. *Racing the beam: The Atari video computer system*. Cambridge, Mass: MIT Press.

Pariser, E. 2012. *The Filter Bubble: How the New Personalized Web is Changing What We Read and How We Think*. London: Penguin Books.
Pasquale, F. 2015. *The Black Box Society. The Secret Algorithms That Control Money and Information*. Cambridge: Harvard University Press.
Passoth, J.-H., and N.J. Rowland. 2010. Actor-Network State: Integrating Actor-Network Theory and State Theory. *International Sociology* 25: 818–841.
Passoth, J.-H., and N. Rowland. 2015. Acting in International Relations? The State Hypothesis, ANT, and Agency. In *Human Nature, Agency and Beyond. Reflecting on the Human Element in World Politics*, ed. D. Jacobi and A. Freyberg-Inan, 286–304. Cambridge: Cambridge University Press.
Plantin, Jean-Christophe, Carl Lagoze, Paul N. Edwards, and Christian Sandvig. 2018. Infrastructure studies meet platform studies in the age of Google and Facebook. *New Media & Society* 20 (1): 293–310.
Rancière, J. 2006. *The Politics of Aesthetics*. London: Continuum.
Ribes, D., and C.P. Lee. 2010. Sociotechnical Studies of Cyberinfrastructure and e-Research: Current Themes and Future Trajectories. *Computer Supported Cooperative Work (CSCW)* 19: 231–244.
Steen, M. 2015. Upon Opening the Black Box and Finding It Full Exploring the Ethics in Design Practices. *Science, Technology & Human Values* 40: 389–420.
Stengers, I. 2000. *The Invention of Modern Science*. Minneapolis: University of Minnesota Press.
———. 2005. Deleuze and Guattari's Last Enigmatic Message. *Angelaki* 10: 151–167.
Striphas, T. 2015. Algorithmic Culture. *European Journal of Cultural Studies* 18: 395–412.
Thaler, R.H., and C.R. Sunstein. 2009. *Nudge: Improving Decisions About Health, Wealth, and Happiness*. New Haven: Yale University Press.
Turing, A.M. 1936. On Computable Numbers, with an Application to the Entscheidungsproblem. *Proceedings of the London Mathematical Society* 2: 230–265.
Vollmer, H. 2007. How to do More with Numbers: Elementary Stakes, Framing, Keying, and the Three-Dimensional Character of Numerical Signs. *Accounting, Organizations and Society* 32: 577–600.
Wajcman, J., and P.K. Jones. 2012. Border Communication: Media Sociology and STS. *Media, Culture & Society* 34: 673–690.

Jan-H. Passoth is Privatdozent at European University Viadrina in Frankfurt (Oder) and Research Group Leader of the Digital/Media/Lab at the Munich Center for Technology in Society (MCTS) at Technische Universität München. His research draws on insights from science and technology studies (STS) and focusses on the use of standardized and interconnected software technologies in popular culture, politics, and the media and the changing role of computer science for contemporary societies.

Part III
TechnoScienceGovernance

Chapter 10
Governing Technoscience in Society: Tracing the Dialectics of Enthusiasm, Ambivalence, and Adjustment

Erik Fisher

10.1 Technoscience in the National Interest

State sponsored programs in nanotechnology, biotechnology, synthetic biology and neuroscience seek to link scientific research to rapid technological innovation that promises widespread social and industrial transformation. Such agendas appeal to governmental interests in promoting national and regional economic growth, competitiveness, security, as well as state legitimacy. U.S. Presidents William J. Clinton, George W. Bush and Barack H. Obama have each given speeches that yoke federal investments in nanoscale research to technological benefits and national interests. Clinton's 2000 announcement of the National Nanotechnology Initiative (NNI) at Cal Tech was itself framed within the larger question of the importance for the state to invest in science and technology (Clinton 2000), while Bush fused post-World War II and post-Cold War science policy rhetoric, describing nanotechnology as a "critical" form of "basic science" in the 2006 State of the Union (Bush 2006). These high-level visions suggest a logic of use-inspired, future-oriented, state-coordinated "big science" that seizes upon possibilities and operates especially well during trying times. Thus, using historical imagery, Obama's 2010 State of the Union portrays technoscience as a chain reaction that springs from crisis to competitiveness thanks to collective willpower and unity:

> Half a century ago, when the Soviets beat us into space with the launch of a satellite called Sputnik, we had no idea how we would beat them to the moon. The science wasn't even there yet. NASA didn't exist. But after investing in better research and education, we didn't just surpass the Soviets; we unleashed a wave of innovation that created new industries and millions of new jobs. This is our generation's Sputnik moment (Obama 2010).

E. Fisher (✉)
School for the Future of Innovation in Society, Arizona State University, Tempe, AZ, USA
e-mail: efisher1@asu.edu

S. Maasen et al. (eds.), *TechnoScienceSociety*, Sociology of the Sciences Yearbook 30, https://doi.org/10.1007/978-3-030-43965-1_10

Obama's speech, delivered only months after his political party was badly beaten in mid-term elections, sought to rally Americans and their elected representatives around his call to "win the future" through investing in scientific research. In the process, it imports a number of characteristics associated with technoscience and late-modern technology into the political realm.

10.2 Technoscientific Language and Logic

In the high-political language of three recent American presidents, science and technology are spontaneously generated, productive of an elaborate future, and crucial to the growth of the state. Such discourse resembles the exuberance of technoscience, which is marked by a political economy of "abundance and excess" (Schwarz and Nordmann 2011) in its lavish promises, investments and trust in the infinite plasticity[1] of both social and material structures. For it suggests that science and technology mutually transform not only one another but also their socio-political context. Thus, the language of technoscience can be seen as performative not only in the shifts it entails in "the self-understanding and epistemic values of science" (Nordmann 2006) but, extending this to broader political contexts, in its orientation of governance practices towards the future; its valorization of design, experimentation, and optimization; and its willingness to tolerate high levels of ignorance (Nordmann, Chap. 2, this volume). This chapter points to such experimental shifts in governance that are imaginatively designed to incorporate both scientific and civic practices into a vision of political life. As we shall see below in the case of the NNI, the technoscientific quest encounters resistance to its own exuberance and responds with adjustments and corrections that result in an even more elaborate image of society.

Resistance to technoscientific excess initially takes form of entrenched distrust and disappointment that emerges out of ambivalence, which can be thought of as kind of risk consciousness. As a pervasive feature of late modern society (Smart 1998), ambivalence is more than a response to enthusiasm; it also becomes a potentially important governance dynamic. In the case of the NNI, governance of nanotechnology becomes linked to a view of technosciencesociety as infinitely malleable. This can be seen, below, in the adjustments that are made to nanotechnology policy in response to the perception of widespread ambivalence. At the same time, the governance of nanotechnology exhibits some of the same features attributed to technoscience and late modern technology, which is marked by a tendency to generate rather than reduce complexity, to rely on principles of self-organization, and to embrace instability (Schmidt, Chap. 3, this volume). Above all, technoscience on this view is characterized by the search for and loss of control (*ibid.*). Accordingly, this chapter shows how such sensibilities as these served to orient the justification,

[1] I am grateful to Alfred Nordmann for suggesting this phrase.

funding, regulation and governance of nanotechnology in society by tracing dialectics of enthusiasm, ambivalence, and adjustment that followed from excessive promotion of technoscience.

10.2.1 Enthusiasm

The role of hype and promising in the promotion of large-scale research and development programs, particularly in their early phases, is well recognized (Borup et al. 2006; Lucivero 2015; Pollock and Williams 2010). In a particularly striking case in point, the visions and agenda used to launch and then establish the NNI were adorned in the futuristic language and imagery of profound social and industrial transformation (e.g., Berube 2006; Macnaghten et al. 2005; McCray 2005; McGrail 2010; Selin 2007). Beginning with the captivating language and inspirational imagery of Clinton's 2000 speech (Anderson 2007), the official and unofficial policy discourse that followed emphasized novel material properties that would usher in enormous economic gains and vast changes in major industries from healthcare to defense. As if these envisioned scientific, technological, and industrial transformations were not enough, they were framed within a larger project of remaking the world that was rooted in the capability to reshape matter at the atomic level (e.g., Crow and Sarewitz 2001). The hype of technoscience was perhaps nowhere more unmistakably on display than in the fact that scientific research at the nanoscale was routinely depicted as leading to the "next industrial revolution" not only in trade journals and the popular press, but also in formal government brochures and reports (Fisher and Mahajan 2006; cf. Kennedy 2008).

Moreover, the entanglement of technoscience promotion and national competitiveness was on full display during three 2003 Congressional hearings on nanotechnology, when promoters—in the attempt to secure legislative authorization for what had so far only been an initiative of the executive branch, albeit one that spanned two presidential administrations—repeatedly dramatized the purportedly tenuous position of American leadership in the global nanotechnology "funding race" (Economist 2004). The U.S. nanotechnology enterprise had since Clinton's announcement grown rapidly to include some twenty-seven funding agencies that chose to affiliate with the NNI, intentionally constructed as a loosely organized cross-institutional coordination mechanism that also sought to facilitate interactions among increasing numbers of industrial and academic researchers, private sector investors, and regulatory agencies. At the same time, in response to the technological momentum that had coalesced in and around the NNI, dozens of national nanotechnology programs had sprung up around the world, a response that members of Congress were repeatedly told posed a threat not only to U.S. economic but also its military competitiveness in the expected trillion-dollar market and context of global leadership (Fisher and Mahajan 2006).

Enacted by Congress in 2003, the *21st Century Nanotechnology Research and Development Act* received strong bipartisan support and authorized a massive funding boost that set the NNI on a trajectory to become "among the largest US civilian technology investments since the Apollo Moon-landing programme" (Lok 2010). By these indications, nanotechnology enthusiasm was tremendously successful as a promotional strategy that interlinked technoscientific advancements, socio-economic benefits, and the highest priorities of the state. At the same time, underlying nanotechnology promotional discourses was an instrumental view of matter—and, as we shall see, of society—as shapeable, malleable, and subject to technoscientific control. Before we can consider this point, we recount the first signs of pubic ambivalence in several notable responses to nanotechnological enthusiasm.

10.2.2 Ambivalence

If heightened expectations helped unify a growing and diverse coalition of supporters and provided a durable platform for steadily increasing funding, they arguably also helped create a more general urgency over the future. Not only did the technoscientific promotion of nanotechnology help spur research investments around the world, it also served to raise and intensify numerous social, ethical, environmental, even existential issues associated with nanotechnology. Reactions to state-endorsed nanotechnology visions, which could border on the utopian and deterministic, in turn produced equally vivid dystopias (Fisher and Mahajan 2006). Science fiction writers (Crichton 2002), scholars (Mnyusiwalla et al. 2003), environmental groups (Arnall 2003; ETC 2003a), political leaders (ETC 2003b) and technology entrepreneurs (Joy 2000) raised provocative and widely discussed scenarios that countered the exuberance of nanotechnology promotion with calls for precaution, relinquishment and moratorium.

The conflicting thoughts and feelings that mark a state of ambivalence are thought to underlie instability and controversy in public perceptions of science and technology more generally as well as widespread distrust of abstract technological systems and dependence upon expertise (Irwin and Michael 2003). Importantly, several prominent reactions to nanohype followed quickly on the heels of the NNI's launch, posing urgent questions about nanotechnological risk, toxicological uncertainty, and regulatory capacity that policy elites quickly scrambled to address. Alongside early evidence of public ambivalence over nanotechnology (cf. Cobb and Macoubrie 2004), policy elites adopted an "affective mode of governance" (Kearnes and Wynne 2007) that acknowledged and sought to take such ambivalence into account. This strategy is striking insofar as it departs from the more familiar "deficit model" of public understanding. As Kearnes and Wynne note, "in contrast to the risk debates that dominated the technological controversies of the late 1990s the vitality of technological innovation is now cast as vulnerable to lack of public confidence and trust in the regulatory and governance structures upon which such innovation depends" (2007: 131).

One indication that ambivalence played a role in NNI policy making is the frequent and repeated use of the affective term societal "concerns" in place of the more usual "perceptions" or "understanding" during this time in commentaries, governmental reports and legislation (Fisher 2019). Repeated use of this language at the time marks a departure from established science and technology policy practices by suggesting a recognition on the part of elite policy actors that public affects and values can play a major role in influencing technoscientific outcomes. A second indication is the abandonment of the more provocative vision of molecular manufacturing that had been originally used to promote the NNI (Peterson 2004). As stated, these and other indications are suggestive of awareness on the part of policy elites of the dynamic relationship between governmental enthusiasm and widespread ambivalence that helped inform the policy adjustments to follow.

However, reflexive awareness on the part of policy makers and, as we shall see, the policy modulations that followed can be seen as taking place within a broader technoscientific frame of imagination, experimentation, and (loss of) control. Curiously, elite ambivalence appears to have played at least as much of a role in NNI policy making as did public ambivalence (Fisher 2019). Second-order forms of ambivalence are evident in the palpable elite concern over public concerns that policy discourse at the time reveals, a situation that was aptly characterized as "nanophobiaphobia" (Rip 2006)[2] and that issues in the "contradictory intent" (Fisher and Mahajan 2006) evident in the policy adjustments we now turn to.

10.2.3 Adjustment

In 2000, the National Science and Technology Council asked the National Science Foundation (NSF) to dedicate a workshop to the "societal implications" of nanotechnology. Citing this workshop, the National Research Council (NRC) published a 2002 report that criticized the NNI's "disappointing" efforts for addressing the societal implications of nanotechnology. Notably, the report tied the "nation's success" in nanotechnology to the NNI's ability to address these implications (National Research Council 2002). This report, in turn, helped frame Congressional hearings and Congressional staff deliberations that eventually led to a novel role for the social sciences to help address these societal implications (Fisher 2019). As noted elsewhere, the Congressional Act broke with tradition by requiring that "ethical, legal, environmental, and other appropriate societal concerns" be "considered during the development of nanotechnology" by providing for societal research, public input and, "integrating research on societal, ethical, and environmental concerns with nanotechnology research and development" (US Congress 2003; cf. Bennett and Sarewitz 2006; Fisher 2005; Fisher and Mahajan 2006).

Against a backdrop of hype and enthusiasm, even though most of the societal issues that were mentioned were not unique to nanotechnology, they nevertheless

[2]A similar situation has been observed in synthetic biology policy discourses, as suggested by the related phrase, "synbiophobiaphobia" (Marris 2015).

received a remarkable level of attention by both promoters and elite decision makers. Moreover, the policy response to these issues was also remarkable as it adapted a prior model of funding social science and humanities for use not only to better understand social and ethical dimensions, but to actually influence and shape nanotechnology directions. As such, this policy response is a departure from the *laissez-faire* linear model of innovation in society and resembles instead a view of science and technology as shaped and modulated by responses and adjustments at multiple levels of governance. Thus, by advancing an alternative model of science-society interactions in which research on public concerns serves to influence nanotechnology research directions, policy makers departed from established models pertaining to public understanding of science (PUS) and ethical, legal and societal implications (ELSI) and did so during a key moment of constitutional fluidity (Jasanoff 2011; Fisher 2011).

In particular, the Congressionally mandated strategies of socio-technical integration and public engagement enrolled scientists, social scientists and public citizens as participants in a broader vision of technosciencesociety: Through structuring the interactions and adjustments between technoscience and society, each of these various actor groups were imagined as helping to "prepare for and adjust to the expected and unexpected social and economic impacts of these new technologies" (National Research Council 2002). This cybernetic mode of governance, in combination with Congressional interest in social research that "influences the direction of ongoing nanotechnology research and development" (House Science Committee 2003), suggest that state policy makers viewed both technoscience and society as having self-organizing properties that simultaneously serve as obstacles of and resources for governance. Not only did such discourses reframe the national interests in technoscience that motivated the NNI in the first place, tempering them with a somewhat more expansive view of its social context, they also reiterated and reinforced them. In short, novel policy developments for more participatory and integrative forms of federally funded research and development originated in elite concern over public ambivalence and were ultimately rooted in governmental enthusiasm over nanotechnology.

10.3 Governing TechnoScienceSociety

The Congressional refashioning of the NNI that occurred at the end of 2003 constituted an inflection of nanotechnology that proceeded from an elaborate vision of the malleability of both science and society. As such, the adjustments to the NNI are indicative of much more than a shift in policy rhetoric of a particular research and development program. By moving away from a linear and deterministic view of technoscience, policy elites tacitly acknowledged previously overlooked dynamic socio-technical interactions and envisioned mechanisms for modulating these interactions. At the same time, policy elites portrayed these dynamic processes and interactions thought to shape technological emergence within a technoscientific frame of government-mediated coordination and control. For not only did the importance of high expectations in the promotion of nanotechnology persist, they were carried over into positive visions of national success through the harmonious embedding of nanotechnology in society as a result of its enthusiastic adoption and dissemination.

A case in point, nanotechnology promoters sought to heighten public and workforce "excitement" in order to ensure a broad base of participation in the envisioned nanotechnology economy (Roco and Bainbridge 2001). As the policy discourse shifted to acknowledging the importance of societal concerns, a broader imaginary of national unity, future preparedness, and socio-technological malleability can be seen to emerge. In the process, we can see a transference of the imaginary of infinite plasticity from matter at the nanoscale to the values and behaviors of society. On the one hand, the new imaginary of socio-technological flexibility and mutual accommodation that emerges can be said to broaden the governance of nanotechnology in society by opening discursive and imaginative spaces for societal concerns and public values to be explicitly brought to bear on scientific and technological decision making (Barben et al. 2008; Fisher 2007; Macnaghten et al. 2005). On the other hand, it is reflective of an attempt to direct public concerns, for instance by recasting them as "needs" to be identified by social experts in order to shape the behavior of an enthusiastic public (Roco and Bainbridge 2001).

Similar to high-level pronouncements by three recent American Presidents, U.S. nanotechnology policy discourse at the turn of the millennium places technoscience at the heart of national interests. Perhaps surprisingly, the imperative to stay the course of technoscientific trajectories in the face of actual or perceived public ambivalence brought policymakers to mandate the influencing of those trajectories through experimental governance mechanisms for public engagement and sociotechnical integration. Their search for control, however, also resulted in a loss of control. For in acknowledging the fluidity of both technological directions and public ambivalence, policy elites simultaneously conceded their lack of ability to directly control that fluidity. As seen by the technoscientific logic underlying the early development of the societal research component of the NNI, state policy makers acknowledged previously overlooked uncertainties, including roles played by non-state actors—namely, publics, scientists and social scientists—in the governance of technosciencesociety. In seeking to experiment with the contingent behavior of publics, researchers and technologies—behavior that could otherwise jeopardize the success of a critical state enterprise—state policy makers also exhibited a significantly higher tolerance for ignorance. For the NNI's novel mechanisms of integration and engagement entail a deeper recognition of unpredictable and uncertain socio-technological dynamic fluctuations than do the conventional and expert-based models of PUS and ELSI.

The simultaneous expansion and loss of state control is reminiscent of the ecology-sovereignty debate, in which the proliferation of international environmental treaties mean that states have come both to acknowledge their own inability to exercise full ecological sovereignty, for instance by admitting that trans-boundary pollution can degrade their own territorially located resources, and to share power with other states and even non-state actors, for instance in the negotiation and implementation of treaties that allow these actors to influence state behavior over what come to be seen as global environmental resources located within its own borders. As Litfin observes, the proliferation of such treaties was initially thought to "undermine the institution of state sovereignty" while it eventually became clear that it actually resulted in "a shift in the practices and norms of sovereignty" (Litfin 1997). Such a shift may indeed have weakened state sovereignty insofar as the state is effectively admitting that it has less control over externally-sourced effects upon its own territorial resources and environmental quality; at the same time, "new international institutions" have in some cases "made some governments more effective problem solvers" (Conca 1994).

Similarly, while the cost of maintaining state control over the increasingly complex and unstable outcomes envisioned for technoscientific investments may appear, paradoxically, to entail a loss of governmental control; the concessions to non-state actors and to the limits of state power over the processes that shape technoscientific directions, namely attitudinal ambivalence and the social construction of technology, suggest increased opportunities for states to achieve their aims—successful technoscientific promotion, production and adoption—through indirectly facilitating the coordination of these processes.

The adjustments to the NNI's societal research program also formed the basis of subsequent efforts by the NNI aimed at "responsible development of nanotechnology." As the NRC report's reliance on the 2000 NSF workshop suggests, insights from scholars in science, technology and society (STS)—whose contributions to that workshop were numerous—appear to have directly and indirectly facilitated novel policy understandings of the complex science-society dynamics that underlie the more innovative aspects of the NNI societal research policy. Additionally, STS scholars have been instrumental in designing and conducting upstream public engagement (Macnaghten et al. 2005), socio-technical integration (Fisher et al. 2015), and other forms of engaged research that otherwise might not have been possible and that in turn has been central in developing frameworks for anticipatory governance (Guston 2014) and responsible innovation (Stilgoe et al. 2013).

As already suggested, while the policy rhetoric and mechanisms of integration and engagement potentially open up possibilities for greater participation by publics and social scientists in governance processes, they were also framed in relation to what could be considered a vision of societal coordination that is suggestive of governmentality. As in the case of state sovereignty, here too we see tensions in the meanings and implications of technosciencesociety governance. For instance, STS scholars have argued that public engagement supports democratic (Wilsdon and Willis 2004) and, conversely, technocratic (Thorpe and Gregory 2010) norms and values. Similarly, socio-technical integration has been upheld as an effective way to enhance reflexivity and responsiveness (Stilgoe et al. 2013; Wynne 2011) yet criticized for potential lack of critical distance and agency (Felt 2014; Valve and

McNally 2013). This has added to the ongoing debate within the science studies community over whether close proximity to technoscientific practices such as that entailed by integrated and engaged research is itself an agent of (Nordmann 2010) or antidote to (Guston 2014) technoscientific hubris.[3]

Such debate within STS is not new and reflects the inherent complexities of engaged scholarship (cf. Hackett and Rhoten 2011; Rip 2009; Schuurbiers 2011; Viseu 2015). For the purposes of the present chapter, it recalls technoscientific entanglements in that it touches upon the relationship of STS to its own object of study and its own actor position within shifting networks of power. Arguably, science studies scholars do not have the luxury of disengagement and, accordingly, have a responsibility to attend to the instrumental, interpretive, and normative dimensions of their work (Jasanoff 2011). Interpretive sensibilities can thus both inform normative commitments and temper the overreach of mere instrumentalism. Furthermore, by refusing to ignore or oversimplify these unavoidable tensions that underlie publicly engaged scholarship and collaborative research, STS scholars are in a rare position to demonstrate the possibility of alternative, more reflexively aware, and socially responsive forms of knowledge, expertise, and technoscience—even as they wrestle with their own ambivalences.

10.4 Conclusion

Drawing on the example of the U.S. National Nanotechnology Initiative, this chapter has developed an understanding of the governance of technoscience in society using the dialectics of enthusiasm, ambivalence, and adjustment. It argues that the exuberance and hype used to promote nanotechnology in the U.S. along with the high political stakes associated with it led to a rare form of elite preoccupation with public acceptance as well as to a critically-informed fluctuation in policy discourse and decision-making. Notably, both of these developments were more nuanced than might otherwise have been expected insofar as they entailed a tacit acknowledgement of previously underappreciated science-technology-society relations and resulted in explicit endorsement of novel forms of socio-technical integration and public engagement. Viewing the early stages of U.S. nanotechnology policy in light of the practices of technoscience accentuates the interrelatedness of exuberance for nanotechnology and openness to shaping science-society dynamics.

Early nanotechnology policy-making was not only characterized by promotional hype but bears clear resemblance to other more fundamental features of technoscience and late-modern technology, including such themes as infinite plasticity, tolerance of ignorance, reliance on self-organization, and the search for and loss of

[3] My own program in socio-technical integration research (STIR; Fisher 2007; Fisher et al. 2015; Fisher and Schuurbiers 2013), which draws inspiration from technology assessment and public engagement and also informs research agendas in both anticipatory governance and responsible innovation, inclines toward the latter view.

control. Exploring these from the standpoint of a dialectic process of socio-technical modulation presents a twofold tale. On one reading, this a story of the ironic loss of control, in which policy makers who embraced the not "there yet" reality of the "next industrial revolution" were thereby led to recognize and accept new sources of ignorance and uncertainty, thus tacitly conceding the lack of state control over broader social dynamics and symbolically accommodating, even empowering, non-state actors by casting them in quasi-governance roles. On the other hand, this accommodation and presumptive empowerment is cast as a form of heroic state coordination and thus suggests a new type of governmentality and control.

In the case of nanotechnology in the U.S., initial exuberance led not only to high-level preoccupation with public concerns but to a symbolically important shift in policy thinking about the relations among science, technology, and society. This included a tacit acknowledgement of traditionally overlooked socio-technical processes and to experimentation with novel mechanisms. By implying more interactive roles for citizens, scientists, and social scientists, policy makers envisioned a dynamic coupling of two malleable sources of governance: public attitudes and technoscientific directions. While this shift in governance is thus suggestive of greater democratic participation, more reflexive understanding of socio-technical interactions, and improved expert responsiveness to public concerns and values, newly envisioned roles were also themselves framed within a broader imaginary that reproduces and reasserts initial technoscience impulses towards control. Closely related to this are ongoing debates among STS scholars who perceive openings and opportunities for their insights and perspectives as well as for their own expertise in designing engagement programs even as they reflect anew on the paradoxes, pitfalls, and power dynamics that these newly enhanced roles suggest (Fisher 2011).

Although the case examined here is limited to one specific research and development program and one particular national setting, it does raise questions that appear to apply across political cultures and forms of technoscience. In an era of increasing mistrust in and rejection of expert institutions, will the embrace of technoscience narratives of excess and transformation continue to inform state policy making? To the extent that it does, it can be expected to further intensify both public and elite forms of ambivalence, amplifying tensions inherent in governing technology in society as narratives of control and loss of control proliferate and reinforce one another. Such amplification would likely spill over into more dynamic interplay between governmental, expert, and public self-conceptions, intensifying efforts to reframe and reengineer the entanglements of state legitimacy, scientific autonomy, and democratic values.

Acknowledgements This material is based upon work supported by the National Science Foundation (grant 1535120).

References

Anderson, B. 2007. Hope for Nanotechnology: Anticipatory Knowledge and the Governance of Affect. *Area* 39 (2): 156–165.

Arnall, A.H. 2003. *Future Technologies, Today's Choices: Nanotechnology, Artificial Intelligence and Robotics; A Technical, Political and Institutional Map of Emerging Technologies*. London: Greenpeace Environmental Trust.

Barben, D., E. Fisher, C. Selin, and D.H. Guston. 2008. Anticipatory Governance of Nanotechnology: Foresight, Engagement, and Integration. In *The Handbook of Science and Technology Studies*, ed. E. Hackett, O. Amsterdamska, M. Lynch, and J. Wajcman, 979–1000. Cambridge: MIT Press.

Bennett, I., and D. Sarewitz. 2006. Too Little, Too Late? Research Policies on the Societal Implications of Nanotechnology in the United States. *Science as Culture* 15 (4): 309–325.

Berube, D.M. 2006. *Nano-Hype: The Truth Behind the Nanotechnology Buzz*. Amherst: Prometheus Books.

Borup, M., et al. 2006. The Sociology of Expectations in Science and Technology. *Technology Analysis & Strategic Management* 18 (3–4): 285–298.

Bush, G.W. 2006. *State of the Union Address*, January 31, 2006.

Clinton, W.J. 2000. *President Clinton's Address to Caltech on Science and Technology*. The White House: Office of the Press Secretary. https://caltechcampuspubs.library.caltech.edu/2676/1/nano_clinton.pdf. Accessed on 20 Mar 2018.

Cobb, M.D., and J. Macoubric. 2004. Public Perceptions About Nanotechnology: Risks, Benefits and Trust. *Journal of Nanoparticle Research* 6 (4): 395–405.

Conca, K. 1994. Rethinking the Ecology-Sovereignty Debate. *Millennium: Journal of International Studies* 23 (3): 701–711.

Crichton, M. 2002. *Prey*. New York: Harper Collins.

Crow, M.M., and D. Sarewitz. 2001. Nanotechnology and Societal Transformation. In *AAAS Science and Technology Policy Yearbook*, ed. A.H. Teich, S.D. Nelson, C. McEnaney, and S.J. Lita, 89–101. Washington, DC: American Association for the Advancement of Science.

Economist, The. 2004. *Small Wonders*, 29 December. https://www.economist.com/node/3494722. Accessed on 20 Mar 2018.

ETC Group. 2003a. *The Big Down: From Genomes to Atoms*. http://www.etcgroup.org/sites/www.etcgroup.org/files/thebigdown.pdf. Accessed on 13 July 2017.

———. 2003b. *Nanotech and the Precautionary Prince*. http://www.etcgroup.org/sites/www.etcgroup.org/files/publication/158/01/precautionprince.pdf. Accessed on 13 July 2017.

Felt, U. 2014. Within, Across and Beyond: Reconsidering the Role of Social Sciences and Humanities in Europe. *Science as Culture* 23 (3): 384–396.

Fisher, E. 2005. Lessons Learned from the Ethical, Legal and Social Implications Program (ELSI): Planning Societal Implications Research for the National Nanotechnology Program. *Technology in Society* 27 (3): 321–328.

———. 2007. Ethnographic Invention: Probing the Capacity of Laboratory Decisions. *NanoEthics* 1 (2): 155–165.

———. 2011. Editorial Overview: Public Science and Technology Scholars: Engaging Whom? *Science and Engineering Ethics* 17 (4): 607–620.

———. 2019. Governing with Ambivalence: The Tentative Origins of Socio-Technical Integration. *Research Policy* 48 (5): 1138–1149.

Fisher, E., and R.L. Mahajan. 2006. Contradictory Intent? US Federal Legislation on Integrating Societal Concerns into Nanotechnology Research and Development. *Science and Public Policy* *33* (1): 5–16.

Fisher, E., and D. Schuurbiers. 2013. Socio-Technical Integration Research: Collaborative Inquiry at the Midstream of Research and Development. In *Early Engagement and New Technologies: Opening up the Laboratory*, 97–110. Dordrecht: Springer.

Fisher, E., et al. 2015. Mapping the Integrative Field: Taking Stock of Socio-Technical Collaborations. *Journal of Responsible Innovation* 2 (1): 39–61.

Guston, D.H. 2014. 'Understanding 'Anticipatory Governance. *Social Studies of Science* 44 (2): 218–242.

Hackett, E., and D. Rhoten. 2011. Engaged, Embedded, Enjoined: Science and Technology Studies in the National Science Foundation. *Science and Engineering Ethics* 17 (4): 823–838.

House Committee on Science. 2003. Report 108–89. S. Boehlert, US House of Representatives, 108th Congress, 1st Session.

Irwin, A., and M. Michael. 2003. *Science, Social Theory & Public Knowledge*. Philadelphia: Open University Press.

Jasanoff, S. 2011. Constitutional Moments in Governing Science and Technology. *Science and Engineering Ethics* 17 (4): 621–638.

Joy, B. 2000. Why the Future Doesn't Need Us, *Wired* 8 (4). https://www.wired.com/2000/04/joy-2/. Accessed on 13 July 2017.

Kearnes, M., and B. Wynne. 2007. On Nanotechnology and Ambivalence: The Politics of Enthusiasm. *NanoEthics* 1 (2): 131–142.

Kennedy, J. 2008. Nanotechnology: The Future is Coming Sooner Than You Think. In *Presenting Futures*, 1–21. Dordrecht: Springer.

Litfin, K.T. 1997. Sovereignty in World Ecopolitics. *Mershon International Studies Review* 41 (2): 167–204.

Lok, C. 2010. Nanotechnology: Small Wonders. *Nature News* 467 (7311): 18–21.

Lucivero, F. 2015. *Ethical Assessments of Emerging Technologies*. New York/Dordrecht/London: Springer.

Macnaghten, P., M.B. Kearnes, and B. Wynne. 2005. Nanotechnology, Governance, and Public Deliberation: What Role for the Social Sciences? *Science Communication* 27 (2): 268–291.

Marris, C. 2015. The Construction of Imaginaries of the Public as a Threat to Synthetic Biology. *Science as Culture* 24 (1): 83–98.

McCray, W.P. 2005. Will Small Be Beautiful? Making Policies for Our Nanotech Future. *History and Technology* 21 (2): 177–203.

McGrail, S. 2010. Nano Dreams and Nightmares: Emerging Technoscience and the Framing and (Re) interpreting of the Future, Present and Past. *Journal of Futures Studies* 14 (4): 23–48.

Mnyusiwalla, A., A.S. Daar, and P.A. Singer. 2003. "Mind the Gap": Science and Ethics in Nanotechnology. *Nanotechnology* 14 (3): R9.

National Research Council; Division of Engineering and Physical Sciences; Committee for the Review of the National Nanotechnology Initiative 2002. 2002. *Small Wonders, Endless Frontier: A Review of the National Nanotechnology Initiative*. Washington, DC: National Academies Press.

Nordmann, A. 2006. Collapse of Distance: Epistemic Strategies of Science and Technoscience. *Danish Yearbook of Philosophy* 41 (7): 7–34.

———. 2010. A Forensics of Wishing: Technology Assessment in the Age of Technoscience. *Poiesis & Praxis* 7 (1–2): 5–15.

Obama, B. 2010. *State of the Union Address*, January 27, 2010.

Peterson, C.L. 2004. Nanotechnology: From Feynman to the Grand Challenge of Molecular Manufacturing. *IEEE Technology and Society Magazine* 23 (4): 9–15.

Pollock, N., and R. Williams. 2010. The Business of Expectations: How Promissory Organizations Shape Technology & Innovation. *Social Studies of Science* 40 (4): 525–548.

Rip, A. 2006. Folk Theories of Nanotechnologists. *Science as Culture* 15 (4): 349–365.

———. 2009. Futures of ELSA. *EMBO Reports* 10 (7): 666–670.

Roco, M.C., and W.S. Bainbridge. 2001. *Societal Implications of Nanoscience and Nanotechnology*. Arlington: National Science Foundation.

Schuurbiers, D. 2011. What Happens in the Lab: Applying Midstream Modulation to Enhance Critical Reflection in the Laboratory. *Science and Engineering Ethics* 17 (4): 769–788.
Schwarz, A., and A. Nordmann. 2011. The Political Economy of Technoscience. In *Science in the Context of Application*, ed. M. Carrier and A. Nordmann, 317–336. Dordrecht: Springer.
Selin, C. 2007. Expectations and the Emergence of Nanotechnology. *Science, Technology & Human Values* 32 (2): 196–220.
Smart, B. 1998. *Facing Modernity: Ambivalence, Reflexivity and Morality*. London: Sage.
Stilgoe, J., R. Owen, and P. Macnaghten. 2013. Developing a Framework for Responsible Innovation. *Research Policy* 42 (9): 1568–1580.
Thorpe, C., and J. Gregory. 2010. Producing the Post-Fordist Public: The Political Economy of Public Engagement with Science. *Science as Culture* 19 (3): 273–301.
US Congress, 21st Century Nanotechnology Research and Development Act of 2003. 2003. Public Law no 108–153, 117 STAT. 1923.
Valve, H., and R. McNally. 2013. Articulating Scientific Practice with PROTEE: STS, Loyalties, and the Limits of Reflexivity. *Science, Technology, & Human Values* 38 (4): 470–491.
Viseu, A. 2015. Caring for Nanotechnology? Being an Integrated Social Scientist. *Social Studies of Science* 45 (5): 642–664.
Wilsdon, J., and R. Willis. 2004. *See-Through Science: Why Public Engagement Needs to Move Upstream*. London: Demos.
Wynne, B. 2011. Lab Work Goes Social, and Vice Versa: Strategising Public Engagement Processes. Commentary on: "What Happens in the Lab Does Not Stay in the Lab: Applying Midstream Modulation to Enhance Critical Reflection in the Laboratory". *Science and Engineering Ethics* 17 (4): 791–800.

Erik Fisher is an Associate Professor in the School for the Future of Innovation in Society at Arizona State University. He studies the multilevel governance of emerging technologies from lab to legislature, focusing on policies and practices for socio-technical integration. He developed the approach of Socio-Technical Integration Research (STIR), which has been applied in over 50 research and innovation organizations on 3 continents. He directs the program in Socio-Technical Integration Research (STIR) and is Editor in Chief of the *Journal of Responsible Innovation*.

Chapter 11
The Politics of Technoscience: From National Visions to Global Problems

David Kaldewey, Daniela Russ, and Julia Schubert

11.1 Introduction[1]

A basic tenet of *Science and Technology Studies* (STS) is that the boundaries between science, technology, and politics are fragile and continually reconfigured. To capture this, scholars conceptualized a "mode 2" of knowledge production (Gibbons et al. 1994); they focused on "hybrids" and "networks" in which science, technology, and politics are not separated but mixed (Latour 1993); and they analyzed the natural and social order as "co-produced" (Jasanoff 2004). The idiom of co-production implies, in a nutshell, that the making of science is "political" (Jasanoff 2004: 21). Given the advancement of technosciences in an increasingly technologized society, Sabine Maasen et al. (Chap. 5, this volume) point to a further reconfiguration of what they call *TechnoScienceSociety*: If technosciences and technologized societies co-design each other, then, evidently technosciences are "political" as well. However, one question remains mostly unaddressed by this framework and STS literature alike: What precisely does it mean to say that (techno)science is political?

[1] We thank the participants of the Yearbook workshop in Munich (November 2015) and two "anonymous" reviewers for their valuable thoughts, ideas, and consistent criticism on earlier versions of this text. The manuscript for this chapter was finalized in September 2016; after that, we only made minor corrections and updates. In the meantime, more elaborated versions of the case studies are to be found in other publications by the authors.

D. Kaldewey (✉) · J. Schubert
Forum Internationale Wissenschaft, University of Bonn, Bonn, Germany
e-mail: kaldewey@uni-bonn.de; j.schubert@uni-bonn.de

D. Russ
Post-Doctoral Researcher, Department of Sociology, University of Toronto, Toronto, Canada
e-mail: daniela.russ@utoronto.ca

S. Maasen et al. (eds.), *TechnoScienceSociety*, Sociology of the Sciences Yearbook 30, https://doi.org/10.1007/978-3-030-43965-1_11

The answer to this question will vary depending on which epistemic community we ask. Mark Brown (2015) uncovered the diverse meanings of "politics" in STS. If one takes heed of adjacent research fields, such as science, technology, and innovation (STI) policy studies, the situation gets even more confusing. The ambiguity of the concept of politics becomes particularly apparent when associated with the corresponding concept of democracy. In STS, notions such as "material democracy," "ontological politics," or "cosmopolitics" have become popular and can be summarized in the idea of "democracy as collective world making" (Brown 2015: 15). In contrast, other disciplines, such as political philosophy, political science, as well as the more policy-oriented strands of STS, adhere to more traditional notions of representative democracy as "collective self-governance," and a corresponding understanding of politics as "purposeful activities that aim for collectively binding decisions in a context of power and conflict" (Brown 2015: 19).

In this article, we do not take sides for either one of these communities nor do we try to present the most adequate concept of politics for STS. Instead, we seek to explore the rationales underlying both "collective world making" and "collective self-governance" in technologized societies. We propose a set of research questions to explain the role of *TechnoSciencePolitics* in *TechnoScienceSociety:* Which visions, values, motives, and presuppositions guide technoscientific knowledge production on the one hand, and the governance of technoscience on the other? Where do these rationales come from? What are the long-term trajectories of these rationales, or, in other words, how does the past shape the development of technoscientific knowledge? Do these rationales remain implicit, operating subconsciously, or can they be explicated discursively, that is, opened up to reflection and adjustment? Finally, who establishes these rationales in the first place, and who makes them explicit? By asking such questions, we avoid defining the "political" analytically. Instead, we conceive of the "political" as a contested semantic space, determined by historically contingent, implicit and explicit rationales. The boundaries of this space are never stable but are constantly reconfigured by scientists, engineers, policymakers, and the public.[2]

In the following sections, we examine the rationales constituting the semantic space of *TechnoSciencePolitics* by means of historical discourse analysis.[3] We focus on two technoscientific discourses, both of which are evolving around specific global problems—one dealing with energy security (Sect. 11.3), the other with climate engineering (Sect. 11.4).[4] As a theoretical starting point, we choose a prominent concept elaborated by STS scholars addressing our basic question of how the co-production of science and politics is guided by historically and culturally

[2] This concept of a semantic space refers partly to insights from constructivist political history: Steinmetz et al. define the political as a "communicative space in history whose boundaries are constantly reconfigured through acts of verbal, visual, and sometimes violent communication" (Steinmetz et al. 2013).

[3] For more methodological details, see Kaldewey et al. (2015).

[4] Both labels, "energy security" and "climate engineering," are contingent and quite controversial. However, for better readability, we renounced putting them in quotes in the following.

contingent rationales: Sheila Jasanoff and Sang-Hyun Kim's notion of sociotechnical imaginaries (Sect. 11.2). However, we make use of this framework under a specific perspective, differing from the approach taken by most of the case studies building on this framework so far. Puzzled by how sociotechnical imaginaries become visible and explicated in practice, we particularly focus on the field of science policy making. We ask how sociotechnical imaginaries mold into the semantic space of *TechnoSciencePolitics*. Our hypothesis is that one way for sociotechnical imaginaries to become politically explicit is the popular reference to "grand challenges." As we will show in the case studies, historical actors translate (implicit) visions encoded in sociotechnical imaginaries into (explicit) problems framed as "grand challenges."

The contribution concludes by asking how an analysis of "grand challenges" can contribute to a refinement of the sociotechnical imaginaries framework. Our findings point to two dimensions that are crucial for understanding how sociotechnical imaginaries translate into actual policies: the interplay between national and global imaginaries, and the oscillation between positively connoted visions and negatively connoted problems.

11.2 Explicating Sociotechnical Imaginaries as "Grand Challenges"

Following Sheila Jasanoff and Sang-Hyun Kim, the co-production of modern societies and technoscience is affected by historically contingent and context-dependent sociotechnical imaginaries. These imaginaries "at once describe attainable futures and prescribe futures that states believe ought to be attained," that is, they encode "collective visions of the good society" (Jasanoff and Kim 2009: 120, 123). Although sociotechnical imaginaries are never "strictly determinative of policy outcomes," they serve as "cultural resources" that implicitly or explicitly shape STI policies (Jasanoff and Kim 2013: 190). Sociotechnical imaginaries are performative and materialize through technology (Jasanoff 2015a: 20). Jasanoff and Kim (2015) bring together a range of case studies illustrating how imaginaries evolve in different national and cultural contexts and around diverse issues such as nuclear power, biotechnology, nanotechnology, information technologies, global security, and global health. Drawing from these case studies, the authors slightly modify and broaden the definition of sociotechnical imaginaries as

> "collectively held, institutionally stabilized, and publicly performed visions of desirable futures, animated by shared understandings of forms of social life and social order attainable through, and supportive of, advances in science and technology" (Jasanoff 2015a: 4).

By referring to abstract notions of "social life" and "social order," Jasanoff and Kim refrain from specifying the concept of politics in their analytical framework. On the one hand, they reject the traditional idea of politics "as consisting simply of purposive, rational action" (Jasanoff 2015a: 7); on the other hand, they likewise criticize approaches that do not pay sufficient attention to power relations (Jasanoff 2015a: 18). While the cases studied within this framework are usually *national* ones, scholars are nonetheless encouraged to go beyond institutionalized STI policies. Coming back to Brown's distinction of the various concepts of politics in STS, the sociotechnical imaginaries framework undermines the distinction between democracy as "collective world making" and democracy as "collective self-governance." Instead, sociotechnical imaginaries serve exactly as the kind of rationales we are interested in: They guide *both* collective world making *and* collective self-governance, and thereby become apparent in technologies *as well as* in political decisions. A more radical interpretation of the framework, therefore, could state that the notion of sociotechnical imaginaries synthesizes both STS and traditional notions of politics. This integrative potential, however, comes at the price of a high level of abstraction.

For STS scholars, sociotechnical imaginaries enable cross-cultural comparisons and open up new perspectives regarding the co-production of science, technology, and society. Eventually, the framework may even help us to assess normatively whether, how, and which imaginaries can be transformative, that is, can become "vehicles for reenvisioning and recalibrating human futures" (Jasanoff 2015a: 27). In other words, Jasanoff imagines sociotechnical imaginaries to take over a function that had once been ascribed to political values, ideas, and ideologies. Consequently, the role of STS scholars can be compared to the political philosophers of the past, fundamentally dealing with the question of the good life and the good society.

In short, at a high level of abstraction, the sociotechnical imaginaries framework provides a general answer to the question of what it means to say "the making of science is political" (Jasanoff 2004: 21). However, it is less evident how the framework connects to the mundane realm of STI policies. Given that sociotechnical imaginaries are often vague and implicit, it is difficult to fathom how they play out in scientists' and other actors' strategies, as well as in political decision-making. How then do sociotechnical imaginaries translate into distinct STI policies? We suggest that sociotechnical imaginaries have to be made *explicit* as actors' categories. To understand this process of *explication*, we examine how the theoretically reconstructed sociotechnical imaginaries become visible in everyday communication and strategic action of scientists, engineers, policy-makers, and the public. Of course, the narratives to be found at this level will not match the complexity of the underlying sociotechnical imaginaries. We assume, however, that those explications and simplifications are influential precisely because they successfully reduce the complexity inherent in culturally embedded sociotechnical imaginaries. Actors build on simplifications in order to transform abstract ideas into practical decisions.

An intriguing example of how sociotechnical imaginaries are explicated in practice is the concept of "grand challenges," which is now widely used in STI policy worldwide (cf. Hicks 2016; Ulnicane 2016; Kaldewey 2018). In the following, we sketch the career of this discourse and illustrate how grand challenges can be understood as translations, concretizations, and explications of sociotechnical imaginaries.

The notion of grand challenges has its origin not in academic discourse but in the trading zones between science and politics. The term "grand challenges" first appeared in US science and technology policy contexts in the late 1980s, indicating the need for federal funding in the field of computational sciences, especially with the aim of building supercomputers (cf. Hicks 2016). However, it was not before the new millennium that the discourse gained strong momentum: A crucial moment was when Bill Gates announced the "Grand Challenges in Global Health" initiative at the World Economic Forum in Davos in 2003 (cf. Enserink 2003). While the grand challenges discourse of the 1980s predominantly focused on US national challenges, the Gates initiative opened up to a global horizon. A grand challenge was now defined as "a specific scientific or technological innovation that would remove a critical barrier to solving an important health problem in the developing world with a high likelihood of global impact and feasibility" (Varmus et al. 2003: 398). A few years later, the idea had become a global rationale for STI policies, as can be seen, for example, in the Obama administration's "Strategy for American Innovation" (White House 2009) or in the current "Horizon 2020" framework of the European Union (European Commission 2011). Beyond these national and supranational STI policy contexts, "grand challenges" have also been taken up by international forums, private foundations, scientific societies, universities, and research institutes (cf. Ulnicane 2016). Many of these actors and organizations produce lists with particularly urgent challenges to communicate priorities in STI policy. In the following section, we explore two prominent cases that are routinely listed and explicitly framed as grand challenges—energy security and global warming.

Evidently, both sociotechnical imaginaries and grand challenges function as rationales that guide "collective world making" and "collective self-governance" in the field of science, technology, and innovation. They influence the way scientists and engineers choose their research problems; they foster or constrain the development of certain research fields; and they systematically privilege or undermine certain methodologies and interdisciplinary collaborations.[5] Furthermore, they influence how science, technology, and innovation are conceived of, modeled, and transformed into an object of governance by policy-makers and other stakeholders. While sociotechnical imaginaries remain implicit and are notoriously difficult to study, discussions about technological problems and challenges are prevalent in policy reports and research proposals. As already indicated, we interpret grand challenges as discursive representations of implicit sociotechnical imaginaries. Of course, this process of making the implicit explicit is not without losses and conflicts. We do not assume that the empirical translations we study were self-evident or necessary. Quite to the contrary, there will always be many ways to translate a sociotechnical imaginary into an actor's category. The practical reframing of general visions as specific "challenges" is contingent and context-dependent.

[5] Several scholars have pointed out that grand challenges "motivate, frame, and communicate research activities" (Winter and Butler 2011: 103), and point to long-term and large-scale research goals. Furthermore, grand challenges are assumed to be defined by various societal stakeholders, thus containing "assumptions about what we value as a society" (Calvert 2013: 474). Evidently, the same could be said about sociotechnical imaginaries.

11.3 The Case of Energy Security

In 2011, the World Energy Council (WEC), one of the earliest providers of international energy data, introduced a new indicator called the "Energy Trilemma Index." This index is supposed to measure countries' policy performances in tackling the three dimensions of the energy problem: energy security, energy equity, and energy sustainability. As Joan MacNaughton, Executive Chair of WEC World Energy Trilemma, puts it in her foreword to the second report,

> "Delivering policies which simultaneously address energy security, universal access to affordable energy services, and environmentally-sensitive production and use of energy is one of the most formidable challenges facing governments—indeed some might argue that it is the most formidable, or even the most important. The World Energy Trilemma report (...) aims to help governments rise to the challenge of tackling this 'trilemma'" (World Energy Council 2012).

Apart from this being "grand challenges" parlance in full swing, two aspects are particularly noteworthy. First, in order for countries to be comparable in a ranking, the indicator presupposes, and thereby states, the existence of a common "energy challenge." Second, the Energy Trilemma Index is one of the first indicators measuring technical data not only on energy systems but also on energy policies. While a systematic comparison of national energy policies of some sort is not entirely new,[6] the Energy Trilemma Index is unique in its avowedly global reach and its form as a quantitative index.

In recent decades, sociologists and anthropologists have taken great interest in understanding quantitative forms of knowledge (Porter 1994; Espeland and Stevens 2008; Heintz 2011) and the standardizing power of rankings and indicators (Porter 1995; Espeland and Sauder 2011; Merry 2011). Indicators bring together the clarity and unambiguity of numbers with a broader theory that puts them in a meaningful context and makes sense of differences and changes over time (Davis et al. 2015). The Energy Trilemma Index, for instance, conveys a meaning of how an equal, sustainable, and secure energy supply looks like, bringing together general theories from developmental and environmental economics. The consequences of indicators are notoriously difficult to capture, and only time will tell whether the Energy Trilemma Index is more than just another project initiated and promoted under a catchy name by a non-governmental international organization. In theory, however, it could well propel the diffusion of a global sociotechnical imaginary (cf. Jasanoff 2015b: 335ff.).

The concept of sociotechnical imaginaries has been repeatedly applied to the study of energy systems, or some of its features (Jasanoff and Kim 2009, 2013,

[6] The attempt to monitor and eventually harmonize national energy policies dates back to the foundation of the Organization of the Petroleum Exporting Countries (OPEC) in 1960, the International Energy Agency (IEA) in 1974, and the Organización Latinoamericana de Energía (OLADE) in 1973, but was at first restricted to countries directly affected by the oil crises. Since then, other regional organizations were set up, such as the ASEAN Centre for Energy and the African Energy Association, producing regular energy reports and portfolios of their member countries.

2015; Tidwell and Smith 2015). According to Jasanoff and Kim (2013: 189 f.), analyzing sociotechnical imaginaries helps to explain the "hidden social dimensions of energy systems." National energy policies, it is argued, differ particularly in their treatment and weighting of technology's risks and benefits—a variation that is seen as an articulation of the various sociotechnical imaginaries involved. Energy systems are conceived as being shaped by sociotechnical imaginaries as well as being "prime repositories of such collective visions" (Jasanoff and Kim 2013: 190).

While studies on the choice of energy sources might indeed tell us something about the diverse perceptions of risk and security, they merely scratch the surface of how radically *social* energy systems actually are. Looking through a historical lens might help here. In an important sense, the problem of a secure energy supply is deeply rooted in more general political, economic, and technological processes of the last 200 years, and has virtually been *co-produced*. These processes are reflected in the concept of energy itself (cf. Mirowski 1989; Smith 1998), but also in the way energy is framed and organized as a national or global "problem." While not entirely denying non-national factors, the concept of sociotechnical imaginaries is clearly designed to highlight national peculiarities, as is apparent in the structure of the case studies building on this framework (Jasanoff and Kim 2015). In the following, we argue that while this focus yields fruitful empirical results in some cases, it does not capture the "hidden social dimensions of energy systems" radically enough. After all, the supply of fuel, fodder, and food was for a long time neither organized within a nation state nor framed as a problem of national security. The idea of energy being a matter of nations and, in turn, the idea of nations exhibiting different energy systems, had yet to be made collectively imaginable. Studying the historical emergence of the problem of energy security and tracing its transformation into a "global challenge" enables a better understanding of how global, national, and local imaginaries come into being and interact with each other.

Stanley Jevons' book *The Coal Question* (1866) might well be the first scientific study on energy security, tying the rise of British coal consumption to the future of the Empire. Although Jevons referred to a very specific situation of the British Empire in the mid-nineteenth century and based his calculations solely on British data, he nonetheless was understood to state a universal economic law. Even before international standardization gained momentum in the early twentieth century,[7] and when little comparable data on coal resources and consumption was available, the occurrence and combustion of coal, as well as the rules of its consumption, were still thought to apply generally. In other words, economic knowledge was imagined as global from very early on. Knowledge on coal in the nineteenth century was embedded in geological, chemical, and physical knowledge, and was thus organized by those disciplines that were understood as producing objective and universal knowledge (cf. Daston and Galison 2007; Heintz 2007). Economics was just about to follow these disciplines in their thrust to formulate universal laws that can be

[7]The forerunner of international standardization, the International Electrotechnical Commission, was founded in 1906.

applied independently of cultural or national contexts (cf. Mitchell 1998, 2002). This indicates an understanding of globalization that is not synonymous with standardization. Standardization can be understood as an actual process of homogenization, for instance, the adoption of a common method or unit of measurement. As such, it is a crucial factor in making scientific knowledge global; but imagining knowledge to be global goes far beyond, and often precedes, de facto standardization. Indeed, it is effective exactly in the fuzzy imagination of energy as a global condition to which humanity is exposed. It is only against this backdrop that "energy systems" can emerge as different ways to relate to a general condition.

It took the First World War to "nationalize" the way we think about energy (and, by the way, the economy as well). Countries with industrialized regions had for a long time produced numbers on several aspects of their energy systems, such as water power resources, the production of coal mines, and the supply and demand of electricity. The information was scattered over several public and private institutions and remained unsystematic—a mirror of the various local factors determining the sources tapped and explored. War changed that. As a term in its own right, "energy security" can be traced back to the First World War, when the supply of fuels for the military became a question of national security (cf. Cherp and Jewell 2011: 202). "The war disrupted prewar flows of fuel, equipment, and people across borders, causing shortages and bottlenecks" (Hausman et al. 2008: 127). Engineers claimed a "rational utilization" of all resources (Hausman et al. 2008: 129) and a previously unseen amount of systematic knowledge on the composition and organization of fuels and electricity was produced.[8]

Occasionally, this meant setting up new institutions dedicated to the study and organization of national fuel supply. In Germany, companies fearing economic losses due to restricted supply chains initially voiced the need for a systematic national organization of resources during war times. This idea was willingly deployed by the government through the foundation of the so-called *Kriegsrohstoffabteilungen* (cf. Krajewski 2006: 204ff.). Fuel and resource supply was hence a problem to be solved by every country taking part in the war, and countries learned from each other in addressing this problem of national security. When sociotechnical imaginaries—for instance, the theories of socialist planning or community economy (*Gemeinwirtschaft*)—were formulated based on war experiences, their audience transcended national publics in a very concrete way. Although this new way of organizing resembled older mercantilist regimes and it was not necessarily linked to transformations of the energy system in terms of new power sources, it certainly changed, and to some extent aligned, the way countries observed and approached the problem of national energy supply. While war had temporarily impeded the flow of energy, it also set off a flow of knowledge on how to organize fuel-dependent societies. By 1938, the majority of the World Power Conference's

[8]At the Fuel Conference of the World Power Conference in 1928, the significance of wartime experience was articulated numerous times, suggesting an influence both in industry development (World Power Conference 1928: 552) and research on fuel composition and classification (World Power Conference 1928: 177, 242).

member countries were able to present data on their "national power economy" along general lines, that is, regarding resources and production. The idea of a world of similar but different national (energy) economies had taken hold.

Nevertheless, for the time being, energy security was enunciated as a problem of national not international concern, and handling it within national boundaries was favored. Wartime strategies were guided by an attempt to become self-sufficient, but at times, they generated new dependencies. When Winston Churchill decided to change the power source of British navy ships from coal to oil, he intended to make them faster than their German counterparts. Being aware of the consequences, he held that the dependency on foreign oil supply could be limited by diversifying the sources: "Safety and certainty in oil lie in variety and variety alone" (quoted by Yergin 2006: 69). To be sure, Churchill's shift of power resources represents only an early step in a more fundamental process. The postwar Marshall plan shifted European energy supply deliberately and radically from coal to oil (cf. Mitchell 2009: 406) and hence created a mutual dependency between Middle Eastern and European countries.

Postwar planning documents show a certain awareness for and reflection on a common problem of resource supply and access, and observers such as John Maynard Keynes and Harry Dexter White argued for a third international institution "alongside the International Monetary Fund and the World Bank, to manage trade in oil and other essential raw materials" (Mitchell 2011: 111 f.).[9] With the Suez Crisis in 1956, and the Oil Crises of the 1960s and early 1970s, two complementary versions of an energy problem emerged: While oil-consuming countries became vulnerable to cuts in oil supply, oil-producing countries experienced shocks of oil price volatility, jeopardizing the balance of their national budgets. Consequently, the Organization of the Petroleum Exporting Countries (OPEC) and the International Energy Agency (IEA) were formed as two complementary international bodies observing, addressing, and managing these problems for the two groups respectively. Under the auspices of the IEA, countries were monitored and ranked according to their energy dependency and efficiency.

Thus, in the wake of the oil crisis, and even more so with the rise of international climate politics, the imagination of energy security has been institutionalized on a global scale. This is not to say that countries will eventually cease to take different decisions on the energy sources they wish to tap and the risks they are willing to take. Increasingly, countries frame their policies as responding to a "global energy challenge," and the Energy Trilemma Index is only a recent facet in a more general process. However, even if the global horizon is not explicitly mentioned and it does not come as a reference to international organizations, national energy infrastructures and policies have always reflected more than a *national* sociotechnical

[9]A forerunner of this idea was Brailsford's claim in *The War of Steel and Gold* (1914) to settle colonial conflicts by including an economic mechanism into the League of Nations that would allow for the common management of resource access. "In other words, the League was to be an economic mechanism to replace, not war between states, but its taproot—the conflict over material resources" (Mitchell 2011: 75).

imaginary. When nations imagine their energy systems, they are equipped with various *global* devices—measures, statistics, and models—making sense and positioning themselves against a global energy environment.

11.4 The Case of Climate Engineering

"All of human life is now lived out not just in the presence of a physically changing climate, but in the new discursive and cultural spaces that have been created by the idea of climate change" (Hulme 2014: ix). In capturing the collective imagination of the modern global public, climate change serves as a powerful sociotechnical imaginary. While the atomic threat dominated the 1960s, and energy and environmental problems arose predominantly in the 1970s, the twenty-first century discovered and framed anthropogenic climate change as the "grand challenge" of our time (WHO 2008; Auswärtiges Amt 2014; NRC 2015). Recently, notions of "climate emergencies" and "tipping points" (cf. Hulme 2014: 21) gave rise to a controversial suggestion, gradually infusing climate policy debates in the twenty-first century, namely the idea to engineer the climate (cf. Royal Society 2009).

Broadly, climate engineering can be defined as the intentional large-scale manipulation of the environment, aiming to prevent symptoms of climate change or remedy its causes (Keith 2000: 245). Within the current scientific debate, we can distinguish two lines of technological concepts. Research on solar radiation management (SRM) aims at tackling the symptoms of climate change and explores the potential of albedo modification or enhancement by cloud seeding or stratospheric aerosol injections (e.g. Govindasamy and Caldeira 2000; Crutzen 2006; Weisenstein and Keith 2015). Carbon dioxide removal (CDR), in contrast, addresses the causes of climate change and assesses options such as enhanced weathering, ocean fertilization, and direct air capture to mitigate climate change (Lal 2004; Keith et al. 2006; Lackner 2009).

Yet, instead of resonating as a positive vision of socio-technical innovation, climate engineering gained political currency as an unappealing, even daunting option of last resort. In our view, a historical perspective on the emergence of climate engineering reveals how the "shared fears of harms" (Jasanoff 2015a: 5) that characterize the climate change imaginary have shifted over time, and eventually brought forth this "bad idea whose time has come" (Kintisch 2010).

First observations of changing and disruptive climatic conditions were made already in the seventeenth century (cf. Parker 2013).[10] At that time, however, observers captured isolated incidents and catastrophic weather events rather than systematic global trends. These extreme weather events were largely attributed to a divine will, as for most of human history the weather had been understood as "the domain of the gods" (Hulme 2014: 12). For early and colonial America, climate change was

[10] Precedent observations were already made in antiquity (Fleming 1998: 137).

a matter of national pride and an essential component of the emerging Republican national ideal (cf. Fleming 1998: 32). The vision was that clearing and cultivating the land would promote a warmer, less variable, and healthier climate (cf. Fleming 1998: 33).

It was not before the late nineteenth century, however, that scholars such as Svante Arrhenius (1896, 1908) and Nils Gustaf Ekholm (1901) began to systematically explore the relation of climate and society—tabulating, charting, mapping, and analyzing their observations and the possibility of global anthropogenic climate change. Observation was further standardized and its geographic scope broadened by the establishment of national weather services in Europe, Russia, and the United States, which later led to international cooperation and global observation systems (cf. Fleming 1998: 41). Based on the invention of meteorological observation systems and statistical analysis, a first scientific picture of climatic conditions became stabilized. This early advancement of climatological knowledge, however, was born out of a deeply fractured and pre-disciplinary field of atmospheric sciences and did not include anticipations of global warming (cf. Fleming 1998: 82; Weart 2008: 7; Edwards 2010: 72). Rather to the contrary—well into the twentieth century, the basic assumption was that human kind would soon experience a new Ice Age.[11]

In this context, emerging insights into the possible human impact on the climate via the burning of fossil fuel sparked hopes, hopes of being able to avert the daunting crisis of a freezing planet (see, e.g., Gramelsberger and Feichter 2011: 16). Significant progress in observational and modeling capacities during the second half of the twentieth century further fueld such hopes to deliberately intervene in and even control the climate. A massive infrastructure of organizations, professionals, programs, and technical equipment stabilized the "quest for perfectly accurate machine forecasts and [...] perfectly accurate data acquisition" as the dominant mode of problematizing climate change for many years to come (Fleming 2010: 192). In this process, the atmosphere had been rendered not only qualitatively describable, but also "calculable" and eventually—this was the hope—controllable (Baker 2017:11). Ironically, at the same time, these scientific developments were essential in formulating human-induced climate change as a societal problem and bring it on the political agenda as such.

The following years witnessed the incremental institutionalization of climate change as a challenge that demanded political action on a global scale. Milestones were the establishment of the Intergovernmental Panel on Climate Change (IPCC) in 1988, and the United Nations Framework Convention on Climate Change (UNFCCC) in 1992. This politicization of climate change at first drowned out luring visions of climate control as a potential response measure. Climate intervention measures were still pursued in academic contexts and occasionally addressed in policy documents. Yet, visions of deliberate climate control lost political steam as climate change became politicized as an issue of environmental safeguarding,

[11] This hypothesis had been established by James Croll or Lord Kelvin (cf. Fleming 2010: 5; Weart 2008: 16).

drowning out the techno-optimism that had shaped the period from the Second World War to the war in Vietnam.

It is only at the dawn of the new millennium, that climate engineering entered the mainstream debate on climate change as a potential response measure to an urgent societal problem. In the early 2000s, newly coined metaphors of climatic emergencies marked a growing sense of urgency in tackling this challenge. The 15th Conference of the Parties (COP) in Copenhagen (2009), however, ended in "failure, rancor and disillusionment" (Hulme 2014: viii). Against this backdrop, suggestions to deliberately engineer the Earth's climate regained political momentum—this time as a potential last resort measure in a hopeless situation. In March 2010, the Asilomar International Conference on Climate Intervention Technologies brought together 165 experts from 15 countries and 6 continents.[12] The conference report states that

> "without aggressive pursuit of a multi-faceted global response strategy to limit and then reverse climate change, the environmental consequences will be severe, with concomitant social costs that are likely to lead to widespread suffering in the most affected regions" (ASOC 2010: 7).

Studying climate engineering as one specific translation of the imaginary of climate change proves particularly instructive when analyzed against the backdrop of the positive/negative distinction implied by this framework. In their first definition of sociotechnical imaginaries, Jasanoff and Kim (2009: 129) emphasized the explanatory relevance of "collective visions of the good society." The authors have modified this selective focus on positive imaginaries by adding the possibility of "resistance against the undesirable" displayed in "collectively held and performed visions of desirable futures" (Jasanoff 2015a: 19).

Fundamentally opposed to positively connoted visions of sustainability, the underlying rationales of climate engineering concepts are propelled by discouraging premonitions of catastrophic climate change. Merging these daunting visions with imagining nature as subject to management and control yields climate engineering as a potential option of last resort. In January 2009, *The Independent* ran a front-page headline "Climate Scientists: It's Time for Plan B," reporting that "the latest technology is needed to save the world from dangerous climate change" (Connor and Green 2009). While national governments are only slowly positioning themselves on the matter, resources are being invested in research and development. European countries, for example, have invested considerable amounts of funds, particularly into accompanying (social scientific, ethical, legal) research.[13]

By means of studying climate engineering as a distinct materialization of the imaginary of climate change, we gain new insights on the historical genesis of both

[12] The conference name and location recalled the 1975 Asilomar Conference on Recombinant DNA (Berg et al. 1975).

[13] See, e.g., the DFG-priority program 1689, established to address risks and side effects of possible climate engineering measures (http://www.spp-climate-engineering.de/schwerpunktprogramm.html).

"collective world making" and "collective self-governance." To make sense of the career of climate engineering, we have to consider, first, daunting visions and, second, a global multi-stakeholder perspective rather than national STI policies. Furthermore, we have to examine how this imaginary is explicated in terms of a "grand challenge," which in this case is prominently employed by scholars, engineers, and policy-makers. Instead of striving for a collective vision of a good society, climate engineering is pursued to address a pivotal problem of our time.

11.5 Conclusion

In conclusion, the cases of climate engineering and energy security present instructive and counter-intuitive examples of "collective world making" and "collective self-governance" in *TechnoScienceSociety*. In both these cases, knowledge production and STI policies are heavily dependent on global developments and problem framings, and at the same time, are surprisingly independent from national political strategies or a favorable public discourse. Coming back to our initial hypothesis that sociotechnical imaginaries become explicit as "grand challenges," we can now reflect on how the historical analysis of such translations, concretizations, and explications in turn contributes to refining the sociotechnical imaginaries framework. Our findings indicate that two aspects regarding the theoretical foundations of the sociotechnical imaginaries framework need recalibration. First, the analysis of energy security and climate change as historically contingent "grand challenges" suggests a reconsideration of how *national* and *global* imaginaries are (co-)produced and (co-)stabilized. Second, by focusing the processes in which sociotechnical imaginaries become explicit, we discover a curious oscillation between (positively connoted) *visions* and (negatively connoted) *problems*. The popular notion of "grand challenges" actually captures this ambiguity very well, as those often are depicted as visionary—particularly, as "compelling enough to capture the public imagination" (cf. Hicks 2016)—as well as problematic, that is, calling for immediate political action and a corresponding adjustment of research agendas.

Regarding the first point, the sociotechnical imaginary framework, by definition, refers to state power and national contexts. While this view is fruitful regarding the possibility of cross-national comparison and critique (Jasanoff and Kim 2009: 120), it draws attention away from the plurality and heterogeneity of other kinds of actors and stakeholders. This approach is particularly surprising as Jasanoff and Kim build on the work of Benedict Anderson (2006) and James C. Scott (1998) when developing their framework of sociotechnical imaginaries. Although both Anderson and Scott focus on the emergence of certain *national* imaginations, their argument goes beyond that by identifying a broader mechanism bringing forth these commonly held imaginations. New media technologies, maps, or statistics are all capable of making people feel like belonging to a group they do not actually know. Following Arjun Appadurai (1996), the whole point of modernity is that new kinds of media and technology generate all kinds of contradicting and overlapping belongings.

Given these works on how collective imaginations are formed, the explicit focus on *national* sociotechnical imaginaries seems somewhat out of place. After all, a myriad of global sociotechnical imaginaries cut across national ones—especially in the field of technology. A case in point is scientific or professional organizations that explicitly address global communities. For example, the National Academy of Engineering's "Grand Challenges of Engineering" report speaks explicitly to an imagined global community of engineers by stating that addressing grand challenges means "to help people and the planet thrive" (NAE 2008). As demonstrated in the case study on energy security, for a long time, there has been an intricate relationship between the global and the national: The construction of globally credible scientific knowledge, measures, and models allows for national sociotechnical imaginaries in the first place. Similarly, climate engineering research and development is geared to an international and highly circumscribed community of experts, centering essentially around people rather than around national programs and institutions.

Finally, and taking up the second aspect, the case studies illustrate how studying "grand challenges" as explications of sociotechnical imaginaries helps to understand the ambiguous rationales underlying and guiding *TechnoSciencePolitics*. While energy security is driven by problems of security, equity, and sustainability, research on climate engineering technologies is promoted by the fear of potentially catastrophic global warming. Drawing on Deborah Stone (1989: 299), defining a problem implies constructing it as amenable to human management, pushing it into the realm of human control. This argument comes close to the (more emphatic) meaning of democracy as both "collective world making" and "collective self-governance." In this sense, the discovery of problems and their specification into distinct challenges yields political power. Visions, here, materialize in urgent challenges, and thus enforce technological developments in *TechnoScienceSociety*. While sociotechnical imaginaries primarily adhere to positive visions of the "good society," our case studies give justice to Jasanoff and Kim's recent expansion of the concept, explicitly including "scientific visions of undesirable futures" (Jasanoff 2015b: 291). Following the "evolutionary pattern" (Rochefort and Cobb 1993) of problem discovery, issue transformation, and the final definition of problems sensibly contributes to capturing the highly ambivalent dynamics of underlying sociotechnical imaginaries. Last but not least, the question of what it means to say that (techno)science is "political" becomes less enigmatic. If visions are transformed into problems and if they ultimately determine research agendas, then exploring and tackling them becomes daily business for scientists, engineers, and policy-makers. An important role for STS scholars, therefore, is not to lose sight of the historical processes by which "grand" and "global challenges" have come into being, and to reflect that such explicated challenges are contingent translations of sociotechnical imaginaries.

References

Auswärtiges Amt. 2014. *Der Klimawandel – eine außenpolitische Herausforderung*. http://www.auswaertiges-amt.de/DE/Aussenpolitik/GlobaleFragen/Klima/. Accessed on 22 Aug 2016.

Anderson, B.R.O. 2006. *Imagined Communities: Reflections on the Origin and Spread of Nationalism*. London/New York: Verso.

Appadurai, A. 1996. *Modernity At Large: Cultural Dimensions of Globalization*. Minneapolis: University of Minnesota Press.

Arrhenius, S. 1896. On the Influence of Carbonic Acid in the Air Upon the Temperature of the Ground. *The London, Edinburgh, and Dublin Philosophical Magazine and Journal of Science* 41 (251): 237–276.

———. 1908. *Worlds in the Making: The Evolution of the Universe*. New York/London: Harper & Brothers.

ASOC, Asilomar Scientific Organizing Committee. 2010. *The Asilomar conference recommendations on principles for research into climate engineering techniques*. Washington DC: Climate Institute.

Baker, Z. 2017. Climate State: Science-State Struggles and the Formation of Climate Science in the US from the 1930s to 1960s. *Social Studies of Science* 47 (6): 861–887.

Berg, P., et al. 1975. Summary Statement of the Asilomar Conference on Recombinant DNA Molecules. *Proceedings of the National Academy of Sciences* 72 (6): 1981–1984.

Brailsford, H.N. 1914. *The War of Steel and Gold: A Study of the Armed Peace*. London: Bell.

Brown, M.B. 2015. Politicizing Science: Conceptions of Politics in Science and Technology Studies. *Social Studies of Science* 45 (1): 3–30.

Calvert, J. 2013. Systems Biology, Big Science and Grand Challenges. *BioSocieties* 8 (4): 466–479.

Cherp, A., and J. Jewell. 2011. The Three Perspectives on Energy Security: Intellectual History, Disciplinary Roots and the Potential for Integration. *Current Opinion in Environmental Sustainability* 3 (4): 202–212.

Connor, S., and C. Green. 2009 Climate Scientists: It's Time for 'Plan B". *The Independent,* January 2.

Crutzen, P.J. 2006. Albedo Enhancement by Stratospheric Sulfur Injections: A Contribution to Resolve a Policy Dilemma? *Climatic Change* 77 (3): 211–220.

Daston, L., and P. Galison. 2007. *Objectivity*. New York/Cambridge, MA: Zone Books.

Davis, K.E., B. Kingsbury, and S.E. Merry. 2015. Introduction: The Local-Global Life of Indicators: Law, Power, and Resistance. In *The Quiet Power of Indicators. Measuring Governance, Corruption, and Rule of Law*, ed. S.E. Merry et al., vol. 1-24. New York: Cambridge University Press.

Edwards, P.N. 2010. *A Vast Machine: Computer Models, Climate Data, and the Politics of Global Warming*. Cambridge, MA: MIT Press.

Ekholm, N. 1901. On the Variations of the Climate of the Geological and Historical Past and Their Causes. *Quarterly Journal of the Royal Meteorological Society* 27 (117): 1–62.

Enserink, M. 2003. Bill Gates Plans a Hit List, With NIH's Help. *Science* 299: 641.

Espeland, W.N., and M. Sauder. 2011. Rankings and Reactivity: How Public Measures Recreate Social Worlds. *The American Journal of Sociology* 113 (1): 1–40.

Espeland, W.N., and M.L. Stevens. 2008. A Sociology of Quantification. *European Journal of Social Science* 49 (3): 401–436.

European Commission. 2011. *Communication from the Commission to the European Parliament, the Council, the European Economic and Social Committee and the Committee of the Regions*. Horizon 2020 – The Framework Programme for Research and Innovation. https://ec.europa.eu/digital-single-market/en/news/communication-commission-european-parliament-council-european-economic-and-social-committee-a-0. Accessed on 19 Sept 2017.

Fleming, J.R. 1998. *Historical Perspectives on Climate Change*. Oxford: Oxford University Press.

———. 2010. *Fixing the Sky: The Checkered History of Weather and Climate Control*. New York: Columbia University Press.

Gibbons, M., et al. 1994. *The New Production of Knowledge: The Dynamics of Science and Research in Contemporary Societies*. London: Sage.

Gramelsberger, G., and J. Feichter (Eds.). 2011. *Climate Change and Policy: The Calculability of Climate Change and the Challenge of Uncertainty*. Heidelberg: Springer.

Govindasamy, B., and K. Caldeira. 2000. Geoengineering Earth's Radiation Balance to Mitigate CO_2-Induced Climate Change. *Geophysical Research Letters* 27 (14): 2141–2144.

Hausman, W.J., P. Hertner, and M. Wilkins. 2008. *Global Electrification. Multinational Enterprise and International Finance in the History of Light and Power, 1878–2007*. Cambridge: Cambridge University Press.

Heintz, B. 2007. Zahlen, Wissen, Objektivität: Wissenschaftssoziologische Perspektiven. In *Zahlenwerk*, ed. A. Mennicken and H. Vollmer, 65–85. Wiesbaden: VS Verlag für Sozialwissenschaften.

———. 2011. Numerische Differenz. Überlegungen zu einer Soziologie des (quantitativen) Vergleichs. *Zeitschrift für Soziologie* 39 (3): 162–181.

Hicks, D. 2016. Grand Challenges in US Science Policy Attempt Policy Innovation. *International Journal of Foresight and Innovation Policy* 11 (1/2/3): 22–42.

House, White. 2009. *A Strategy for American Innovation. Driving Towards Sustainable Growth and Quality Jobs*. Washington, DC: US Executive Office of the President, National Economic Council, and Office of Science and Technology Policy.

Hulme, M. 2014. *Can Science Fix Climate Change? A Case Against Climate Engineering*. Chichester: Wiley.

Jasanoff, S., ed. 2004. *States of Knowledge. The Co-Production of Science and Social Order*. London/New York: Routledge.

———. 2015a. Future Imperfect: Science, Technology, and the Imaginations of Modernity. In *Dreamscapes of Modernity: Sociotechnical Imaginaries and the Fabrication of Power*, ed. S. Jasanoff and S.-H. Kim, 1–33. Chicago: University of Chicago Press.

———. 2015b. Imagined and Invented Worlds. In *Dreamscapes of Modernity: Sociotechnical Imaginaries and the Fabrication of Power*, ed. S. Jasanoff and S.-H. Kim, 321–341. Chicago: University of Chicago Press.

Jasanoff, S., and S.-H. Kim. 2009. Containing the Atom: Sociotechnical Imaginaries and Nuclear Power in the United States and South Korea. *Minerva* 47 (2): 119–146.

———. 2013. Sociotechnical Imaginaries and National Energy Policies. *Science as Culture* 22 (2): 189–196.

———, eds. 2015. *Dreamscapes of Modernity: Sociotechnical Imaginaries and the Fabrication of Power*. Chicago: University of Chicago Press.

Jevons, W.S. 1866. *The Coal Question*. London: MacMillan & Co.

Kaldewey, D. 2018. The Grand Challenges Discourse: Transforming Identity Work in Science and Science Policy. *Minerva* 56 (2): 161–182.

Kaldewey, D., D. Russ, and J. Schubert. 2015. Following the Problems. Das Programm der Nachwuchsforschergruppe 'Entdeckung, Erforschung und Bearbeitung gesellschaftlicher Großprobleme. *FIW Working Paper* No. 2.

Keith, D. 2000. Geoengineering the Climate: History and Prospect. *Annual Review of Energy and the Environment* 25 (1): 245–284.

Keith, D., M. Ha-Duong, and J.K. Stolaroff. 2006. Climate Strategy with CO2 Capture from the Air. *Climatic Change* 74 (1–3): 17–45.

Kintisch, E. 2010. *Hack the Planet: Science's Best Hope – Or Worst Nightmare – For Averting Climate Catastrophe*. Hoboken: Wiley.

Krajewski, M. 2006. *Restlosigkeit. Weltprojekte um 1900*. Frankfurt am Main: Fischer.

Lackner, K.S. 2009. Capture of Carbon Dioxide from Ambient Air. *The European Physical Journal Special Topic* 176 (1): 93–106.

Lal, R. 2004. Soil Carbon Sequestration to Mitigate Climate Change. *Geoderma* 123 (1): 1–22.

Latour, B. 1993. *We Have Never Been Modern*. Cambridge, MA: Harvard University Press.

Merry, S.E. 2011. Measuring the World. Indicators, Human Rights, and Global Governance. *Current Anthropology* 5 (33): 83–95.

Mirowski, P. 1989. *More Heat than Light: Economics as Social Physics, Physics as Nature's Economics*. Cambridge/New York: Cambridge University Press.

Mitchell, T. 1998. Fixing the Economy. *Cultural Studies* 12 (1): 82–101.

———. 2002. *Rule of Experts: Egypt, Techno-Politics, Modernity*. Berkeley: University of California Press.

———. 2009. Carbon Democracy. *Economy and Society* 38 (3): 399–432.

———. 2011. *Carbon Democracy: Political Power in the Age of Oil*. London/New York: Verso.

NAE, National Academy of Engineering. 2008. *Grand Challenges for Engineering*. http://www.engineeringchallenges.org/File.aspx?id=11574&v=ba24e2ed. Accessed on 24 Aug 2016.

NRC, National Research Council. 2015. *Climate Intervention: Carbon Dioxide Removal and Reliable Sequestration*. Washington, DC: The National Academies Press.

Parker, G. 2013. *Global Crisis: War, Climate Change and Catastrophe in the Seventeenth Century*. New Haven: Yale University Press.

Porter, T.M. 1994. Making Things Quantitative. *Science in Context* 7 (3): 389–407.

———. 1995. *Trust in Numbers: The Pursuit of Objectivity in Science and Public Life*. Princeton: Princeton University Press.

Rochefort, D.A., and R.W. Cobb. 1993. Problem Definition, Agenda Access, and Policy Choice. *Policy Studies Journal* 21 (1): 56–71.

Royal Society. 2009. *Geoengineering the Climate: Science, Governance and Uncertainty*. London: The Royal Society.

Scott, J.C. 1998. *Seeing Like a State: How Certain Schemes to Improve the Human Condition Have Failed*. New Haven: Yale University Press.

Smith, C. 1998. *The Science of Energy: A Cultural History of Energy Physics in Victorian Britain*. Chicago: University of Chicago Press.

Steinmetz, W., I. Gilcher-Holtey, and H.-G. Haupt, eds. 2013. *Writing Political History Today*. Campus: Frankfurt am Main/New York.

Stone, D.A. 1989. Causal Stories and the Formation of Policy Agendas. *Political Science Quarterly* 104 (2): 281–300.

Tidwell, A.S.D., and J.M. Smith. 2015. Morals, Materials, and Technoscience The Energy Security Imaginary in the United States. *Science Technology & Human Values* 40 (5): 687–711.

Ulnicane, I. 2016. Grand challenges' Concept: A Return of the 'Big Ideas' in Science, Technology, and Innovation Policy? *International Journal of Foresight and Innovation Policyll* 11 (1/2/3): 5–21.

Varmus, H., et al. 2003. Grand Challenges in Global Health. *Science* 302: 398–399.

Weart, S. 2008. *The Discovery of Global Warming*. Cambridge: Harvard University Press.

Weisenstein, D.K., and D. Keith. 2015. Solar Geoengineering Using Solid Aerosol in the Stratosphere. *Atmospheric Chemistry and Physics Discussions* 15 (8): 11799–11851.

WHO, World Health Organization. 2008. *Message from WHO Director-General*. http://www.who.int/world-health-day/dg_message/en/. Accessed on 22 Aug 2016.

Winter, S.J., and B.S. Butler. 2011. Creating Bigger Problems. Grand Challenges as Boundary Objects and the Legitimacy of the Information Systems Field. *Journal of Information Technology* 26: 99–108.

World Energy Council. 2012. *World Energy Trilemma: Time to Get Real – The Case for Sustainable Energy Policy*. London: World Energy Council.

World Power Conference. 1928. *The Transactions of the Fuel Conference of the World Power Conference*. Vol. 1. London: Percy Lund, Humphries & Co.

Yergin, D. 2006. Ensuring Energy Security. *Foreign Affairs* 85 (2): 69–82.

David Kaldewey is Professor for Science Studies and Science Policy at the University of Bonn.

Daniela Russ is a Historical Sociologist at Bielefeld University, currently working on a project "Globalizing Energy: Engineers and the Making of the Global Energy Economy (1880–1990)."

Julia Schubert is a Research Associate in the field of science studies at the University of Bonn, where she has been working on a project "Engineering the Climate: Science, Politics, and the Career of a Plan B."

Chapter 12
TechnoSecuritySociety: Catastrophic Futures, Pre-emptive Security & Mass Surveillance

Michael Nagenborg and Jutta Weber

12.1 From Technopolis to TechnoScienceSociety

"[T]echnology itself is a political phenomenon. A crucial turning point comes when one is able to acknowledge that modern technics ... now legislates the conditions of human experience. New technologies are institutional structures within an evolving constitution that gives shape to a new polity, the technopolis in which we do increasingly live. For the most part, this constitution still evolves with little public scrutiny or debate. Shielded by the conviction that technology is neutral and tool-like a whole new order is built ..." (Winner 1978^2: 323 f.)

In 1977, Langdon Winner sketched a rough picture of society being reconfigured into a technopolis respectively TechnoScienceSociety which is increasingly driven by radical sociotechnical changes and the reconfiguration of science and technology themselves; this conversion of society continues to proceed, even 40 years later, with little public awareness. In 1985, STS scholar Donna Haraway described the new character of an emerging 'high-tech culture' explicitly as "an emerging system of world order analogous in its novelty and scope to that created by industrial capitalism; we are living through a movement from an organic, industrial society to a polymorphous, information system – from all work to all play, a deadly game." (Haraway 1991/1985: 161). In 1987, Bruno Latour popularized the term 'technoscience' to indicate the fusion of science and technology as well as the messy networks linking research and development, industry and society[1].

[1] "I will use the word technoscience from now on, to describe all the elements tied to the scientific contents no matter how dirty unexpected or foreign they seem" (Latour 1987: 174).

M. Nagenborg (✉)
University of Twente, Enschede, Netherlands
e-mail: m.h.nagenborg@utwente.nl

J. Weber
Paderborn University, Paderborn, Germany

S. Maasen et al. (eds.), *TechnoScienceSociety*, Sociology of the Sciences Yearbook 30, https://doi.org/10.1007/978-3-030-43965-1_12

However, until the end of the twentieth century, few scholars in the humanities or social sciences engaged with and analysed the close intermingling of science, technology and society. It was the emerging interdisciplinary field of (cultural studies of) science and technology studies that noted from the early 1980s onwards that (techno)'science *is* culture' (Franklin and McNeil 1991; Haraway 1985a).

Despite manifold ethical, STS or technology assessment discourses, the understanding of technology as a force that formats society, as a decisive life form and as a pervading medium in everyday life is still rarely recognized, as dominant neoliberal thought repeatedly suggests the autonomy and responsibility of the human subject. Only in short historical moments, such as the Fukushima nuclear disaster in March 2011, does a discussion arise on the character of contemporary sociotechnologies.

The disclosure of mass surveillance by data analyst Edward Snowden in 2013 could have served as another starting point to open a broader debate on the nexus of society and technology, especially since Snowden revealed the degree to which state agencies tapped into the constant data streams within the private sector. The possibilities of ubiquitous digital interconnectivity, big data storage, and enhanced data mining tools – a part of the 'polymorphous, information system' (Haraway 1991/1985) – make the political character of technology more than obvious. We are actively involved in many public debates on how to 'tame' data monopolists like Google or control our intelligence agencies but are rarely involved in public debates about digital 'technology as culture',[2] or this phenomenon's intrinsic logic and consequences.

The epochal claim of technoscience includes a profound sociopolitical, economic, and biopolitical reconfiguration of society or a "New World Order" (Haraway 1991/1985) that comes, beneath others, with the dismantling of the welfare state, neoliberalization, 'turbocapitalism', globalization, and the massive deregulation of many societal realms. These developments began in the second half of the twentieth century and are accompanied by profound epistemological and ontological changes as well as the transformation of norms and values in the age of technosciences (including computer science, robotics, biomedicine, nano- and neuroscience). The 'ontological politics' of technoscience(s) as well as knowledge/power relations get thoroughly reconfigured and accompanied by an extensive production of hybrids (Haraway 1991/1985; Latour 1993, etc.). Concepts such as 'nature', 'body', 'technology' or 'subjectivity' gain new meanings. Today, science is no longer a valuable project of knowledge acquisition that is inherent to progress by applying technological insights towards practical solutions but rather an entrepreneurial and pragmatic project wherein technology is the main driver in developing 'innovations' for new markets and (sometimes) specific societal problems. Until the 1980s, theory-based science was understood as the precondition for engineering to turn scientific insights into technical artefacts, to design technical systems and to develop useful

---

[2] See Cultural Studies of Technoscience: i.a. Franklin and McNeal (1991); Haraway (1985a, b, 1997); Reid and Traweek (2000); Suchman (1987).

applications. In the age of technoscience (Weber 2003; Nordmann 2004, 2010; Forman 2007), science and technology have indiscriminately fused, and the dominant narrative is no longer the one of knowledge and progress but rather of application and innovation.

In this chapter, we will demonstrate the need to take this transformation seriously in studying security and surveillance architectures in contemporary (post-)democracies. It is astonishing that few STS scholars have studied the role of security respectively technosecurity (Weber 2011) in our TechnoScienceSociety.[3] How do the digital interception of mass communications, smart CCTV, body scanners, biometric passports, electronic border systems, etc., shape our technopolis? How do these elements reconfigure our understanding of our polity? What do the extensive implementation and ubiquity of surveillance architectures with increasingly opaque and presumably autonomous security technologies mean for democracy?

12.2 A Post-Newtonian Rationality

One of the outstanding features of technosecurity is its 'Post-Newtonian' rationality, its interest in unpredictability, chance, emergence, processes of becoming, and accordingly delimiting forms of behaviour and thought (Hagner and Hörl 2008; Hempel et al. 2010, etc.). This rationality builds on the exploitation of contingent processes to make complex, non-deterministic systems controllable and to find new solutions to problems 'we do not (yet) really understand' (Holland 1992). Thus, contemporary technosecurity practices, such as preventive and pre-emptive risk management, need to be understood in the context of contemporary technoscience, where unpredictability and the unknown are turned into an integral factor of control by the systematized exploitation of processes of trial and error (Weber 2010). Technosciences no longer focus on the intrinsic properties of organisms or the objective description of universal laws. Evolution via tinkering, the processes of trial and error, search heuristics and post-processing have become important tools for constructing complex, dynamic and adaptive systems. This post-Newtonian techno-rationality uses processes of imitation and imagination to resource the unpredictable and to find possibilities for exploiting surplus processes in a technical way (Haraway 1985a, b, 1997; Hayles 2003; Weber 2003, 2010). It is not a mere coincidence that technosecurity follows the same 'Post-Newtonian' rationale.

The background to this development is the feeling of being at risk that has become frequently manifested in Western societies. Ulrich Beck already described in 1986 the rise of the 'risk society' (1986) in which potential threats, such as nuclear disasters or global warming, induced by accelerated technoscientific processes, become no longer calculable, quantifiable and predictable. Anthony Giddens

[3] There are, of course, notable exceptions besides the scholars mentioned in the text, for example, Monahan (2006, 2010). Still, it is fair to claim that (techno-)security remains an underresearched topic in STS.

pointed to the growing preoccupation of Western societies with their future(s), thereby generating the feeling of risk (1999): the future is perceived as catastrophic (Aradau and van Munster 2007; Horn 2014), which might not be too surprising a perception in a world in which societies are becoming progressively heterogeneous while personal and social relations and controls are weakened (Knorr-Cetina 1997). Consequently, risk discourses are increasingly enlarged: "… the structural demand for knowledge relating to risk becomes insatiable. As well because the accumulation of such knowledge adds awareness to new sources of risk, the risk-knowledge process gains its own internal momentum" (O'Malley 1999: 139). Risk discourses not only address health, natural, and technological disasters but also terrorism, organized crime, and illegal immigration. While security was traditionally achieved primarily via the empirical identification and assessment of threats framed by a TBC causal logic (Aradau et al. 2008), it is now reconfigured in the logic of a predictive maximum technosecurity based on a hyper pro-activity. As risks are increasingly experienced as limitless, demands for pre-emptive technosecurity measures are spreading (Aradau and van Munster 2007; Kaufmann 2011): Security has become governed by a precautionary logic warning of the "dangerousness of the future" (Aradau et al. 2008: 148) and by attempts to manage contingency and unpredictability. As risks are increasingly experienced as limitless, demands for pre-emptive technosecurity measures as well as social media networks are spreading (Amoore and De Goede 2008; Grusin 2010; Kaufmann 2011).

12.3 Technosecurity as Culture

Technosecurity is highly technology-oriented and driven by the Post-Newtonian rationality that is genuine to contemporary technoscience culture. Security today is less regarded as a social, political or ecological issue and more conceptualized as maximum security "obtainable by high-tech, total surveillance" (Brown 2006: 24). Technology becomes the silver bullet for security issues (Aas et al. 2009; Marx 2001). Security and surveillance technologies converge and are used to monitor, track, search and profile almost every realm of society – from the economy, politics, and the military to everyday life. CCTV, RFID chips, drones or scanners are used to search for 'terrorists', monitor sport events, grant access to ATMs, and control employees.

Simultaneously, social media has made monitoring an everyday event: millions of people are using Facebook, as well as locative media such as Foursquare, jogging trackers, and apps, to track the cell phone of a partner (Andrejevic 2007; Kaplan 2006). In parallel, huge amounts of data are collected, sorted and processed not only by state authorities (military, police, intelligence agencies) but also by private companies, such as Google and Facebook, for predictive and future analysis. The hope is to gather new information, discover hidden patterns and 'connect the dots' in the hope of pre-empting future political and economic risks. Meanwhile, online users are sharing data, tracking the movements of others, or being tracked themselves.

Some seem to enjoy the comforting gaze of the other or the self-reflective loops of social monitoring. Surveillance technologies are not only operated top-down by state authorities but also used in diverse interactive ways.

Emphasizing the 'globalization of surveillance' in the aftermath of 9/11 and the techno-fetishism of recent military theory, Armand Mattelart coined the term 'techno-security' to signify "(t)he exclusively technological approach to intelligence gathering, at the expense of human intelligence" (Mattelart 2010: 138). We suggest expanding the concept of technosecurity, because Mattelart works primarily with a predominantly top-down, institutional approach with a focus on military technology. Hence, his particular perspective does not allow him to make a stronger connection to everyday practices and thus culture.

In this context, culture in general and technosecurity culture in particular are understood as multifaceted, dynamic sociopolitical practices with a broad variety of agents and actors. Technosecurity culture is conceptualized as a heterogeneous, embodied and complex process in which not only states and other authorities but also software, concepts, machines, and humans participate in the production of meanings, standards, categories and norms. Christopher Daase highlighted the fruitfulness of the concept of "security culture" (Daase 2012). Traditionally used with regard to safety issues in technology assessment, this concept was redefined by him as a framework to understand the reconfiguration of security in the course of the profound sociopolitical changes of the last few decades. Framing security as culture makes it possible to focus not only on institutional actors, such as the military or the police, but also to enable a deeper understanding of how security governs policy and everyday life. We share Daase's interest in everyday culture, but we also need to acknowledge and overcome his neglect of the central role of technology in security discourses and practices. As we noted earlier, it is crucial to analyse technology as an integral part of politics and to understand it as culture. Framing a theory of technosecurity culture will allow us to develop a more encompassing understanding of how security governs our lives.

12.4 The Hyper-Pro-Activity of Security

If we had to single out a decisive characteristic of the understanding of" security" in the age of technoscience that differs from the understanding of "security" in earlier times, the move from reactive to proactive ways of producing security would be a good candidate.

Currently, security is about "doing something" (Molotch 2012). Whenever something bad or evil has occurred, the questions arise, "How could we let this happen? Why didn't we do something to prevent this?" Hence, whoever takes (or has to take) responsibility for the perceived failure of providing and maintaining security has a strong preference to say that she or he actually did *something*. This points to the larger context of a culture in which we find it hard to accept that we are not in control of everything, for example (Capital-N) "Nature" (Böhme 2012; Shaklar 1990).

Earthquakes are the prime example here: Seneca was one of the most prominent authors writing on *securitas* in the ancient world. However, in *Naturales quaestiones* (book 4), he asked, "But if the earth itself stirs up destruction, what refuge or help can we look for?" Ever since the 1755 Lisbon earthquake, Western societies have seemed to be less and less inclined to accept this stoic stance. Since security currently "is about preventing adverse consequences from the intentional and unwarranted actions of others" (Schneier 2003: 11), we are confronted with hyper-pro-activity due to uncertainty about the seemingly unlimited number of potential threats that the future holds.

Technologies play an important part in this line of thinking. They are perceived as the means to address those threats by rendering the future actionable (Anderson 2010a, b) and providing safeguards against potential threats. At the same time, technologies are seen as potential threats themselves, as well as targets that need to be protected. As Langdon Winner (2010) has put it: "The horror of the World Trade Centre attack was that the power of two wonders of modern technology—the skyscraper and the jet airliner—came crashing together causing the carefully contained power of both systems to be released in catastrophic explosion, inferno and collapse." (Winner 2010: 166) However, "... the ultimate fear driving public and private policies in the post–9/11 era, is an awareness that seemingly secure, reliable structures of contemporary civilization are, taken together, an elaborate house of cards. The collapse of the Twin Towers foreshadows other techno-social disasters too numerous to list, and perhaps the collapse of society as a whole, possibilities that now seem to justify the most urgent, ultimately violent measures." (Winner 2010: 167).

While we will not focus on the vulnerabilities created by our dependency on technological systems, it is worth noting that Winner, towards the end of the lengthy quote, engages in what Grusin (2010) refers to as *premediation* by actively contributing to mapping out potential — and in this case undesirable — futures. This premediation is especially prominent when he describes how the airplanes that were crashed into the Twin Towers were flying over his home and speculates about how the pilots could have maximized the damage by aiming for the "nuclear reactors at the Indian Point electrical power plant approximately 60 miles south. Since these facilities were not designed to withstand a direct hit by an airliner, targeting them might have caused catastrophic failure, and possibly a core melt down as the fuel sank into the mud and water of the Hudson River." (Winner 2010: 157). Here, Winner (2010) clearly follows a model that has been a prominent feature of mass media since 9/11: even when confronted with a horrific attack, there is still the tendency to think about even something more horrifying that we need to prevent. One more example of a "journalistic celebration of premediation" from van Goede's paper is as follows: "Imagine your most unthinkable nightmare of the next terrorist attack. Now try to imagine something even worse." (Goede 2008: 156). The role of imagination in both critical and journalistic writings is revealing in that it points to the degree to which future is made present (Anderson 2010a) in contemporary security discourses. However, before we turn towards the link between security and futurity, let us briefly explore the general trend from re-active towards pro-active security.

To a certain degree, security has always been about taking certain measures to prevent future acts. For example, Thomas Hobbes in *De Cive* (first published in 1642) provides his readers with the following list of practices that point to the preventive nature of different security measures: "We see that all commonwealths, even if they are at peace with their neighbours, still defend their borders with garrisons of soldiers, their cities with walls, gates and guards. What would be the point if they had nothing to fear from their neighbours? Even within commonwealths, where there are laws and penalties set against wrongdoers, individual citizens do not travel without a weapon to defend themselves or go to bed without barring their doors against their fellow citizens and even locking their chests and boxes against their servants in the house. "(Hobbes 1998: 10) Countries and towns guard themselves against *potential* attacks, laws are enforced to punish *potential* offenders, and individuals arm themselves and use locks and keys to protect what they regard as their own property against *potential* intruders and perpetrators. In this light, it is tempting to say that security technologies (such as locks and keys) have also been thought of as the means to prevent future acts.

We also need to account for at least two more recent developments in liberal societies: the idea of the state's monopoly on the legitimate use of physical force and the idea of the security of the individual. Both developments lead to a re-active style of policing, where on the one hand the state is thought of as the sole actor who may use force to protect its citizens and their properties, while on the other hand the liberal state has to ensure the individual's security by limiting its own power. A relevant example of this line of reasoning can be found in the ruling of the U.S. Supreme Court in Wheaton v. Peters (33 U.S. 591, 634 (1834)), which states the following: "The defendant asks nothing – wants nothing, but to be let alone until it can be shown that he has violated the rights of others." This is the first time that the "right to be alone" was introduced, and this concept later famously provided the basis for Warren and Brandeis to call for a "right to privacy." (Standler 1997). There is a recognizable practical challenge in upholding the "right to be let alone": after all, how can the state show that the rights of others have been violated, if the state is not allowed to search for the evidence? However, the idea is to limit the power of the state by allowing the state and its agents to act only *after* obtaining evidence of something that already has occurred. The contemporary shift from re-active to pro-active forms of policing is thus a clear departure from the traditional liberal understanding of the state, where the state's power is limited (in theory, at least) to reacting to crimes and offenses already committed.

The break with the traditional liberal understanding of the state is less obvious when we account for the emergence of *social security* in the nineteenth century, which needs to be understood, at least in part, as a *prophylactic* measure to prevent individuals from becoming criminals (and to decrease the risk of social unrest and political opposition). However, Tobias Singelnstein and Peer Stolle noted in their monograph on the "Sicherheitsgesellschaft" (Security Society) (2011) that we need to make a clear distinction between the *prophylactic* approach to crime in the past and the *pre-emptive* approach of today: the former aimed at preventing individuals from becoming criminals, while the latter aims at preventing crimes before they are committed. This pre-emptive approach can be achieved through the use of legal instruments by labelling as criminal activities those that are perceived to be steps taken in preparation for a crime, such as joining certain organizations, travelling to suspicious locations, looking for dangerous information on the Internet, etc. However, technologies also contribute to the shift from re-active to pro-active security.

A prominent example of this shift is *predictive policing*: "The innovative predictive-policing model moves law enforcement from focusing on what occurred to focusing on what will occur and how to effectively deploy resources in front of crime, thereby changing outcomes. With new technology, new business processes, and new algorithms, predictive policing is based on directed, information-based patrol; rapid response, supported by fact-based prepositioning of assets; and proactive, intelligence-based tactics, strategy, and policy. The analytic methods used in the predictive-policing model surface particular times and locations predicted to be associated with an increased likelihood of crime." (Beck and McCue 2009). While it is not surprising that in the current age of digitalization, policing has become data-driven and "intelligence led" (Ratcliffe 2012), we need to be aware of the fundamental shift from re-active to pro-active policing, wherein police officers are assigned to tasks based on the probability of future crimes.

A concrete example of "Predictive Crime Fighting" is IBM's "Blue CRUSH," where "CRUSH" stands for "Criminal Reduction Utilizing Statistical History" (IBM 2011). According to IBM's promotional material, "a predictive model that incorporates fresh crime data from sources that range from the [police department's] records management system to video cameras monitoring events on the street" is at core of this system that allows the department, in the words of the Director of Police Services in Memphis, "to shift officers to a particular ward, on a particular day, right down to the shift level. It's a bit like a chess match and it's enabling us to make arrests we never could have before." (IBM 2011). The reference to playing a game of chess nicely illustrates the role of a playful attitude in providing technosecurity.

It's worth noting that the full name of the IBM system points to the long-standing use of statistics in modern policing. Indeed, as Ian Hacking (1990) has shown in great detail, the emergence of the modern understanding of policing went hand in hand with the rise of statistics in the nineteenth century. However, we also need to note the incorporation of multiple data sources into one unifying platform, which is presented as a tool to enable decision-making based on real time data. Therefore, predictive policing systems also need to be understood as the offspring of the "Semi-Automatic Ground Environment" (SAGE) air defence project and other real-time virtual control systems that were developed in the 1950s. According to Patrick Crogan (2011), "the success (however imagined) of the system rested on the effectiveness of its advance mapping of the real environment's potential eventuality … and the execution of a controlling, pre-emptive gesture based on that mapping." (Crogan 2011: 11). Works like Crogan's historical account on the emergence of technoculture highlight the need to closely study the role of technology in what, at first glance, might be perceived as an epistemological shift in which the border between factual knowledge about past events and the probability of future events has become blurred. Staying with the example of the SAGE system, what we see here is the representation of factual knowledge (the mapping of the environment) and predicted events ("potential eventuality") on one flat screen, where no distinction between the ontological differences (actual/potential) is being made. Thus, we would argue that the rise of the idea of predictive policing does not indicate that police agencies all over the world have suddenly started to believe in predicting the future. Rather, the discussion on the feasibility of predictive policing needs to be taken as an indication of the role of technology in shaping our everyday security practices.

Predictive policing turns into an imminent part of technosecurity culture because not only is it increasingly embedded into policing systems all around the world, from the US to China, but its underlying technologies are also increasingly becoming part of everyday culture. Predictive policing is no longer only a subject of popular culture, as was paradigmatically portrayed in Steven Spielberg's *Minority Report*.

Jordan Crandall (2005) provides us with an elaborated account of the internal logic of media technologies and argues that the ideas of real-time tracking of behaviour, scenario-techniques, and distributed, interactive simulations (of behaviour) are grounded in the cybernetic conception of today's surveillance (and more generally information) technologies. These new technologies of tracking and simulation enable, at least partially, the impression that it is possible to control future actions—or even the future as a whole—via a ubiquitous panoptical system (Crandall 2005).

This new epistemic culture of data-processing, planning and control is no longer based on representation, causality and objectivity (Haraway 1985a, b; Weber 2016) but rather on recombination, correlation and resourcing the unpredictable, thereby using systematized and automatized processes of trial and error and tinkering. One example of this new orientation is data mining, which is central to predictive policing (but also to so-called counter-terrorism operations). Data mining is conducted based on flexible databases with structured and unstructured data that can be searched systematically using advanced data mining algorithms. Huge piles of data are searched and clustered to produce patterns of correlations between data and thus to 'discover knowledge in databases' (Hildebrandt and Gutwirth 2008; Kitchin 2014). The discovery approach does not rest on the idea of correlations based on causal relationships but rather assumes that a (possible) past correlation will appear again at some point in the future. "With smart applications ... the target is to collect and aggregate as much data as possible, in order to mine them for relevant patterns that allow the profiler to anticipate future behaviours. The hiding of data in fact diminishes the (so-called) 'intelligence' of the applications" (Gutwirth and Hildebrandt 2010: 7). Data mining is a tinkering approach that is set in place to produce endless re-/combinations in the hope of pre-empting unwanted future actions. It is only within this epistemic framework that tracking and simulation have become key in our technoscience and technosecurity culture. We experience a shift in focus from the traditional measurement of humans and nature towards the projection and control of behaviour in organic, technical and social systems.

Another kind of technology that embodies the move towards pro-active security are so-called "Smart CCTV systems" that aim to recognize so-called 'abnormal' behaviour. Again, the novel quality of the technology seems to be obvious at first glance. At the same time, the emergence of "Smart CCTV" is unsurprising if we consider "Smart CCTV" to be the logical next step in the development from classic CCTV systems to automated systems, which began with plate recognition systems and led to face recognition systems in the 2000s (Brey 2004; Introna and Wood 2004). Such face recognition systems can be broadly connected to the history and ideology of biometric measurements in the nineteenth century (Lyon 2008; Hacking 1990) as well as the history of biometrical devices developed since the 1950s (Wayman 2007). However, Smart CCTV systems for abnormal behaviour detection no longer operate on a given representation of a specific person but come with the promise of identifying what does not belong and is at odds with everyday activities. They are no longer designed to re-identify and re-discover what is known (e.g., by matching the biometrical data captured by the camera with those stored in a database) but are designed to detect what is not known yet. Smart CCTV systems of this kind are no longer looking for previously registered offenders but are watching out for 'suspicious' behaviour. This phenomenon in turn increases the pressure on everybody under surveillance to behave in 'non-suspicious' ways. This phenomenon might lead into truly Kafkaesque situations because there is no way to distinguish between regular and Smart CCTV systems, no way to know how the data is being processed, and hence, no way to know what kind of behaviour is considered to be 'non-suspicious'.[4]

Both predictive policing and smart CCTV technologies are therefore good examples of the move towards proactive security measures. Additionally, both technologies offer a historical reading that allows them to be framed as the next logical step in an ongoing development. However, we should not allow ourselves to be misled by the continuation of the technological development. We also need to note and analyse the frictions embodied in these technologies.

The necessity of continual analysis also holds true for other security practices, which we may refer to as *anticipatory actions* (Anderson 2010a): "In relation to terrorism, climate change and trans-species epidemics, acting in advance of the future is an integral, yet taken-for-granted, part of liberal-democratic life. … [Bombs] are dropped, birds are tracked, and carbon is traded on the basis of what has not and may never happen: the future" (Anderson 2010a: 777). Anderson (2010a, b) offers two noteworthy examples of how the future is rendered actionable in technoscientific practices: the "Infectious Disease Catastrophe Model" (2010a) and RAND's TableTop strategic games (Anderson 2010b). According to Anderson (2010a), Risk Management Solutions' 'Infectious Disease Catastrophe Model' "generates a stochastic event set of, approximately, 2000 possible pandemics. The

[4]One may take some relief in the fact that the providers of security are very much interested in minimizing the number of false-positives. However, for a person marked as suspicious, the basic uncertainty remains. The individual may even not be aware that she is being confronted by security providers because of an alarm triggered by a technological system.

possible geographies of the pandemics vary from one another on the basis of infectiousness and lethality of a virus, spatial and temporal location of an outbreak, pandemic lifecycle, and countermeasures. Each 'possible pandemic' is generated through standard metrics for counting and tracking the geographies of actually existing pandemics. These include virology, epidemiology, case studies of past epidemics, and diagnostic pandemic surveillance data" (Anderson 2010a: 784).

Again, it is very tempting to focus on historical continuation in the reasoning behind the design and employment of such models, which are currently employed to predict various kinds of "low probability-high impact" events, "including hurricanes, flooding, infectious diseases and terrorism" (Anderson 2010a: 784). After all, epidemics have been studied using probabilistic means since the seventeenth century (Hacking 2006). However, what sets the 'Infectious Disease Catastrophe Model' and other modelling solutions apart from other established forms of risk management (Power 2004) is the shift from *probability* to *possibility*. The models can also be understood as a kind of premediation in that they follow the same logic of imagining something even worse when confronted with already existing threats, only in this case it is numbers being used rather than the text that newspapers use to premediate events and identify the next, even more dangerous threat.

Anderson's second example of how the future is rendered actionable in techno-scientific practices is the tabletop strategic games 'played' in 2004, in which participants acted out their response to the detonation of a nuclear weapon at the Port of Long Beach: "The advent of the event is imagined to disrupt the connections and circulations that make up life in the Los Angeles basin. People die and are injured from flash burns, radiation poisoning, flying debris, and traffic accidents. Infrastructure is damaged, causing power outages, and the loss of telecommunications. Global equity markets plummet. All commercial air traffic is halted. Millions flee the LA basin. What else will occur after the advent of event is, however, unpredictable" (Anderson 2010b: 227). However, the organizers ask, "In the weeks and months after the attack, what would the longer-term economic implications be?" (RAND, quoted by Anderson 2010b: 227). Neither exercises as such these nor the use of tabletop games are new developments. However, in comparison to, for example, the original *Kriegsspiel* used to train Prussian military officers in the nineteenth century, contemporary tabletop exercises serve multiple purposes: they continue to be training sessions and "drills for habituated response" but are also "tests of existing response capability; audits of emergency plans … as well as laboratories to generate knowledge of future disruptive events" (Anderson 2010b: 230). Thus, tabletop exercises also need to be understood as a means of knowledge production. The question is not whether the participants make the correct moves. Rather, the participants' creative engagement in the course of the fictional events contributes to establishing knowledge about a potential event, which is considered to be "unpredictable".

12.5 Technosecurity Technologies

While societies are preoccupied with their possible catastrophic futures and trying to gather knowledge about possible future events to pre-empt and manage possible risks, technosciences, such as computer science and Artificial Intelligence, are inventing ways to exploit the unpredictable and the processes of inventing new solutions to tame the unpredictable.

Accordingly, contemporary security technologies in civil security (as well as warfare discourses and practices) build on systematized practices of imagination, playing/tinkering, and mining of (often highly unlikely) possibilities, practices in which the unpredictable becomes a central resource for the management and control of open, dynamic systems that are made productive in the arenas of policing and killing. The new approach also rests on an ontology that turns the suspect/enemy into a possibilistic system that can—most prominently in the military context—be found, fixed, (finished), exploited and analysed (F3EA) with the help of information superiority based on real-time tracking, data mining, and an omnipotent sensoric control that is tested not only in warfare simulations and games but also in computational counterinsurgency and law enforcement (see Belcher 2013; González 2015; Mayer and Weber under review). This ontology perfectly demonstrates the idea and logic of post-processing and is driven by the hope of "finding, tracking, and targeting virtually in real time any significant element moving on the face of the earth" (Crandall 2005).

At the same time, security, surveillance, and killing technologies configure a "regime of technologically enhanced identification techniques" (Ruppert 2009: 4) and build on data collections of huge populations "not only to monitor certain targets in real time but also to be able to retrace any individual's itinerary of relations if in the meantime this has become of interest" (Chamayou 2015: 7). Ideally, any person in a war/riot/mega-event zone[5] who is categorized as a possibilistic risk (not only a concrete threat) to the dominant order is eliminated, disposed or excluded. Elsewhere, so-called risk populations are identified and registered. Social unrest is hoped to be pre-empted by taking suspects in custody. Technologies of identification, such as data mining, profiling, biometrics, and big data analytics, are governed by a productive technorationality that generates non-representational and non-objective but highly productive knowledge on the basis of imagination (Salter 2008; Amoore and de Goede 2008), speculation and an epistemology of (semi-)automated tinkering. Predictions are produced and profiles constructed, creating a new reality by projecting past 'reality' into the future using parameters and values that prefer specific categories and especially highlight specific correlations and not others.

The precondition for this new knowledge/epistemic regime of information processing, planning and control is not only a new ontology but an epistemological

[5] One example of the application of these techniques would be the arrests of peaceful demonstrators at the Olympic Games in London and their ban from the Olympic arena, which implies a suspension of their civil rights.

shift from the representational (Haraway 1985a, b; Pickering 2002) towards an optimized logic of control that exploits the unpredictable with the help of systematized tinkering. Only then can virtuality, real-time tracking and simulation take over the command. The traditional measurement of humans and nature is substituted by the projection and control of human and non-human behaviour in organic, technological and social systems.

12.6 Conclusion

What seems to be the crucial epistemological and ontological difference between contemporary technosecurity compared to modern forms of security is the focus of the former on the possibilistic. Contemporary technosecurity seeks to premediate any possible future event, even highly unlikely ones, in the fear that they might turn catastrophic. This approach fits into an epistemic regime, a post-Newtonian rationality that builds on the exploitation of the unpredictable, on the unknown, on automatized practices of tinkering, and on mapping our world as completely as possible by collecting and mining any available data.

Our findings demonstrate the need to build a stronger connection between current technosecurity discourses and practices and the phenomena of technosciences at large. To do so means, on the one hand, to avoid focusing on the role of specific technologies within a specific institutional setting alone (as Mattelart's 2010 study on the military did) and to take seriously the role of technology within the contemporary "security culture" (Daase 2012) on the other hand. We need to take seriously the idea of technosecurity *as* Culture. Therefore, our contribution should be understood as an invitation for scholars of techno-science to bring their knowledge and skills to the study of security and surveillance architectures.

References

Aas, K.F., H.O. Gundhus, and H.M. Lomell, eds. 2009. *Technologies of InSecurity: The Surveillance of Everyday Life*. Abingdon: Routledge-Cavendish.

Amoore, L., and M. de Goede, eds. 2008. *Risk and the War on Terror*. New York: Routledge.

Anderson, B. 2010a. Preemption, Precaution, Preparedness: Anticipatory Action and Future Geographies. *Progress in Human Geography* 34 (6): 777–798.

———. 2010b. Security and the Future: Anticipating the Event of Terror. *Geoforum* 41 (2): 227–235.

Andrejevic, M. 2007. *ISpy: Surveillance and Power in the interactive era*. Lawrence: University Press of Kansas.

Aradau, C., and R. Van Munster. 2007. Governing Terrorism Through Risk: Taking Precautions, (Un)Knowing the Future. *European Journal of International Relations* 13 (1): 89–115.

Aradau, C., L. Lobo-Guerrero, and R. van Munster. 2008. Security, Technologies of Risk, and the Political: Guest Editors' Introduction. *Security Dialogue* 39: 147–154.

Beck, U. 1986. *Risikogesellschaft: Auf dem Weg in eine andere Moderne*. Suhrkamp: Frankfurt am Main.
Beck, C., and C. McCue. 2009. Predictive Policing: What Can We Learn from Wal-Mart and Amazon about Fighting Crime in a Recession? *Police Chief* 76 (11): 18.
Belcher, O. 2013. *The Afterlives of Counterinsurgency: Postcolonialism, Military Social Science, and Afghanistan 2006–2012*. PhD thesis. https://circle.ubc.ca/bitstream/handle/2429/45520/ubc_2014_spring_belcher_oliver.pdf?sequence=5. Accessed 30 Dec 2015
Bohme, G. 2012. *Invasive Technification*. Trans. C. Shingleton. London: Bloomsbury.
Brey, P. 2004. Ethical Aspects of Facial Recognition Systems in Public Places. *Journal of Information Communication and Ethics in Society* 2 (2): 97–109.
Brown, F. 2006. Rethinking the role of surveillance studies in the critical political economy of communication. IAMCR Prize in Memory of Dallas W. Smythe. http://www.msu.ac.zw/elearning/material/1330622850Curran%20and%20Gurevitch.pdf. Accessed on 17 Feb 2013.
Chamayou, G. 2015. *A Theory of the Drone*. Trans. Janet Lloyd. New York: New Press.
Crandall, J. 2005. *Operational Media*. http://www.ctheory.net/articles.aspx?id=441. Accessed on 17 Feb 2013.
Crogan, P. 2011. Gameplay Mode. In *War, Simulation, and Technoculture*. Minneapolis/London: University of Minnesota Press.
Daase, C. 2012. Sicherheitskultur als interdisziplinäres Forschungsprogramm. In *Sicherheitskultur. Soziale und Politische Praktiken der Gefahrenabwehr*, ed. C. Daase, P. Offermann, and V. Rauer, 23–44. Frankfurt am Main/New York: Campus.
De Goede, M. 2008. Beyond Risk: Premediation and the post-9/11 Security Imagination. *Security Dialogue* 39 (2–3): 155–176.
Forman, P. 2007. The Primacy of Science in Modernity, of Technology in Postmodernity, and of Ideology in the History of Technology. *History and Technology* 23 (1–2): 1–152.
Franklin, S., and M. McNeil. 1991. Science and technology: Questions for cultural studies and feminism. In *Off-Centre. Feminsm and Cultural Studies*, ed. S. Franklin, C. Lury, and J. Stacey, 129–146. London/New York: HarperCollins.
Giddens, A. 1999. Risk and Responsibility. *Modern Law Review* 62 (1): 1–10.
González, R.J. 2015. Seeing into Hearts and Minds: Part 2. 'Big Data', Algorithms, and Computational Counterinsurgency. *Anthropology Today* 31 (4): 13–18.
Grusin, R. 2010. *Premediation: Affect and Mediality After 9/11*. New York: Palgrave Macmillan.
Gutwirth, S., and M. Hildebrandt. 2010. Some Caveats on Profiling. In *Data Protection in a Profiled World*, ed. S. Gutwirth, Y. Poullet, and P. de Hert, 31–41. Dordrecht: Springer.
Hacking, I. 1990. *The Taming of Chance*. Cambridge: Cambridge University Press.
———. 2006. *The Emergence of Probability*. Cambridge: Cambridge University Press.
Hagner, M., and E. Hörl (eds.), 2008. Die Transformation des Humanen. In *Zur Kulturgeschichte der Kybernetik*. Frankfurt am Main: Suhrkamp.
Haraway, D. 1991/1985. A cyborg manifest: Science, technology, and socialist-feminism in the late twentieth century. In *Simians, Cyborgs, and Women: The Reinvention of Nature*, ed. D. Haraway, 149–182. London: Routledge. First published, Haraway, D. 1985. Manifesto for cyborgs: Science, technology, and socialist feminism in the 1980s. Socialist Review 80:65–108.
Haraway, D. 1985a. Manifesto for Cyborgs: Science, Technology, and Socialist Feminism in the 1980s. *Socialist Review* 80: 65–108.
———. 1985b. A Cyborg Manifest: Science, Technology, and Socialist-Feminism in the Late Twentieth Century. In *Simians, Cyborgs, and Women: the Reinvention of Nature*, ed. D. Haraway, 149–182. London: Routledge.
———. 1997. Modest Witness@Second Millenium. In *Female Man Meets Oncomouse: Feminism and Technoscience*. New York: Routledge Chapman & Hall.
Hayles, N.K. 2003. Computing the Human. In *Turbulente Körper, Soziale Maschinen: Feministische Studien zur Technowissenschaftskultur*, ed. J. Weber and C. Bath, 99–118. Wiesbaden: VS Verlag für Sozialwissenschaften.

Hempel, L., S. Krasman, and U. Bröckling, eds. 2010. *Sichtbarkeitsregime: Überwachung, Sicherheit und Privatheit im 21 Jahrhundert. Leviathan Sonderheft 25/2010*. Wiesbaden: VS Verlag für Sozialwissenschaften.

Hildebrandt, M., and S. Gutwirth. 2008. *Profiling the European Citizen. Cross Disciplinary Perspectives*. Dordrecht: Springer.

Hobbes, T. 1998. *On the Citizen*. Ed. and Trans. R. Tuck, and M. Silverthrone. Cambridge: Cambridge University Press.

Holland, J.H. 1992. Genetic Algorithms Computer Programs That 'Evolve' in Ways That Resemble Natural Selection Can Solve Complex Problems Even Their Creators do Not Fully Understand. *Scientific American* 267: 66–72.

Horn, E. 2014. *Zukunft als Katastrophe*. Frankfurt am Main: Fischer Verlag.

IBM. 2011. *Memphis PD: Keeping Ahead of Criminals by Finding the Hot Spots*. https://www.ibm.com/smarterplanet/us/en/leadership/memphispd/assets/pdf/IBM_MemphisPD.pdf. Accessed on 23 July 2015.

Introna, L.D., and D. Wood. 2004. Picturing Algorithmic Surveillance: The Politics of Facial Recognition Systems. *Surveillance and Society* 2 (2–3): 177–198.

Kaplan, C. 2006. Precision Targets: GPS and the Militarization of U.S. Consumer Identity. *American Quarterly* 58 (3): 693–714.

Kaufmann, S. 2011. Zivile Sicherheit: Vom Aufstieg eines Topos. In *Sichtbarkeitsregime. Überwachung, Sicherheit und Privatheit im 21. Jahrhundert*, ed. L. Hempel, S. Krasmann, and U. Bröckling, 101–123. Wiesbaden: VS Verlag für Sozialwissenschaften.

Kitchin, R. 2014. *The Data Revolution. Big Data, Open Data, Data Infrastructures & Their Consequences*. Los Angeles: Sage.

Knorr-Cetina, K. 1997. Sociality with Objects: Social Relations in Postsocial Knowledge Societies. *Theory, Culture and Society* 14 (4): 1–30.

Latour, B. 1987. *Science in Action*. Milton Keynes: Open University Press.

———. 1993. *We Have Never Been Modern*. Cambridge, MA: Harvard University Press.

Lyon, D. 2008. Biometrics, Identification and Surveillance. *Bioethics* 22 (9): 499–508.

Marx, G.T. 2001. Technology and Social Control: The Search for the Illusive Silver Bullet. *International Encyclopedia of the Social and Behavioral Sciences*. http://web.mit.edu/gtmarx/www/techandsocial.html. Accessed on 17 Feb 2013.

Mattelart, A. 2010. *The Globalization of Surveillance: The Origin of the Securitarian Order*. Cambridge: Polity Press.

Mayer, K., and J. Weber. Under review. From Optimizing Military Operations to Targeting Terrorist Networks: Social Network analysis in Data-Driven Warfare. *Science, Technology and Human Values*.

Molotch, H. 2012. *Against Security: How We Go Wrong at Airports, Subways, and Other Sites of Ambiguous Danger*. Princeton: Princeton University Press.

Monahan, T. 2006. Surveillance and Security. In *Technological Politics and Power in Everyday life*. New York/London: Routledge.

———. 2010. *Surveillance in the Time of Insecurity*. New Brunswick: Rutgers University Press.

Nordmann, A. 2004. Was ist TechnoWissenschaft - Zum Wandel der Wissenschaftskultur am Beispiel von Nanoforschung und Bionik. In *Bionik – Neue Forschungsergebnisse aus Natur-, Ingenieur- und Geisteswissenschaften*, ed. T. Rossmann and C. Tropea. Berlin: Springer.

———. 2010. A Forensics of Wishing: Technology Assessment in the Age of Technoscience. *Poiesis & Praxis* 7 (1–2): 5–15.

O'Malley, P. 1999. Governmentality and the Risk Society. *Economy and Society* 28 (1): 138–148.

Pickering, A. 2002. Cybernetics and the Mangle: Ashby, Beer and Pask. *Social Studies of Science* 32 (3): 413–437.

Power, M. 2004. *The Risk Management of Everything. Rethinking the Politics of Uncertainty*. London: Demos Publications.

Ratcliffe, J.H. 2012. *Intelligence-Led Policing*. London: Routledge.

Reid, R., and S. Traweek, eds. 2000. *Doing Science and Culture. How Cultural and Interdisciplinary Studies Are Changing the Way We Look at Science and Medicine*. New York/London: Routledge.

Ruppert, E.S. 2009. *Number Regimes. From Census to Metrics*. Cresc Working Paper Series [Working paper No. 68]. http://hummedia.manchester.ac.uk/institutes/cresc/workingpapers/wp68.pdf. Accessed on 19 Sept 2017.
Salter, M. 2008. Risk and Imagination in the War on Terror. In *Risk and the War on Terror*, ed. L. Amoore and M. de Goede, 233–246. New York/London: Routledge.
Schneier, B. 2003. *Beyond Fear: Thinking Sensibly about Security in an Uncertain World*. New York: Copernicus Books.
Shklar, J.N. 1990. *The Faces of Injustice*. New Haven: Yale University Press.
Singelnstein, T., and P. Stolle. 2011. *Die Sicherheitsgesellschaft: Soziale Kontrolle im 21. Jahrhundert*. Wiesbaden: VS Verlag.
Standler, R.B. 1997. *Privacy Law in the USA*. http://www.rbs2.com/privacy.htm. Accessed on 30 Dec 2016.
Suchman, L. 1987. *Plans and Situated Actions: The Problem of Human-Machine Communication*. Cambridge, MA: Cambridge University Press.
Wayman, J.L. 2007. The Scientific Development of Biometrics over the Last 40 Years. In *The History of Information Security: A Comprehensive Handbook*, ed. K. de Leeuw and J. Bergstra, 263–276. Amsterdam: Elsevier.
Weber, J. 2003. *Umkämpfte Bedeutungen: Naturkonzepte im Zeitalter der Technoscience*. Frankfurt am Main/New York: Campus.
———. 2010. Making Worlds: Epistemological, Ontological and Political Dimensions of Technoscience. *Poiesis und Praxis* 7 (1–2): 17–36.
———. 2011. Techno-Security, Risk and the Militarization of Everyday Life. In *The Computational Turn: Past, Presents, Futures? Proceedings of the International Association for Computing and Philosophy, MV Wissenschaft*, ed. C. Ess and R. Hagengruber, 193–200. Münster: Aarhus University.
———. 2016. Keep Adding. On Kill Lists, Drone Warfare and the Politics of Databases. *Environment and Planning D: Society and Space* 34 (1): 107–125.
Winner, L. 1978. Autonomous Technology. In *Technics-Out-Of-Control as a Theme in Political Thought*. Cambridge, MA/London: MIT Press.
———. 2010. Trust and Terror: The Vulnerability of Complex Socio-Technical Systems. *Science as Culture* 13 (2): 155–172.

Michael Nagenborg is Assistant Professor for Philosophy of Technology in the Philosophy Department of the University of Twente. His current research focus is the interplay between technologies and the city.

Jutta Weber is a STS Scholar, Philosopher of Technology, and Professor for Media Sociology at the University of Paderborn, Germany. Her research focuses on computational technoscience culture(s), asking how and for whom the nonhuman/human actors work. She has been Visiting Professor at several universities including Uppsala (Sweden), Vienna (Austria), and Twente (the Netherlands). Her recent publications include "Degendering the Driver", a special section with an introduction to "Autonomous Driving and the Transformation of Car Cultures" in a special issue of *Transfers: Interdisciplinary Journal of Mobility Studies* (8(1) 2018, edited together with Fabian Kröger); "Tracking and Targeting: Sociotechnologies of Identification" in a special issue of *Science, Technology, & Human Values* (42 (6) 2017, edited together with Karolina Follis and Lucy Suchman); and "Keep Adding. Kill Lists, Drone Warfare and the Politics of Databases" in *Environment and Planning D: Society and Space* (February 2016 vol. 34 no. 1, 107–125; see also www.juttaweber.eu).

Chapter 13
Mobilizing the Emergence of Phronetic TechnoScienceSocieties: Low-Carbon E-Mobility in China

David Tyfield

13.1 Introduction – 'What Do I Do Next?'

A surge of new terms denotes the emerging birth pangs of a new era. TechnoScienceSociety (TSS), Anthropocene (Pálsson et al. 2013), metamorphosis (Beck 2015), etc.… All hope to capture a significant global discontinuity of unprecedentedly far-reaching *ontological* human-constructed novelty and thus of equally unprecedented responsibility; responsibility for all this innovation and even for "reality itself".

'TechnoScienceSociety' aims to capture something specific about this and regarding the importance of STS, and the new directions STS must take, in responding to this imperative. As I understand it, TSS connotes the condition of intra-active and mutually implicated crises, specifically of each of 'technology', 'science' and (the familiar and conventional mechanisms of public, socio-political and administrative governance that are) 'society' considered *individually* vis-à-vis a world in which they are being brought into multiple and deepening but novel, inseparable and as-yet ungovernable *conjunction* (= 'TSS') while *also* generating novel insoluble problems in the process. Moreover, these crises for each of 'society', 'technology' and 'science' concern both their productivity on their own terms (e.g. for science, of new knowledge) *and* their capacity for problem-solving of contemporary real world issues, which as intrinsic and extrinsic conditions together construct their public legitimacy.[1] This suggests the global turbulence may be understood as a

[1] See chapters by Schmidt (Chap. 3) and Reinhardt (Chap. 4), and Nordmann (Chap. 2), Fisher (Chap. 10) and Maasen (Chap. 5) respectively.

D. Tyfield (✉)
Lancaster Environment Centre, Lancaster University, Bailrigg, Lancaster, UK

Joint Institute for the Environment (JIE), Guangzhou, China
e-mail: d.tyfield@lancaster.ac.uk

S. Maasen et al. (eds.), *TechnoScienceSociety*, Sociology of the Sciences Yearbook 30, https://doi.org/10.1007/978-3-030-43965-1_13

game of 'catch-up' regarding forms of government (in the Foucauldian sense of the conduct of conduct) and politics up to the task of governing and fostering this new world, i.e. of government *by* TechnoScienceSociety *of* TechnoScienceSociety itself. This is thus a challenge of emergence, and one that places 'science' and 'technology' – or better research & innovation, or better still their conjunction as TechnoScienceSociety – at the heart of contemporary issues of government and politics.

But this is not just a matter of redecorating the edifice of our old politics with the new furniture of contemporary ideas and issues. Too much has changed and is changing for that approach to work – and, indeed, not to exacerbate present system failures. For TSS is a new *ontological* predicament and one that is profoundly challenging at present in both the speed and *slowness* of its emergence. In short, there is no shortage of evidence today that TSS is emerging primarily in deeply troubling forms.

"What can be done?" thus becomes the primary issue for TSS, as new practical and political challenge of how to respond, and respond prospectively and constructively, in the face of such irreducible complexity and uncertainty. For without prospect of rational mastery what emerges clearly is the irreducibly practical nature of the imperative, still firmly in place, of knowledgeably guiding such complex systems, a knowing that is now only possible situated *within* such systems and through the active process of interaction with them. In this pragmatised, situated, processual and limited perspective of knowledge, then, we are no longer discussing adequate means for the definitive achievement of clear ends, abandoning also the subject-object dualism underpinning this dichotomy. Rather we have practices of learning that build up deepening capacities of skilful, complexity-attentive judgement, which in turn underpins those practices. 'Transition' and subsequent 'maintenance', e.g. of the government of TSS, are thus seen to be simply different temporal gazes on the same process at different stages of its fruition.

TSS thus spells the wholesale reorganization of the constellation of related concepts that together make up the very common-sense meaning of 'politics' and 'government', not least of which today are knowledge, science, technology and society, all of which are currently founded on essentially literal and dualistic, not pragmatic and processual, grounds. We will argue here that grappling with TSS and a *politics* of TSS demands a shift broadly from a universalist (if possibly critical) epistemo-politics of critique or criticism to an onto-politics of situated practical wisdom (*phronesis*). Important pointers in this direction come from existing work in actor network theory (ANT)-inspired STS. But while useful starting points, this work is also insufficient in some key regards, to the point that we must now move beyond it. In particular, the need for a shift to a situated, processual and practically engaged perspective applies no less to discussions such as this in STS. We here trace out some of the argument for, and consequences of, that move.

The argument proceeds as follows. First we will consider the after-ANT argument of ontological politics and how this demands a break with the familiar epistemological and 'one world' politics still dominating contemporary discourse, mainstream and critical. Then we consider some of the key problems with this conception of ontological politics vis-à-vis the predicament of an emerging TSS. This includes a brief discussion of an alternative perspective, of complex power/knowledge systems (CP/KS) within a phronetic onto-politics. Finally, we illustrate the arguments by analysing, using this CP/KS and onto-political perspective, a key case study of contemporary TSS: the ongoing attempts of innovation towards a transition in urban mobility system in China.

13.2 Ontological Politics and TechnoScienceSociety

The predicament of TSS, and its skilful government, is greatly illuminated by ANT and the post-ANT discussion of ontological politics. ANT provides thick empirical description of heterogeneous assemblages of humans and non-humans, thereby showing how *knowledge* too is an enacted achievement of such processes of effortful construction, not the rational uncovering of *a priori* truths. This is thus a non-teleological and empirically-grounded account of an ontology of multiple concrete actor-networks for which some apparent closure, for instance into what we call 'facts', 'structures' or 'identities', can only be achieved in actual practice, and so is necessarily unknowable abstractly and in advance. Such picture clearly resonates strongly with precisely the political predicament of TSS, regarding both the abstract challenge of complexity and unknowability, and the more concrete one of firmly grasping the extraordinary ontological productivity of the endeavour ANT fuses into 'techno-science'.

But it also follows from this account that there are not just different *perspectives* on a common, singular reality, as is generally supposed, but rather *multiple realities*. This shift not only undermines the clear distinction between knowledge and reality. It also necessarily has profound implications for politics, including the politics of STS, given the presupposed importance of the category of knowledge in contemporary default understandings of both 'politics' and the 'agency' involved in it.

STS started in the 1960s/70s as a form of science and technology criticism (if not critique), following Kuhn's historicizing of the process of scientific knowledge construction together with contemporaneous political and normative concerns about specific scientific developments. This was based on a philosophical triad: ontological realism, epistemic relativism and a conception of critical social science as guidance and illumination to political praxis, holding actual S&T to explanatory and normative account vis-à-vis the (single) reality. Let us call this epistemo-politics: a now *post*-Enlightenment, *post*-positivist politics built on and against the still-dominant common-sense presumptions of reason, reality and progressive social Enlightenment of which science itself is usually such an important standard-bearer.

As the empirical fecundity of less realist approaches became apparent through strong social constructivist and then ANT research, however, this edifice became increasingly shaky. But the major blows to epistemo-politics really come from ('after'-)ANT at the turn of the century onwards; through attention to the growing socio-political centrality and complexity of *political* (and not just scientific) controversies centrally involving S&T issues, and the growing evidence of the inadequacy of the 'many views on one world' approach. Symptomatically, these are inadequacies of explanation only *secondarily*, and primarily inadequacies in politics and practical understanding: e.g. how environmental controversies prove so resilient to 'the facts', this then raising questions regarding '*which* facts?' or rather '*whose* facts?' and thence 'whose *objects* and cosmologies/ontologies?'

This led to formulation of a picture of never-ending jostling amongst multiple lived realities and ontologies of what *is* in the world (Latour 2004), including both human and *non*-human agencies (the latter including both 'nature' and 'technology') and with the latter often vastly outnumbering and/or outweighing the former in any given actor-network; a move that further heightens the relevance of ANT to discussions of TSS given the characteristic proliferation in the latter condition of precisely such human-techno-natural assemblages. This conception is thus clearly set against the more orthodox presumption of political contestation being amongst rationally cognizing (human) agents by means of argument, supposedly moving to rational consensus. These theoretical developments were then propelled by political objection and experience regarding the residual and politically-troubling epistemic superiority implicit in the 'many views, one world' approach of epistemo-politics. The result is 'ontological politics'.

The classic statement of 'ontological politics' is Mol (1999). Tracing ANT's shift in emphasis from analysis of the relation between knowledge/science and reality to that between reality and politics, Mol uses the example of anaemia to show how ANT reveals multiple realities not just multiple perspectives. Anaemia, here, is not a singular reality manifest in three settings ('pathophysiological', 'statistical' and 'clinical') that may sometimes disagree with each other; but actually three different, if overlapping, phenomena that are then constructed, more or less successfully, – and then commonly treated or mistakenly 'known' as – singular. For (after-)ANT, therefore, anaemia is *nothing but* the contingent assemblage of diverse phenomena, human and non-human, technoscientific and somatic, that in any given particular case achieves a diagnosis and subsequent treatment as 'anaemia'; and there are *three* of these realities that go under the same name.

This account thus presents the essential qualitative ontological sameness of 'technology', 'science' and 'society', rather than their *a priori* difference, and hence affords accounts of their mutual and novel implication, conjunction and interaction that speak directly to the challenge of (understanding) TSS. Moreover, regarding a new politics of TSS, then, the most obvious questions that arise concern both the multiplicity and contingency of realities constructed with technoscientific activity. On the one hand, *other* realities could (have) be(en) constructed as dominant instead of those that currently prevail. Different understandings of the world and different socio-political positionings would then be privileged (as 'natural', 'real' and/or 'rational'). On the other, the realities that have actually been constructed thereby contingently enable and disable some of the humans and non-humans all interdependently involved in the construction of that world. While this may not matter particularly regarding anaemia, in the case of infrastructure mega-projects the symmetrical attention to humans and non-humans reveals how rivers and streams, for instance, or the planetary ecology as a whole are recruited and disciplined in such processes, and likely with little attention to their own integrity as agents (Rowland and Passoth 2014). This is thus to highlight crucial political issues for TSS that are otherwise occluded.

But things are then immediately complicated further by ANT. In that any given heterogeneous, enacted assemblage is not the result of any identifiable agential choice. Indeed, even the subjects and their knowledge(s) are as much *results* of this construction. Against a founding premise of epistemo-politics – the pre-eminence of rational agency –, therefore, a world of multiple, possibly novel and/or proliferating technoscience-mediated realities calls for a concept of politics that is substantially alien to incumbent common-sense: not just in terms of being concerned with (as object) the very substance of particular realities, rather than contending perspectives on and explanations of it; but also in terms of it not being primarily conducted through and regarding such cognitive positions (as subject and means). And yet it is recognisably and undeniably politics nonetheless given the ways in which different assemblages privilege and penalize both differentially and contingently.

For Mol, multiple worlds and their inseparable knowledges give rise to an 'ontological politics' of multiple dispersed and empirical recalcitrances, hence one *conducted* ontologically. But also, true to ANT, one without (definitive) agency, directability or direction. And without which, in reflexive coherence, nothing much informative can be said in the abstract. *It all depends* – on what/who/how the specific issue is formed and enacted. Ontological politics thus becomes defined as a process of *interference* of what seem *post hoc* to have emerged as contingent 'choices' of one 'world' over another. This also implies a new and different relation with (what still gets called) 'social science': an 'associational' analysis (Latour 2005), tracing the empirical connections in particular cases. Hence not serving up totalizing and explanatory critique but rather new, detailed, effortfully-constructed descriptions that may *themselves directly* be ontopolitical interferences, opening up given common-senses of what/who matters (ontologically) and does not. This is thus a politics as permanent meta-political process of iconoclasm and stirring up regarding *specific* issues and, in the process, thereby constantly reproblematizing 'one-worldism' and the seemingly resilient human compulsion to treat as singular that which (ANT claims) is multiple.

Finally, then, this is not just a non- (indeed anti-) utopian politics, not motivated by and condensed around a Grand Vision; but also one that does not even *claim or expect*, much less strive for, relatively predictable outcomes from its political engagements. Regarding the key question of TSS regarding prospective practical guidance, therefore, it appears that an ANT ontological politics can offer only limited assistance. At most, this approach affords *post hoc* grounds for political strategic judgement as criticism (not critique) from the thick description of existing closures; and/or a counsel to the epistemically modest goal of a programme of unrelenting Baconian experimental tormenting of given realities-cum-common-senses to *see* what these interferences achieve and to prevent one-worldist common-senses from solidifying in the first place. At the least, however, a post-ANT approach can offer no guidance at all, as in the case of future actor-networks that are, by definition, inaccessible to thick description as not yet actual; or, more abstractly and ethically, in the turn instead to an orientation of care or concern that simply embraces the total bafflement of rational mind implicit in complex, concrete network realities and their ceaseless and unmasterable rising and falling, and so chooses to take an undifferentiated positive affective gaze to all things and eventualities.

Such responses are understandable and not unpersuasive. Certainly, an ANT ontological politics brings out particularly clearly the self-defeating irrelevance of a post-Enlightenment epistemo-politics in that the more such world mastery is attempted the more it simply encounters (and itself elicits) further (possibly novel) recalcitrances, in a treadmill that has, with the emergence of TSS, reached planetary proportions (Latour 2013). Positively, therefore, this approach highlights the need for a paradigm shift in common-sense, abandoning the realist concepts of modernity – not least 'technology', 'science', 'society' – in search of new and empirically-grounded ones of their overlapping construction into manifold techno-science-social worlds.

As such, ontological politics also captures crucial aspects of the ethical/normative challenge of TSS, in that it essentially grasps the extraordinary ontological productivity of contemporary technoscience and the responsibility this brings with it. This includes showing clearly how contemporary default understandings stand in the way of assuming that responsibility, by soothing us with false reassurances about the necessary progressiveness to human mastery implicit in 'science', 'technology' and 'society', there presupposed irreducibly in their very meaning. In short, ANT ontological politics brings to life how these concepts are not only not essentially good, but that they are not even essentially anodyne. To the contrary, they are potentially (and have proved in fact to be) massively destructive, particularly of 'other' humans and non-humans (and their cosmologies), and especially where they are indeed vested with such unreflective 'progressive' and tacitly imperialistic commitments. If this does not strike us into over-awed, ethically reflective silence, what will?

13.3 From Ontological Politics to Phronetic Onto-Politics

This is a compelling new vista of how profoundly 'politics' must be rethought before we can grapple with the multiple realities being continually constructed in the technosciencesociety that ANT also effectively describes. But for all these important ways in which it illuminates the predicament of TSS, ANT ontological politics still does not address the key pragmatic challenge discussed above, which can specified now even more clearly as: if and how it is possible to conceive of politics within an account of reality of complex assemblages of human and non-human agents, and to do so in a forward-looking way that can inform (re-)constructive *practice* in real-time, not just as retrospective accounts in *thought* of what has already happened?[2]

Phrased thus, however, the political priority of TSS emerges as different to that of after-ANT: no longer (just, or even primarily) the problematic proliferation of novel technoscientific-mediated worlds not held sufficiently to account, as specific, manifest issues and their evident contestations and even injustices; but now also the problematic *slowness* of the emergence of *better* such worlds. Meanwhile, the current relentless emergence of TSS presents deepening and accelerating positive feedback loops of innovation focused on facilitating ever-more individualized consumerism, corporate-financial appropriation and corporate-state violence, engendering worsening environmental, geopolitical and socio-economic crises. These dynamics must be actively displaced not merely destroyed. In short, the political problem of the present is as much, if not primarily, one of the urgent need for *construction* of new sociotechnical assemblages. And where these must be *both* successfully expedited (vs. incumbent system lock-ins) *and* somehow qualitatively 'guided' in terms of the specific social futures to which they are giving rise; an intrinsically political question.

To be sure, the tools and insights of an ANT-inspired perspective certainly could, and have, furnished significant insights regarding such a politics (e.g. Passoth and Rowland 2010; Barry 2013). But what we may crudely characterize (with my apologies) as a 'pure' ANT approach tends not to, since – and to the extent that – it is framed as an intellectual endeavour of the observer-analyst. Conversely, we here argue for an approach that builds on the insights of ANT but does so while fully embracing the key challenge of TSS and hence rebases the analysis itself in an explicitly pragmatic, processual and strategic approach that is directly engaged in the problem field that it is studying.[3] This is provided by a turn towards *phronesis* – or situated, strategic and power-aware wisdom – whereupon the ANT ontological politics analysis remains an important step but no longer the only, nor the final, one; and in a process of engagement with TSS that not only illuminates TSS better but also, in doing so, *is* directly the cultivation of the wise government of complex systems that it is seeking to move towards. Instead of wasting the potential contribution of ANT to TSS, therefore, it is harnessed, but only by taking a key step beyond ANT towards its regrounding in an essentially pragmatic, engaged project.

What is meant by 'phronesis'? Phronesis is the third form of knowledge outlined by Aristotle, alongside *techne* and *episteme*. It is a practical wisdom that is irreducibly situated and the basis of skilful judgement relevant to the specific and unique situation in play. As such, it is also always the presupposed basis for the successful deployment of *techne* and *episteme*, as in the tacit skills and embodiments of the master plumber or laboratory scientist. Yet, while the latter two are utterly

[2] I thank one of the anonymous referees for helping with this formulation.

[3] By 'strategic' I mean not means/end calculative reasoning but the more circumscribed and itself processualized, complexified sense of intelligent, productive but limited self-conscious guidance of conduct implicit in Foucault's discussions of power/knowledge and 'government' as the conduct of conduct in a world of uncertainty.

familiar – the work of consultants and 'disinterested' academics respectively –, phronesis has been systematically ignored, or even disparaged, in the modern period. Today, however, in a moment of the exceptional overlapping crises of the emergence of TSS, both of these familiar forms of knowledge are self-evidently inadequate in the face of these wicked problems (as 'technology' and 'science' respectively) yet we have just as much need for knowledgeable guidance as ever, if not more so. What is needed, thus, is precisely such situated, engaged and politically-attentive wisdom as a cultivated, practical virtue – a way of life (Hadot 1995) that also thereby includes an implicit knowledge-attentive ethics that is itself also processual, situated and pragmatic.

To this Aristotelean account work of Flyvbjerg et al. (2013) adds an important Foucauldian twist, following Foucault's later work on power. Here, power is treated not in familiar juridico-discursive terms as that which is held by A over B and is illegitimate until it is rationally consented to. Instead, power is essentially productive, dispersed and normatively ambivalent – dangerous not bad – and constitutive of social phenomena in terms of the power *relations* between people and things, and thereby also of those people and things themselves as *relata*. Moreover, amongst these things are concepts and knowledges, as key mediations of this process of co-production, so that we have now constitutively strategic power/knowledges. Returning to phronesis, then, we can update the still-quite-rationalist Aristotelean account of phronesis to take into account the dynamic, complex constitution of things through power/knowledges. Phronesis is thus not just a superior form of reasoning (itself supreme, as for Aristotle), but also the crucial process through which oneself *and others* may be brought to insightful strategic reflection upon how they are situated, and so constituted, within dynamic power/knowledge relations in real-time.

A post-Foucauldian phronesis thus also yields a framework for analysis of TSS phenomena that is both distinctly resonant with ANT and yet also importantly different. Instead of heterogeneous assemblages of empirical things, we explore reality as composed of complex, dynamic systems of power/knowledge relations and technologies, or complex power/knowledge systems (CP/KS) (see Fig. 13.1) (Tyfield et al. 2015a, b).

This redefines ontological politics as an onto-politics (*Cf* epistemo-politics) in which, following ANT's lead, both ontology and politics are redefined in active negotiation with each other. An ontology of complex, dynamic systems of power/knowledge relations is akin to ANT's in several key respects: (power-)knowledge is still incorporated within and as an achievement of specific enactments and complex, concrete condensations of human-with-non-human things; the contingency and power-distributive effects of specific systems that prevail is preserved; hence there is no all-powerful and unquestioned/able presumption of a 'single reality' that can do all kinds of (usually tacit) political (and epistemic) work, as opposed to the priority of the productive, strategic effects of knowledge claims in possibly constructing systems that tend to *make* their claims 'real'; and thus the agents, knowledges and realities are both irreducibly dynamic achievements *and* strategic or political, in formation and effect.

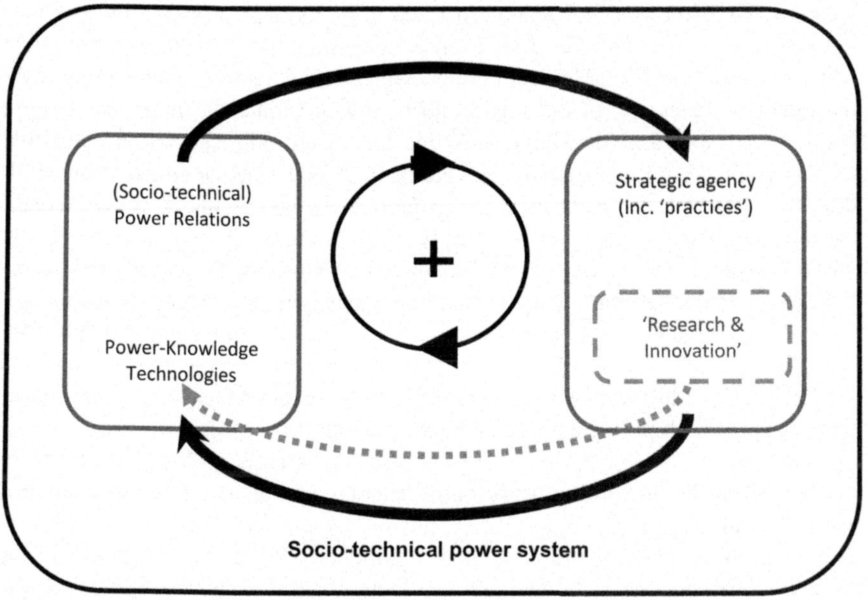

Fig. 13.1 Complex power/knowledge systems

But it is also crucially different too: the systems are emergent, yielding a synchronic emergent powers materialism (Bhaskar 1998) that exceeds a rigorously-policed flat ontology (to the extent it *is* thus policed in ANT studies); so too, the constituent elements of systems are *also* emergent as relata; hence relationality is thorough-going, rather than limited to the relations amongst empirically-given things and so capable of yielding qualitatively new, as-yet-non-actual 'things'; and this relationality is taken as irreducibly strategic, constitutively oriented to its own preservation in any given case. We are thus working with shifting relational assemblages, but as emergent, complex, dynamic systems, not ontologically flat mixtures (nor, of course, realist 'structures').

The mutually compatible combination of an ontology constituted by power/knowledge relations and an analysis that is phronetic in approach and motivation thus completely transforms the purpose of such studies as well. The goal is no longer to analyse existing closures and perhaps be enlightened by their evident (political) contingency, so changing minds (of the analyst/reader) and orientating them instead to concern. But rather it is to examine possibly tendential processes of ongoing attempted closure in order to illuminate *for participating actors* their plausible prospective strategic positioning in the systems taking shape, and its interdependence on positioned others (human and non-human). The latter, however, *is itself* precisely the process of 'wise' government of complex TSS systems, albeit no doubt initially just embryonic.

For this 'onto-politics', therefore, ontology and reality is irreducibly and constitutively political, being *constituted* always in mediation by relations and technologies of power/knowledge. To be is to be strategic. Onto-politics is thus arguably 'onto-' in a more profound sense, concerning the substance of being real *per se*, than is ANT's ontological politics, which seeks instead (only!) to highlight the hitherto neglected but essential contribution of (non-)human agencies to the contingent enabling of some humans. While conversely, politics is reconceptualised as always *primarily* a matter of the rolling (re)production, in strategic material-discursive practice, of systems that are co-produced with specific power/knowledge relations and technologies that *are* 'reality', and the forms of living together – who with whom and on what terms – that they enable and disable. Politics is thus indeed, with ANT, a never-ending process of strategic jockeying (incorporating both competition and collaboration, compulsion and consensus, argument and (possibly everyday) practice in interaction of human and non-human things), but also and primarily of world-*production* conducted in ontic-enacting practice of everyday life, not (just) high-political argument and strategizing.

This thus preserves but also entirely transforms what is 'political' about onto-politics, just as it transforms what is 'onto-' about it, in mutual redefinition. Notably, against ontological politics, 'politics' remains distinctively human and emergent only from the irreducible presence in such human-non-human assemblages of the strategizing, cognizing *zoon politikon* that is the human in pursuit of their specific understanding of a 'better world'; a strategic process (in the deflated Foucauldian sense described above) that is fundamentally about and conducted through the ontologically-constitutive power/knowledge relations of systems-selves. It is onto-*politics* only insofar as it *also* involves the power/knowledge relations that are distinctive of, and constitutive of the emergence of, humans, but whose primary character is their ontological productivity, not argument, contestation and/or coercion. Such jockeying as constitutes ontological politics is more simply accounted for as 'onto*geny*'.

Yet this is also *not* just the traditional notion of politics as a sphere pertaining to humans alone. For it does not just 'pertain' to humans in the sense of 'applying' to humans alone, even as it does presuppose their presence, somewhere, in the assemblage under discussion.[4] Indeed, now constituted through relations with others, human and non-human, and their emergence in turn to greater systems, the concept formally known as 'human agency', as supposedly self-subsistent and answerable only to itself, is seen to be the emergent product of always concrete inter-dependent relations. This is thus to bring human agency profoundly and permanently into question. This question, however, is one specifically for the self-reflective human agent; and one, moreover, that is again answerable only in reflective practice, based in an essentially pragmatic orientation of study of reflective, strategic interaction with the world.

This shift to such a practical, phronetic basis is also arguably immanent in ANT ontological politics itself. For an ANT ontological politics perspective implies the

[4]For instance, note how this onto-politics affords a spectrum of emergent strategicness as differences in degree not kind, e.g. for higher mammals.

inclusion of the (ANT) analyst themselves in both such a world and the uncomfortable predicament of the axiological imperative confronting irreducible uncertainty to the exact same extent that it is itself a compelling analysis. But this immediately subjects the very distinction of analyst-agent to the same process of deconstruction as ANT has achieved for the subject-object dualism more broadly. While this may not entail the total collapse of one onto the other (see below), it certainly does demand the profound reframing of ANT and/or ontological political analysis itself as primarily characterized by precisely the same epistemic shift that it counsels for knowledge and science more broadly, i.e. towards a pragmatic, processual orientation itself within, and so working (messily) with, complex, dynamic unknowable actor-networks.

To admit this essentially pragmatic motivation and drive to, and immanent in, a post-ANT analysis, however, is also thereby dramatically to reframe the entire enterprise, as in the shift to phronetic CP/KS analysis. For it means that the essentially analytical question of ANT regarding "what is 'X'?" – "what actually *is* the manifest human-non-human assemblage and its current (here, now) closure that we unreflectingly refer to as 'X'?" – is neither the first nor the last question implicit in that very train of thought. Rather the motivation for even asking "what is 'X'?" in the first place arises for a (human) agent and from their encounter with a problem in the world. Moreover, this will likely be a difficult problem not answerable through the default first attempt at its resolution, in which that thinking agent identifies the problem as (caused by) 'X' and then proceeds to attempt to deal with 'X' in practice. Of course, this very often works, since our understanding of the world is sufficiently intelligent that we do indeed know the world as it is amenable to our given understanding – an understanding, after all, emergent from and in part constitutive of that world.

But, as ANT would expect, it is also sometimes not the case that this approach works, and especially now regarding issues of TSS, since these are by definition realities constituted in ways that transgress our unreflective default understanding (i.e. 'our' as early twenty-first-century Westerners) in terms of 'science', 'technology' and 'society' (and 'nature'). In these instances, and others besides, therefore we need to ask the question of ANT ontological politics (or CP/KS) and adopt the productive disciplines of its conceptual poverty and empirical 'flat' ontology in order to attempt to construct more insightful conceptual innovations that, hopefully, can 'see' X and its world more clearly.[5]

But what is meant by "more clearly"? How is it judged? The answer, of course, is in a further step beyond in which that revised ontological understanding – concrete and abstract – is *tested* in a new attempt to resolve the practical problem with which we started, to practical standards of satisfaction. Having rethought and reopened 'X', in other words, we are impelled to return to test and see what can now be done. Of course, the strategic predicament of the limited constitutively-situated (human) agent remains unchanged as regards confronting a world of real but

[5] I thank an anonymous referee from the idea of productive "conceptual poverty".

limitlessly unknowable recalcitrances of others. This new post-ANT-analysis (or CP/KS-analysis) understanding, thus, does not reinstate the empty promise of 'now' knowing the world (around 'us') completely and comprehensively, with the concomitant and all-important promise of its potential mastery (also 'for and by us'). But neither are we simply back where we started as regards our total ignorance of the recalcitrance that is currently frustrating us, even as we still interact with the world in significant (indeed, overwhelming) ignorance as to its 'actual' constitution.

Rather the analysis has made two crucial changes, in terms of what we have learned *and* what we are now open to and capable of learning, going forward. Simply going through this thought process we (will) have already learned much in both concrete and abstract about the practical problem field at hand. Concrete insights cannot be specified here, of course, since they pertain only to the specific issue under consideration (though see below for an example). But abstractly we have learnt that the 'objective' reality we are confronting is not, in fact, an 'object' at all – as supposedly fixed, monolithic, self-subsistent and substantial – but is constitutively relational, complex, dynamic and inchoate, as we are ourselves. And so we have also become newly capable of learning – indeed, primed to learn further and be ourselves changed by – how this given reality specifically manifests each of these characteristics, through our ongoing experimental encounters, in practice, with world(s), thereby cultivating a newly informed and other-attentive phronetic stance.

Returning to practice in this final step, however, raises one final problem for ontological politics and the need for its reframing. For, back in an explicitly practice-oriented position ourselves, we find immediately that the opening of 'X', however much it may be insightful in its own right, fails in itself to assist us with deciding on the practical question of "what do I do next?" To answer this question, which presents itself to us interminably (including as the motivation for ANT itself, as we have seen), is immediately and inescapably both: to demand some active reclosure, by that thinking self, of what 'X' is such that 'I' can now interact with it on the new, revised basis; and so, in turn, to adopt a position that, in practice, presupposes both the ontological monism of self and world and hence the epistemic dualism of subject-object. Such closure is thus productive, prospective, active and dangerous, and done by 'me'; not post hoc, already achieved by others (vis-à-vis the analytical gaze) and normatively appraisable as such. It is the latter, however, that can be analysed by ANT and is implicit in its ontological politics. Yet it is the former that is needed to be able to return finally to the practical problem and to its (possible, possibly temporary) closure in turn.

This thus leads to arguably the fundamental philosophical difference between an ontological politics and onto-politics. The watchword of the former is to stand against the cardinal sin of one-worldism, as an error not just of epistemic judgement but of ethical and (ontological) political domination. On this account, one worldism – the positing of the twinned commitments of ontological monism and subject-object epistemic dualism – is precisely and directly to deny the foundational insights of such ontological politics that worlds are multiple and composed of contingent non-human and human (Western and non-Western) assemblages in which the agency of the (Western, male etc.…) human is actually utterly dwarfed by that of the ‘others’ it thereby comes to dominate. One-worldism is thus intrinsically and totally opposed to ontological politics and vice versa.

Similarly, then, ontological *politics* can then be conceptualized in terms of a symmetrical contestation and negotiation between any number or quality of agencies that do not have to – indeed, should not – be accorded the rational, means-end calculative powers that would always privilege the (Western…) human over and to the detriment of all other agents. To the contrary, what is needed, it is insisted, is a new conception of politics, and the agency implicit in it as a ‘politics’, that systematically redresses the imbalance away from the vastly greater but no less ‘real’ or ethically worthy collection of non- or ‘other’ human agencies that one-worldism denies.

To be sure, this is a compelling ethical call and one all the more important to heed carefully amidst the emergence of all the novel human-non-human agencies of global, cosmopolitized, planetary TSS and its problems of emergence. But the problem is that it is only possible to stick righteously attached to this position insofar as one remains in the role of an observer-analyst, which is thereby to reinstantiate the subject-object dualism and hence, not just to not contribute to the practical construction of TSS, but actively to participate in its continuing frustration.

In order to respond to this call, then, we must do so in practice. And this demands resuming a practical, engaged perspective, of ‘our’ agency in the world, situated *within* the complex heterogeneous assemblages and *as* ourselves; i.e. reflective, self-conscious and righteous (Haidt 2015) beings and hence as beings who spontaneously, relentlessly and unstoppably posit ontological monism and subject-object dualism. The option before us, then, is no longer one between retaining a regressive ontological monism or embracing an enlightened ontological politics – and both as explicit, argued position – but of practically deploying ontological monism either unreflectively or reflectively, the latter being in the light of permanently discomforting insights regarding its fundamental groundlessness and normative dangerousness.

But this shift in perspective is much easier to take if the entire thought process has been conceptualized, from the outset, in the fundamentally pragmatic terms of phronesis and CPKS analysis. For, from this perspective, ontological monism is no longer an irreparable and inexcusable mistake (of judgement). Rather it is an inescapable, world-productive and constitutive concomitant of distinctively human, cognizing agency – i.e. that is expressive of the real emergence of the world of humans in the *human*-non-human systems of relations from which that species being itself arises – but also an active ‘doing’ that can nonetheless be done more or less skilfully and with more or less situated strategic wisdom, or phronesis. Again, therefore, we converge finally on the cultivation of phronesis, which becomes both the challenge and its own solution, as in the emergence of (self-)government of TSS.

Finally, then, this shift in perspective then illuminates the significant shift in our understanding of the conjunction of a complex heterogeneous ontology and a politics as practice of possibly *making the world better* that still includes, but has fundamentally resituated, familiar epistemo-politics in ways that can now be pursued in phronetic practice (albeit therefore fallibly and falteringly). An ontological politics attempts to do away with and sideline epistemo-politics by theoretical fiat, arguing that it is mistaken and wrong while supposedly replacing it, even as it affords no alternative from of prospective guidance. By contrast, onto-politics subsumes and reconstructs epistemo-politics as an enduring condition of being human, but one that can be pursued differently and better to the extent it is based in *practices* of phronesis.

To be sure, an ANT ontological politics analysis can help significantly in this process in that, as *itself* a practice, it no doubt can and does cultivate, as embedded intuition (Cf Haidt 2015), the presumption that our ignorance of the world(s) we encounter massively outweighs our knowledge of them, and that those realities are precisely complex, dynamic, relational, systemic, inchoate etc.… This can go towards transforming how it is we interact in and with the world. But these changes come to nought unless and until they are brought back to changes effective in practical, complex-system-situated life and its productive encounters between mutually mysterious selves and worlds, including in STS research. Again, therefore, we see that an ANT ontological politics is only itself brought to fruition in a step beyond itself.

Here, then, the world is still (and inescapably) encountered as an object by a subject, but where our very sense – perhaps only cognitively at first, but possibly more deeply over time – of these terms are themselves loosened, dynamized and brought closer together in ways that actively move away from subject-object dualism and ontological monism as practiced ontological commitments that unceasingly tend to the creation of world in their (split) image. Phronesis and CPKS thus definitely does not simply lead us back to a familiar politics and dualistic human agency even as these remain apparently in place. For the cluster of attendant and constitutively relational (intensional and inter-dependently defined) concepts surrounding that familiar constellation of epistemo-political concepts has been entirely transformed, both in form and in substance, respectively: now as self-consciously constitutively relational, complex, systemic, dynamic and pragmatic in form; and substantively inseparable from novel concepts of phronesis and its cultivation, productive power/knowledge relations and emergent systems thereof, government and its grounding in *self*-government and, of course, TSS.

In short, going beyond ANT ontological politics, we need a programme of *phronetic* analysis of the emerging co-production of technoscience *innovations* and forms of TSS *government*. For this will then move towards a productive onto-politics that can itself enable and instantiate practical (cosmo-political) projects of 'transition', or world creation and reconstruction, and their ongoing wise government.

Returning to STS research, then, this is thus a programme that can *only* be delivered by way of concrete, situated and engaged analysis, not abstract theorizing. To illustrate it at work, therefore, we turn to a key global challenge of emerging TSS: low-carbon urban mobility transition in China. Our question here is: how (when, by what/whom (Savransky 2013)) is urban mobility changing and in co-production with what emerging trajectories of socio-political regimes? To repeat, though, this is addressed with a view to practically informing diverse forms of onto-political strategic redirection, *not* to 'get the answer right'.

13.4 E-Mobility in China: The Politics of Who 'I' (and 'China') May Become

A key contemporary site of the emergence of an onto-political TechnoScienceSociety is low-carbon innovation in urban mobility in China (Tyfield et al. 2015a, b). It is an important case study for many reasons, regarding each of these terms.

First, as 'low-carbon' and 'innovation' it relates directly to contemporary (techno-science-societal) efforts to forge new ecologically sustainable forms of life in response to the multiple environmental global risks reaped by/in the emergence of TSS to date; as well, therefore, as the new predicaments of the Anthropocene, cast in official Chinese policy in the slogan of 'Ecological civilization'. 'Innovation' is also, of course, a fetishized policy buzzword demanding critical attention by STS (Godin 2006). Then as 'urban' it achieves particular significance with the twenty-first century emergence of humanity as a predominately urban species; with cities as key sites of ecological footprint and innovation; and with cities (especially in the 'global South') as key sites of what 'living together' will actually be like for the majority in the twenty-first century. Urban *mobility* is, in turn, a crucial element of all of the three previous terms, as well as a central social practice and socio-technical and political system constitutive of (emerging) techno-science-societies (Urry 2007). Finally, in China we find how each of these, and other major issues besides, attains global superlatives; including as the world's largest (and still rapidly growing) car market, the world's top emitter of GHGs on an absolute basis (with *per capita* emissions also now overtaking those of the EU), and site of the largest and fastest (ongoing) urbanization in history that will greatly shape the global environmental impact of future humanity.

In this context of titanic structures and movements, the decarbonisation and cleaning (and decongestion) of Chinese urban mobility is an increasingly pressing imperative – for Chinese society itself, as well as the planet. As regards the practical question motivating study of this case, therefore, we are concerned with not just expediting but qualitatively shaping 'low-carbon transition' in Chinese urban mobility, such that (techno-science-)social futures they instantiate are not just green (for some) but also equitable, empowering and convivial, i.e. politically enabling for the broadest possible group of human and non-human agents.

Certainly, there can be no doubt that the intense interest in e-mobility in China is profoundly political. For instance, the coming of e-mobility is seen as a unique industrial opportunity by the Chinese government of breaking into the vaunted global citadel of the global automotive majors by 'overtaking around the corner', through taking the lead in the plug-in *electric* car (EV). This is then manifest in perhaps the world's most ambitious national industrial policy, of a technology-push policy orthodoxy, supporting the EV, and applied to an industry that is still substantially state-owned within a country that supposedly has especially strong authoritarian levers for top-down initiatives. Surely, then, China should be a ferment of new EV-based innovation, leading the world, and saving the planet to boot?

In fact, there is no such leadership in evidence. Rather, for all its support by government (across many levels), Chinese mobility today primarily evidences not accelerating low-carbon transition, but a deepening internal combustion engine (ICE) automobility system, at extraordinary pace and scale. Even to the extent EV sales are growing, they remain comparatively minuscule – a sales surge in late 2015 perhaps reaching 1.5% of the comparable annual sales of ICE cars, let alone the total number on Chinese roads and still maybe 25% below government targets (Bloomberg 2015). This certainly augurs no meaningful transition at system level at current rates for many decades. These sales are also utterly dependent on governmental subsidies, the continuation of which has recently been brought into question as widespread fraud has emerged (in turn, also raising questions about how many of the 'record breaking' 2015 sales figures were real EVs (Yang 2016)).

"Why, or rather how, are things working out this way?" thus becomes the tricky ("what is 'X'?") problem of conceptual innovation for a CP/KS analysis. But it is not the only one posed by contemporary low-carbon e-mobility in innovation and mainstream policy understandings. Another, of equal significance, arises from the extraordinary – and starkly contrasting – success of the the e-bike or electric two-wheeler (E2W). Over 200 million of these now ply China's roads (more than *ICE* cars, let alone EVs), the vast majority of which are themselves Chinese brands. This is thus an existing global industrial success story that also has the potential to place Chinese innovation at the global core of a new, affordable, lighter-footprint and 'liveable' model of urban auto-mobility of China's burgeoning mega-cities (and those across the global South more generally). Yet far from embracing this opportunity, we find instead only its containment, if not outright rejection, by government, with E2Ws banned in many big cities and a major, nationally-coordinated clamp-down on them in spring 2016. This, therefore, raises precisely the same question, where the stasis of both the EV and of the E2W contributes to the overall weakness of low-carbon transition more generally.

The central relevance of the foregoing discussion regarding onto-politics comes through in this case study in three key aspects. First, the key point emerging from these two concrete problematics is that while low-carbon transition is fervently desired, what is currently systematically lacking – and missed by policy (analysis) – is a clear and present constellation of agencies that may be identified as the primary producers and beneficiaries of such a transition. In other words, the key question, that can only be answered in concrete (as ANT would counsel), of "*who* and *what* is – or will power – Chinese e-mobility transition?" (Tyfield 2014) is not only generally overlooked but also currently without answer. Yet the practical, strategic predicament compels precisely furnishing some (faltering, incomplete, initial) sense of *which* closure, *for* and *by* whom/what. Through the lens of phronetic CPKS analysis, in short, we see that the fundamental question for expediting low-carbon transition is the fundamentally **onto-political** and prospective one of "*which* qualitatively new, **constitutively strategic** agencies will be co-produced (**will come to exist**) with, and in the process of, e-mobility transition?"[6] This is thus a question that can be addressed neither by a conventional analysis of political manoeuvring and/or transition, working with and assuming only existing agencies, nor an ANT analysis, which makes no such assumption but then struggles to address the prospective question. Rather we must trace embryonic emergent dynamic feedback loops amongst artefacts, situated everyday practices amidst environmental risks, subjectivities and stratifications, cityscapes, regulations and high-level innovation policies and investment strategies in the possible emergence of qualitatively novel complex systems of power/knowledge relations.

Secondly, though, an onto-political analysis is needed because the problem field is not only thereby irreducibly onto-political (i.e. concerning what irreducibly political beings exist) but is also experienced as such in everyday life with an intensity that is uncommon, especially against a presumed Western 'normality'. This has three key aspects. First is the particular lived relevance in everyday life of the question above but translated into the first person as the essentially onto-political question of "Who will I (myself and my country, as Chinese) become?" This question presents itself as banal and everyday amidst a rapidly changing and individualizing Chinese society that is widely appreciated by those living within it to be dynamic, pitiless, still in-formation and littered with intense existential human-non-human-assembled dangers regarding, e.g., food safety and air quality (Yan 2010; Ren 2013; Han and Shim 2010). Here, in other words, it is a common-sense, lived reality that *everything* – one's very future being – is potentially at stake, conditioning a profound competition of social status and equally profound distrust of socio-technical-natural risks.

Moreover, while not, of course, necessarily conversant with the *avant garde* heights of mobilities theory (Urry 2007), mobility choices nonetheless are appreciated to be highly significant in this regard, reflecting how strongly (personal positions within) mobility systems condition and are co-produced with 'who we are' (Kaufmann 2010; Sheller and Urry 2016). Such concerns play out, for instance, in the intense Chinese consumer identity politics of 'car' ownership, with an entrenched common-sense preference for the existential security represented by a private, foreign-built (hence high technical quality), big and expensive (ICE) car: as in the now famous catch-phrase of a young female game-show contestant who would rather 'cry in the back of a BMW, than laugh on the back of a bicycle'.

Yet, and secondly, this is far from being a settled, sedimenting and stable common-sense. Rather contemporary Chinese e-mobility innovation reveals the essential onto-political contestation of a set of mobility systems all of which are currently 'monsters', i.e. precisely politically controversial in their very existence.

[6] Where the terms in bold and underlined respectively go together.

This would include the massively incumbent ICE model just described, since this is palpably and permanently beyond the reach of a huge proportion of the population, and simply has no prospect of attaining to the aspired American levels of penetration without catastrophic planetary (i.e. onto-political) environmental and economic consequences.

Alongside the rapacious monster of late twentieth century ICE auto-mobility, gorging itself not just on fossil fuels, but also land and clean air, however, the monstrousness of the EV and the E2W (both as emergent mobility systems) is even clearer, albeit more 'hopeful' (Cf Mokyr 1991; Law 1991). On the one hand, then, we find what may be informatively caricatured as the revolted rejection of the (systems of the) EV as a hideous parody of the 'real' internal combustion engine (ICE) car; a Frankenstein… in green lipstick. Faced with the deep-seated ontological concerns for one's future security and very identity, the 'monstrous' but still expensive EV offers only inadequate and uncertain (onto-political) boost to one's status to compensate for its new (onto-political) frustrations and risks, as in persistent anxieties about charging and/or being stranded (e.g. on a congested ring-road), financial risks of rapid depreciation and/or personal hazards (e.g. high-profile battery explosions and a widespread fear of 'radiation' from the batteries).

Yet on the other, for all its unquestionable demand, the E2W too is a monster. It inhabits a grey area both legally – in terms of regulations of its use (e.g. licence/registration required?) and production (e.g. standards) –, and regarding (industrial) policy – at best ignored as a potential Chinese global industrial champion, but also, at worst, penalized off the roads. This monstrousness must also be understood in onto-political terms. Crucial here is the larger context of the E2W's production and use. China is a highly dynamic society shaped by two key processes that are together amongst the most illustrative examples of the exceptional conjunction of deepening economic liberalism ('Reform & Opening Up' to globalization) and continued one-party-state government: respectively, migration from the countryside to the burgeoning industries of China's cities and the authoritarian biopolitics of the household registration system (*hukou*), that ties every citizen primarily to one location (classified as 'rural' or 'urban') for access to social services.

Together, this has created two clear and formally-defined tiers of society with only limited social movement between them, but with massive numbers of 'rural' citizens now in cities. This stratification is then performed and reproduced not just in terms of different employment but also – in the full gamut of human-non-human assemblages constitutive of the difference in lives – in dress, gait and language and in urban zoning, standards of housing and conspicuous consumer choices, not least of which is an auto-mobile vehicle: a car (and a 'good' car) being an expensive acquisition out of reach to the majority as against an affordable E2W. The onto-politics of individualised competitiveness regarding identity and possible projected life-course ("who will I become?") thus become particularly heightened, crystallizing in the key symbol of one's access to mobility – both physical and, thence, social (*Cf* Kaufmann 2010).

In this context, then, the proliferation of E2Ws – the growth of this monstrous new enabled agency – is primarily viewed by many, including government at different levels, not as a fantastic opportunity to shape a new, Chinese-dominated and China-appropriate (given the density of cities) e-mobility system, but as an existential 'security threat', with bans of them in several major cities. This takes many forms. At its most mundane it refers to the new menace of silent, quick, dangerous mobility, compounded by the unruliness of its use (on and off pavements, the wrong way down roads, skipping red lights, maybe veering unpredictably across traffic). Undeniably underlying and mediating this security discourse, however, is the imaginary of e-bike users as low 'quality' (*suzhi*) (Anagnost 2004) citizens, vividly and readily exemplified in the transparent division of urban society into residents and migrants ('peasant workers'). While systemically-essential demotic mobility, underpinning China's continuing (if slowing) economic 'miracle', therefore, the e-bike also enables the 'wrong' people regarding the key form of contemporary government that is 'govern-mobility' (Bærenholdt 2013) – the people no-one aspires to become but rather fear (themselves falling back among), especially those constituencies currently most politically enabled. Similarly, against a widespread understanding of car-based automobility as the acme of modernity, from a Chinese state perspective of national industrial profile and urban landscape visible to the imagined rich-world tourist, the e-bike is a symbol of a backward, 'developing' country, not one to be cultivated as the acme of twenty-first century green city living.

The result is the dynamic monster of the E2W (system): unruly, unfamiliar, dangerous and growing, yet also unregulated and unregulatable, let alone amenable to being banned or eliminated. But amid such widespread disquiet, or even governmental rejection, neither does the E2W threaten actual system change in competition with the parallel protean growth of the incumbent ICE car system. In short, as things stand, the answer to the question of "who is (the new agency co-produced with the transition to) e-mobility?" is a set of monsters that are monstrous precisely in the uncertain but potentially existential onto-political threat they pose to existing agencies. To be perfectly clear, therefore, the landscape of Chinese e-mobility transition in 2016 presents *no* clear options that could be embraced by a (more) rational politics (critical and bottom-up and/or managerial and top-down). Instead, as an essentially onto-political arena, the strategic question of "what will ultimately prevail and how?" offers itself to the analysis only in terms of tracing the complex strategic relations and interactions of the present, and the (possibly technological) mediations thereof, and how these could productively play out into the construction of entirely new mobility systems-cum-sociopolitical relations; and with this as lived predicament, not just theoretical conundrum.

This thus points to the third crucial way in which this case is fundamentally onto-political, namely how it is contested in actual practice and is thereby being brought into being. The constraints in contemporary China regarding explicit epistemo-political contestation compound this, though issues of mobility, such as parking or clean air (Geall 2013) are in fact amongst the issues regarding which citizens are freest. But arguably more important are the no less intense, but banal, everyday and material contestation of (yet-to-be-settled) settlements of the power/knowledge relations of the ever-emerging green mobility systems (e.g. practices of human-non-human encounters with charging infrastructures, weathers and clothing, pavements or traffic signals or the movements of other vehicles…).

The use of hundreds of millions of E2Ws every day in such humdrum but strategically productive ways by people *just living their lives* (vs. radical epistemo-politics), however, adds up to the persistent and consequential consternation of the top-down regime, feeding its equally existential determination to shape the very existence of the population – the subjectivities and socio-technical relations constitutive of a future China – in ways that preserve its monopoly of state power. The result, therefore, is the dynamic of contemptuous rejection of the E2W by the party-state, despite its seeming promise from the perspective of both expedited low-carbon transition and even techno-nationalist political economy, *and* its resilient persistence. Conversely, the EV has persisted to date, despite its widespread onto-political rejection and consequent lack of demand, because of the onto-political project it bears for the party-state in the construction of a prosperous, green and globally-competitive … and CCP-led… China.

The key dynamic of contemporary e-mobility innovation, thus, is the dynamic clash between existing (ontopolitical) constituencies (and power/knowledge relations) and the emergent socio-technical monsters they support regarding "who/what 'I' will become" in the context of an *urgence* of rapid and individualizing social and planetary-environmental change, with potentially everything to lose and/or gain; and one in which mobility (physical and socio-economic), and their parallel emergence of socio-technical mobility systems and subjectivities, occupies a central role. To the extent any clear conclusions can be drawn from this constitutively strategic process, therefore, what seems clearest is that both are converging on the increasingly self-conscious and deliberate attempt actively to shape the meaning of the greatest prize available in this contest, namely the very *form and meaning* of the burgeoning Chinese 'middle class'. For it is this still-as-yet-undefined agency in particular that appears to be the answer to the question of "who is Chinese e-mobility transition?" – as the future monster but now *domesticated* (Zuev and Tyfield 2016) – since it is in claiming this label as constitutively connected in everyday common-sense with a specific model of e-mobility that such a model stands to amass the greatest power momentum towards its successful emergence (Tyfield et al. 2015a, b).

As regards some meaningful practical guidance that can arise from this analysis, therefore, it directs us to consideration of the multiple ways in which 'middle class'-ness is being co-produced today with such e-mobility innovation, and the multiple possible (essentially political and strategic) future mobility systems to which this could lead. This is thus to reach the final way in which onto-politics is crucial to this analysis, namely in terms of presenting just such emergent futures, e.g. as a set of plausible scenarios, to multiple agents and stakeholders with whom they are then discussed, refined and republicized to stimulate yet broader strategic – phronetic – reflection. In the case of this particular research, this was done through a bilingual workshop held in Shenzhen in March 2016 to over 40 Chinese stakeholders – from government, business (both 'big' and 'start-up'), NGOs and civil society, and academia (both staff and students).[7]

Of course, this is hardly the only, or necessarily the best, way to do this, though. Rather, the very framing of situated strategic and informative/informed interventions in concrete public TSS issues demands, in itself, forms of situated practical wisdom for judgement regarding what ways present themselves as most promising, given the issue, issue-public and incumbent power/knowledge relations at hand. This may take the form of familiar public, perhaps even polemical, interventions stimulating public discussion about issues hitherto not seen or overlooked. But it may not. And indeed, in the case of low-carbon urban mobility in China, given current limitations on free speech (i.e. epistemo-politics), this was hardly a viable option.

What is clear from this case study, though, is how the lens of ontopolitics illuminates every stage of its analysis, from conception of its motivation to framing of its goals via methodological choice to theoretical/substantive approach and insights. As regards the latter, for instance, this is not just a familiar case of post-socialist identity politics since what it precisely is not is a matter of 'catch-up' to a given and stable socio-political model, this being exactly what is essentially in question and for planetary reasons. Nor it is just a familiar case of authoritarian politics failing to contain a bottom-up overspill and mushrooming appetite for (TSS-enabled realization of) greater individual autonomy, since it is also and essentially a matter of how everyday *onto-political* practices by that very same individualizing but security-thirsty citizenry serve to perform, reinstantiate and possibly transform the power relations of the party-state, as with E2Ws. And nor, of course, is it just a familiar case of authoritarian state-capitalist 'power' getting its wicked way, since the EV is struggling, and for ontopolitical reasons. Rather, it is a case that is amenable only to the onto-political gaze, and thereby, in turn, demanding of insightful phronetic engagement not correct, arms-length analysis.

13.5 Conclusion – STS as Midwife of TSS?

How does this approach furnish (the beginnings of) an answer to what a STS for and in TSS might look like, doing reconstitutive strategic and ethical work? Space forbids all but the briefest and programmatic presentations of an entire research programme. But the outlines are clear. For the example above illuminates a series of existing *problems* but where these are also *openings* and indeed *responsibilities* for STS, as well as providing the capacity to respond to them, regarding both orientation and ways to do so.

[7] These scenarios are also now available at: http://steps-centre.org/2016/blog/four-scenarios-of-future-urban-e-mobility-in-china/ (accessed on September 20, 2017).

First, the meso-level dynamic of the emerging Chinese middle classes and its ongoing active shaping not only highlights likely embryonic trajectories of innovation-cum-socio-political change. But it *also* allows us to place ourselves imaginatively amongst this emergent power-knowledge system and ask about who will be the winners and losers in this emerging TechnoScienceSociety and how. This then allows us (i.e. you, reader) to ask questions about where society may be going *now*, making 'present', as lively political issues, potentially constitutive socio-political divisions of future TechnoScienceSociety that are currently just emergent. And with that comes the responsibility to do something about it. Yet, since what is described are problems of emerging ontopolitics, and the ongoing and uncertain constitution of futures, this also crucially qualifies the register of the analysis in ways that are then potentially empowering: both politically, viz. the possibility of strategic action framed by this possible future and at a time when these (political) realities have not yet taken definitive shape; and epistemically, viz. shedding the impossible burden of getting the (single) future 'right'. An ANT ontological politics, unless based in an engaged programme, tends not to be able to assist with such prospective guidance. A CP/KS phronesis, by contrast, asks: "*Which* 'one world' is being constructed? For and by whom? Could it be otherwise? and, How?" (*Cf* Flyvbjerg et al. 2013). Here, then, STS both forges and grasps a strategic window and opportunity for intervention by way of the power/knowledge technologies that *it itself* is also in the business of producing: such as those that emerge from a CP/KS analysis.

Secondly, though, what is to guide this process of (modest but committed) strategic intervention? The answer lies in the foregoing discussion of complex power/knowledge systems and the vision that emerges from that of the phronetic society: multiple practices of situated practical wisdom, governing the multiple and growing techno-science-social innovations, and thereby generating – through a learning-by-doing, not as trial-and-error but rather as a building-by-exerting-oneself-to-be – strategically wise self-government of the complex systems, thereby *constituting* the emergence of technosciencesociety habitable for humanity. But this is also inseparable from the parallel emergence of its 'critical' (but also constitutive) politics, to which STS is crucial. For here STS itself is self-consciously reframed – from the outset of any specific concrete analysis – as new paradigm of 'knowledge production' as virtue practice of situated strategic wisdom. STS thus becomes a professionalized practice of cultivation of power/knowledge technologies for maximal strategic reflection and ethical reflexivity amongst as broad a human and non-human public as possible on issues of techno-science and innovation and their respective implications for who and what *is* (becoming) and how. In short, as stewards and physicians (perhaps with our own Hippocratic oath) of the complex, equitable and democratic government of technosciencesociety.

In short, if ontological politics comes 'after ANT', then – to answer Mol's (1999: 87) seminal question – perhaps we can say (or hope) that, amidst the turbulent emergence or birth pangs of TSS, phronetic onto-politics comes after that.

References

Anagnost, A. 2004. The Corporeal Politics of Quality (*Suzhi*). *Public Culture* 16: 189–208.

Bærenholdt, J.O. 2013. Governmobility: The Power of Mobilities. *Mobilities* 8 (1): 20–34.

Barry, A. 2013. The Translation Zone: Between Actor-Network Theory and International Relations. *Millennium: Journal of International Studies* 41 (3): 413–429.

Bhaskar, R. 1998. *The Possibility of Naturalism* (2nd edn). London: Routledge.

Beck, U. 2015. Emancipatory Catastrophism: What Does it Mean to Climate Change and Risk Society? *Current Sociology* 63 (1): 75–88.

Bloomberg. 2015. China Makes 100 Billion Yuan Bet on Electric Vehicles. *Automotive News China*. Available at: http://www.autonewschina.com/en/article.asp?id=14169. Accessed on 18 Dec 2015.

Flyvbjerg, B., T. Landmann, and S. Schramm, eds. 2013. *Real Social Science*. Cambridge: Cambridge University Press.

Geall, S., ed. 2013. *China and the Environment*. London: Zed.

Godin, B. 2006. The Linear Model of Innovation: The Historical Construction of an Analytical Framework. *Science, Technology & Human Values* 31 (6): 639–667.

Hadot, P. 1995. *Philosophy as a Way of Life: Spiritual Exercises from Socrates to Foucault*. Oxford: Blackwell.

Haidt, J. 2015. *The Righteous Mind*. London: Penguin.

Han, S.J., and Y.H. Shim. 2010. Redefining Second Modernity for East Asia: A Critical Assessment. *British Journal of Sociology* 61 (3): 465–488.

Kaufmann, V. 2010. *Rethinking the City*. London: Routledge.

Latour, B. 2004. Whose Cosmos, Which Cosmopolitics? Comments on the Peace Terms of Ulrich Beck. *Common Knowledge* 10 (3): 450–462.

———. 2005. *Reassembling the Social*. Oxford: Oxford University Press.

———. 2013. Facing Gaia: Six Lectures on the Political Theology of Nature, the Gifford Lectures on Natural Religion Edinburgh, February 18–28. http://www.bruno-latour.fr/sites/default/files/downloads/GIFFORD-ASSEMBLED.pdf. Accessed 19 Sept 2017.

Law, J., ed. 1991. *A Sociology of Monsters*. London: Routledge.

Mokyr, J. 1991. *The Lever of Riches*. Oxford: Oxford University Press.

Mol, A. 1999. Ontological Politics: A Word and Some Questions. *The Sociological Review* 47 (S1): 74–89.

Pálsson, G., B. Szerszynski, S. Sörlin, J. Marks, B. Avril, C. Crumley, H. Hackmann, et al. 2013. Reconceptualizing the "Anthropos" in the Anthropocene: Integrating the Social Sciences and Humanities in Global Environmental Change Research. *Environmental Science & Policy* 28: 3–13.

Passoth, J.-H., and N.J. Rowland. 2010. Actor-Network State: Integrating Actor-Network Theory and State Theory. *International Sociology* 25 (6): 818–841.

Ren, H. 2013. *The Middle Class in Neoliberal China: Governing Risk, Life-building, and Themed Spaces*. Abingdon/New York: Routledge.

Rowland, N.J., and J.-H. Passoth. 2014. Infrastructure and the State in Science and Technology Studies. *Social Studies of Science* 45 (1): 137–145.

Savransky, M. 2013. Worlds in the Making: Social Sciences and the Ontopolitics of Knowledge. *Postcolonial Studies* 15 (3): 351–368.

Sheller, M., and J. Urry. 2016. Mobilizing the New Mobilities Paradigm. *Applied Mobilities* 1 (1): 10–25.

Tyfield, D. 2014. Putting the Power in 'Socio-Technical Regimes'–E-Mobility Transition in China as Political Process. *Mobilities* 9 (4): 585–603.

Tyfield, D., A. Ely, and S. Geall. 2015a. Low-Carbon Innovation in China: From Overlooked Opportunities and Challenges to Transitions in Power Relations and Practices. *Sustainable Development* 23 (4): 206–216.

Tyfield, D., D. Zuev, P. Li, and J. Urry. 2015b. 'Low Carbon Innovation in Chinese Urban Mobility: Prospects, Politics and Practices', *STEPS Centre Working Paper 71*, Brighton: STEPS Centre.

Urry, J. 2007. *Mobilities*. Cambridge: Polity.

Yan, Y. 2010. The Chinese Path to Individualization. *British Journal of Sociology* 61 (3): 489–512.

Yang, J. 2016. 'How Beijing's EV Plan Came Unglued – And What To Do About It' *Automotive News China*. http://www.autonewschina.com/en/article.asp?id=14289. Accessed 2 Feb 2016.

Zuev, D., and D. Tyfield. 2016. 'Beauty… and/or the Beast: Domesticating the Electric Two-Wheeler', presentation to *Low-Carbon China: Emerging Phenomena and New Questions for Innovation Governance*, 18th/19th January 2016, Tsinghua University, Beijing.

David Tyfield is a Reader in Environmental Innovation and Sociology at the Lancaster Environment Centre, Lancaster University, and Executive Director of the Joint Institute for Environmental Research and Education (JIE), Guangzhou, China. His research focuses on the interaction of political economy, social change, and developments in science, technology, and innovation, with a particular focus on issues of low-carbon transition in China, especially urban e-mobility. He was Principal Investigator for a UK ESRC-funded project "Low Carbon Innovation in China: Prospects, Politics and Practices" (2013–2017) and is Author of *Liberalism 2.0 and the Rise of China: Global Crisis, Innovation and Urban Mobility* (2018, Routledge).

Part IV
TechnoSciencePublics

Chapter 14
Technoscientific Citizenship in Citizen Science. Assembling Crowds for Biomedical Research

Sascha Dickel

What does it mean to be a citizen in a society shaped by technoscience? This question has gained prominence in political theory and the social sciences alike. While many authors have proposed normative frameworks that try to define what "good" (scientific or technological) citizenship should be like (Barney 2007; Feenberg 2011; Frankenfeld 1992; Ribble et al. 2004), some STS scholars and sociologists have taken a reflexive and empirical approach to the problem of citizenship. They ask how citizenship is performed, enacted, and transformed (Bora and Hausendorf 2006; Irwin 2001; Jasanoff 2004; Rose 2000; Felt 2003). In late-modern society, one important feature of enacting citizenship is *participating* in activities that are expected to contribute to the societal good (Jasanoff 2004; Sutter 2004).

"Citizen science" is a contemporary variant of participatory citizenship. The term was coined independently by Alan Irwin (1995) and Rick Bonney (1996). While Irwin used the notion of citizen science to address issues of scientific responsibility in democratic societies, Bonney conceived citizen science as a "practice in which people [...] contribute observations or efforts to the scientific enterprise" (Eitzel et al. 2017: 6). The latter understanding has gained particular prominence in recent years. Under the umbrella term of citizen science, thousands of participatory projects have aspired to translate the work of millions of contributors into scientific outcomes (Bonney et al. 2014). Science policy actors have embraced the term citizen science and integrated it into their open science and public engagement agendas (see, e.g., Open Science Policy Platform 2018). Digital information and communication technologies are the main drivers of this recent boom of participation in research (Franzoni and Sauermann 2014).

Meanwhile, the social sciences have just begun to study the implications of citizen science (Franzoni and Sauermann 2014; Riesch and Potter 2014; Dickel and Franzen 2016; Strasser et al. 2018). Recently, there has been an increase in

S. Dickel (✉)
Institute of Sociology, Johannes Gutenberg University Mainz, Mainz, Germany
e-mail: dickel@uni-mainz.de

S. Maasen et al. (eds.), *TechnoScienceSociety*, Sociology of the Sciences Yearbook 30, https://doi.org/10.1007/978-3-030-43965-1_14

scholarly attention paid to the *terminology* of citizen science: What does the term "citizen" in citizen science actually mean, and how does it differ from other terms that are used to describe the role of contributors, like "volunteer" or "amateur" (Edwards 2013; Eitzel et al. 2017)? Eitzel et al. suggest that "[c]itizen science as participatory data collection [...] has existed for a long time without specific descriptive terminology. Members of the public have collected observations of nature for hundreds of years in such fields as archaeology, astronomy, and natural history" (Eitzel et al. 2017: 6). Might it be that citizen science is therefore just a new term for amateur practices that were always a part of science, but which have been made invisible due to the professionalization of research? Wooley et al. propose that language *matters* and that the term "citizen" in citizen science needs to be carefully considered. The notion of citizenship is always connected to normative understandings of rights and duties as well as to visions of the societal good (Woolley et al. 2016: 13). Hence, citizen science projects address people not only as "epistemic subjects" (Kasperowski and Hillman 2018) but also as *normative subjects.*

Following these reflexive approaches, my chapter investigates citizen science by studying how two biomedical citizen science projects, Cell Slider and EteRNA, address potential participants. How is the "social positioning" (Bora and Hausendorf 2006) of citizens performed in these cases?

In the first section of this chapter, I discuss citizen science as a mode of digital participation and "prosuming". The second section is devoted to the two case studies. The third section relates my findings to a broader discussion of citizen science and participation in contemporary democracies.

14.1 Participation as Digital Prosuming

Citizen science as "crowd science" (Franzoni and Sauermann 2014) is part of a digital transformation of the public sphere. This transformation is frequently linked to ideas of empowerment and democratization. In particular, the emergence of interactive web interfaces – often labeled "Web 2.0" – is conceived as a revolutionary shift towards a new culture of involvement and collaboration (Beer 2009). These visions reproduce a utopia of *ubiquitous participation* that has accompanied almost every introduction of new media since Bertolt Brecht's theory of radio in 1932 (Dickel and Schrape 2017).

Digital participation enables practices of cooperation and collaboration beyond professional and organizational boundaries (Reichwald and Piller 2006; Tapscott and Williams 2007; Benkler 2006). These practices have recently been described as "prosuming" (Hellmann 2010; Toffler 1989), a term originally coined by Alvin Toffler (1989). Toffler predicted a "third wave" of civilization (after the agricultural and industrial revolutions), in which the gap between the spheres of production and consumption would be closed:

> In short, whether we look at self-help movements, do-it-yourself trends, or new production technologies, we find the same shift toward a much closer involvement of the consumer in production. In such a world, conventional distinctions between producer and consumer vanish. The 'outsider' becomes an 'insider' (Toffler 1989: 275).

Although Toffler's civilizational shift has not happened, the notion of the prosumer has become a popular way of describing user activities in digital ecologies (Hellmann 2010). As prosumers, individuals are expected to transcend their roles of users and consumers (Kelly 2005).

Critical studies of participatory web cultures have provided reasons to doubt the "rhetoric of democratization" (Beer 2009: 986) that is prominent in these discourses. It has become apparent that the web can hardly be conceived as a global village where citizens come together to engage in participatory activities. It can rather be understood as a distributed socio-technical apparatus that serves as a platform for various kinds of participatory "social machines" (Shadbolt et al. 2013; Smart and Shadbolt 2014) and complex technological assemblages that rely on the active participation of human agents in technological infrastructures, alongside artificial agents. Examples of such machines are social media platforms (e.g. Facebook, Twitter, and LinkedIn) as well as interfaces that enable peer-to-peer collaboration (e.g. Wikipedia or Linux), crowdworking (e.g. Amazon's Mechanical Turk), or crowdfunding (e.g. Kickstarter). Along with the development of digital technologies of participation, the notion of the prosumer is increasingly being applied to contexts of *knowledge production*. Narratives of the "wisdom of crowds" (Surowiecki 2004) and an accompanying decline in the relevance of exclusive expert knowledge have become commonplace in the discourses of "TechnoScienceSociety" (see the introduction to this volume).

Citizen science extends participatory knowledge production by technological means to the field of science, questioning the distinction between professional scientists and laypeople, and suggesting a new contract between science and the public. Within the 'post-war social contract' (Jasanoff 2004: 227) scientists were often regarded as a distinctive truth class, sharply separated from ordinary citizens. As a professionalized social sphere, science established itself as an exclusive system for knowledge production, offering laypeople and amateur scientists hardly any opportunities to include their non-certified expertise (Collins and Evans 2002). This exclusive social order of science has been questioned in recent decades, leading to intensified forms of science communication as well as exercises in public engagement.

However, while public *participation in decision making*, in the case of critical technological innovations, has become an established feature of contemporary science and technology policy (Bogner 2012; Wehling 2012), public *participation in research* was long considered futile and counterproductive because of the gap between expert and lay knowledge. This view was also endorsed by social scientists who embraced the intuitionalist tradition of science studies: Weingart claims that "[s]cience *for* the people [...] may in a loose sense be realistic; science *by* the people is not" (Weingart 1982: 55). Cases with very specific socio-epistemic conditions like the "construction of lay expertise" (Epstein 1995) in the case of AIDS activism seemed to be the exceptions that prove the rule.

The aim of citizen science is to prove the rule wrong. It does not primarily involve people in the mode of deliberative governance but as contributors to research, often enabled by digital infrastructures and mobile devices. Citizen science is currently gaining a lot of attention from policy makers, scientists, civil society, and the media. The public debate is driven by great expectations: Citizen science is presented as a new mode of participatory knowledge production that might democratize science and constitute a new approach to address society's grand challenges (Prainsack 2014).

Contemporary science policy discourses present citizen science as a tool of knowledge production as well as a tool for increasing scientific literacy. According to the narratives of citizen science, scientific research may become a public matter (again): "What was once a novel idea – lay people engaging in the scientific enterprise – is becoming mainstream" (Bonney et al. 2015: 12). While the scientific enterprise of modernity was inherently linked to the scientific profession of certified experts, the socio-epistemic regime of citizen science aims to open research to non-professionals (Dickel and Franzen 2015) and to blur the boundaries of production and consumption: "By using new digital tools openly, we are changing the traditional research culture: citizens are now both producers and consumers of the scientific outcomes required to address today's global challenges" (Serrano 2013).

This quote from Serrano, a coordinator of an EU funded project on citizen science, shows how citizen science reproduces the discourse of "prosuming" that is so prevalent in contemporary society. However, if citizen science is indeed just another case of digital "prosuming" – this time within the sphere of research – why is it not just called volunteer science or amateur science? In which ways are people *positioned as citizens* in citizen science and how does this relate to the design of the digital interfaces that are installed to assemble large crowds of participants?

14.2 The Role of Citizens in Crowdsourced Biomedical Research

To find preliminary answers to these questions, I investigate two cases of citizen science, the biomedical projects Cell Slider and EteRNA. Cell Slider was launched in 2012 by Cancer Research UK, a cancer research and awareness organization. It ended 2015. It encouraged participants to take part in cancer research by analyzing pictures from breast cancer trials. Cell Slider was part of Zooniverse, one of the most important platforms for digital citizen science. Zooniverse grew from the popular crowd science project Galaxy Zoo, a project devoted to classifying galaxies by means of digital mass participation. Today, Zooniverse hosts more than 100 projects, "ranging from the humanities and social sciences to physics and astronomy"

(Kasperowski and Hillman 2018: 8). Several scientists and developers from different research institutions administer the platform. Zooniverse describes itself as

> the world's largest and most popular platform for people-powered research. This research is made possible by volunteers – hundreds of thousands of people around the world who come together to assist professional researchers. Our goal is to enable research that would not be possible, or practical, otherwise. Zooniverse research results in new discoveries, datasets useful to the wider research community, and many publications. At the Zooniverse, anyone can be a researcher. You don't need any specialised background, training, or expertise to participate in any Zooniverse projects. We make it easy for anyone to contribute to real academic research, on their own computer, at their own convenience (Zooniverse 2018).

Cell Slider embraced this idea from Zooniverse that no particular expertise was needed to work as a citizen scientist. Thus, Cell Slider was a paradigmatic example of "involvement made easy" (Marres 2012: 62). Once a user had joined the Cell Slider crowd, his or her job was to classify images from breast cancer trials. On each picture, users targeted cancer cells, which differed in size and shape from blood and tissue cells. Each analysis consisted of a sequence of predefined steps. When a new picture appeared, users needed to decide which types of cells were shown in the picture. Afterwards, they counted the irregular (cancer) cells and estimated the percentage of stained yellow cancer cells in the given sample, as well as the brightness of the cell core.

EteRNA is an ongoing citizen science project led by scientists at Carnegie Mellon University and Stanford University that takes the form of an online game. EteRNA asks users to solve increasingly complex problems related to the folding of RNA molecules. In doing so, the project "crowdsources the computationally challenging tasks of RNA design and structure prediction" (EteRNA Wiki 2014). EteRNA presents itself as a series of puzzles in which users design RNA molecules in an online gaming environment. Before users engage in research, they unlock several "tools" in a tutorial, which takes a lot of time. EteRNA's puzzles become more and more challenging. After the initial stage of the game, participants may take part in so-called "labs". In these "labs", players submit their designs to the Stanford researchers, who manufacture and test the virtual molecules. EteRNA also allows users to participate in chats to discuss both their designs and the further development of the project itself. EteRNA also offers them the possibility of contributing to academic papers. In recent years, several articles have been released in which EteRNA players appear as authors (Lee et al. 2014; Anderson-Lee et al. 2016; Wellington-Oguri et al. 2018).

Cell Slider and EteRNA are examples of crowdsourcing. Crowdsourcing can be defined as an open call to a mass of individuals who are asked to perform a task that would normally be performed by employees (Howe 2010). Crowdsourcing is usually enabled by online platforms and networks, which allow large groups of potential volunteers to participate. The use of online tools makes participation possible for individuals in different places, at different times, and with different backgrounds. Digital citizen science uses crowdsourcing technologies to implement a new division of scientific labor, which has been labeled "crowd science" (Franzoni and

Sauermann 2014). Assembled crowds are expected to work towards a certain goal: in the case of Cell Slider and EteRNA, to contribute to biomedical research.

The crowdsourcing approach of the two projects offers people the option to participate while sitting in front of a computer screen. Both use crowdsourcing for biomedical research. In addition to these similarities, the projects differ in several aspects. Cell Slider is an already finished project, whereas EteRNA is still ongoing. The tasks that individuals had to fulfill by taking part in Cell Slider were not very complex and quite limited. The responsibilities of EteRNA users are considerably more challenging and open-ended. I will show that despite these differences, both projects reconfigure participation and citizenship in a similar way: as *responsible* and *functional participation*.[1]

14.2.1 How Cell Slider Addresses Potential Participants

When you visited the Cell Slider homepage (Cancer Research UK 2015a) you were greeted by the following welcome address (Fig. 14.1):

Let us consider this message more closely. The homepage welcomes users with a cry for help: "Research is killing cancer, but we need your help!" This appeal addresses the user as an actor responsible for the acceleration of scientific progress. It depicts science itself as a subject in a war against a dire enemy that must be "killed". The legitimacy of the act of killing is not put into question. The weapon used to kill the enemy is "research". In this appeal, science does not enter the scene as basic research. It is not framed as an endeavor to test theories or to understand unknown phenomena, but as a political agent in a deadly conflict.

"Research", so we learn, is currently *in the process* of killing its enemy, cancer. Note that it is not questioned *if* research will be successful. It all seems to be a

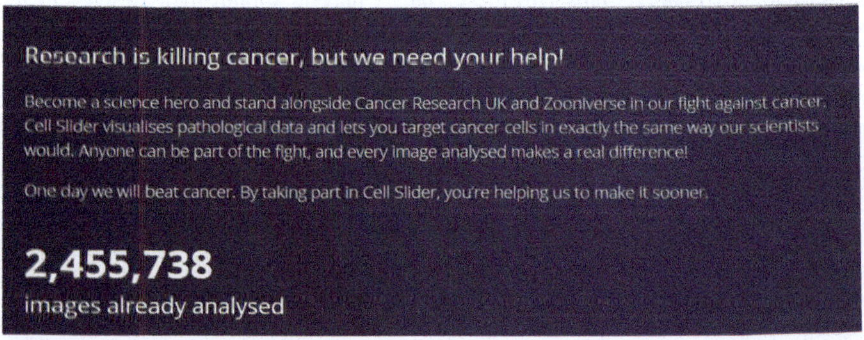

Fig. 14.1 Cell Slider, welcome message

[1]For the methodological background of the following reconstructive analysis, see Bora and Hausendorf 2006 and Wenninger 2015.

question of *when*. The final result of the fight is fixed in advance, suggesting a teleological structure of the conflict: it is the *fate* of research to kill cancer. This assumed teleology weaves the participation of individuals into a fateful chain of events. In order for "research" to fulfill its destiny, *you* must join the fight. The reader of this message is addressed like a soldier who must help a personalized abstract entity (usually, in cases of war: the nation, here: research) to fight a common enemy who needs to be killed in the context of a just and fateful war.

The next lines of the welcome address confirm this interpretation: "Become a science hero and stand alongside Cancer Research UK and Zooniverse in our fight against cancer". The language of war and fate persists. The website suggests the possibility of a profound transformation of the user: the transformation into a "science hero". Heroes are agents of fate, destined to perform extraordinary deeds in pursuit of a greater good. It is their action (or: inaction) that makes all the difference in a conflict or quest. The "we" in the cry for help is also specified in the quoted sentence above. "We" refers to Cancer Research UK and Zooniverse. These organizations are already involved in the fight against cancer, but they seem to need the hero to bring the battle to its destined conclusion. By positioning the user as a hero in a fight against a universally accepted evil force, he or she is made personally responsible in an emphatic way. The welcome address tells the user that his or her heroic participation is needed so that research can fulfill its destiny to kill cancer.

But how exactly can the user participate in this fight? What is the hero's actual role? The following phrase answers these questions: "Cell Slider visualizes pathological data and lets you target cancer cells in exactly the same way our scientists would". Cell Slider is now itself positioned as an actor – a technological actor that empowers users. It allows them to master a specific task "*in exactly the same way our scientists would*". Cell Slider does not promise users they will become full scientists themselves, but that they – together with the Cell Slider technology – will be able to act like one in a clearly defined area: targeting cancer cells.

We now also gain a clearer picture of the heroism promised. Heroes perform extraordinary deeds. However, what is extraordinary in targeting cancer cells? One would expect that heroes can do deeds that regular soldiers cannot, but this is not implied here. Instead the task of the user is toimitate the practice of actors who are already involved in the fight against cancer. The key in understanding this communicative offer lies in a relational interpretation of the *extraordinary*. Extraordinary events take place outside the everyday practice of subjects. The heroism that users are promised is that they are leaving the world of their everyday life behind to enter the practice of the research. This practice, which for professional scientists is not extraordinary at all, is extraordinary only for laypeople. Thus, the user is not a hero compared to a professional scientist, but compared to all those who do not accept Cell Slider's offer and stay inside the boundaries of their life worlds as laypersons.

The technology of the website asks users to imitate the practice of professional scientists. In doing so, they could do something that is extraordinary for laypeople: taking personal responsibility in the epic fight against cancer. If they refrain from doing so, however, if they reject the role of the hero, they are also responsible for the persistence of the evil force of cancer, because "[a]nyone can be part of the fight".

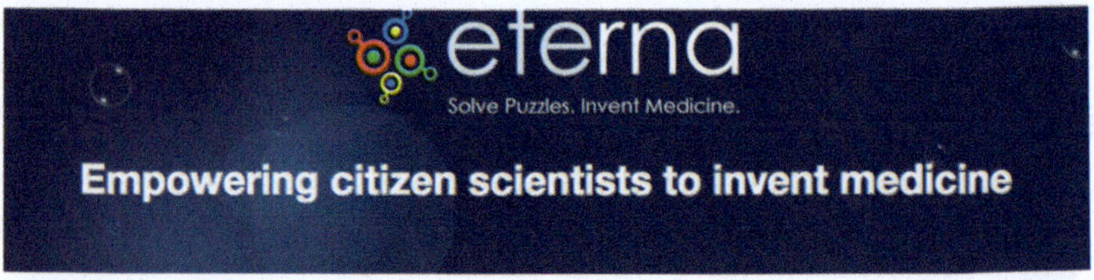

Fig. 14.2 ETERNA, home

This responsibilization is reinforced by the easiness of the task. Anyone can do it. Every single instance of participation matters, because "every imaged analyzed makes a real difference". Technology makes it possible to be a hero without any heroic effort – so why don't *you* help? It is implied that research would be slower (in killing cancer) without the participation of citizen scientists. The last sentence of the welcome message reads as follows: "One day we will beat cancer. By taking part in Cell Slider you're helping us to make it sooner". Any unnecessary cancer-related death is now the personal responsibility of the reader.

14.2.2 How EteRNA Addresses Potential Participants

The project EteRNA reveals a different discourse[2] that does not rely so much on a language of war and heroism, but which is characterized by "playful hyper- responsibility" (Andersen and Knudsen 2016). The project homepage (EteRNA 2018) defines this ambivalent social positioning. Let us first look at the website's title (Fig. 14.2).

Beneath the name of the project, we find a brief description: "Solve Puzzles. Invent Medicine". One the one hand, solving puzzles is an intellectual activity, fitting for the domain of science. One the other, it is also a playful activity that one might perform in the context of a game. It can be inferred that this ambivalence is constitutive for gamified citizen science projects like EteRNA. The description suggests that solving puzzles is an undertaking that leads to the invention of medicines. This not only implies a linear model of innovation (from research to applications) but also suggests that the invention of medicine can be achieved by doing actual research *in the mode of playful activities*.

The slogan "Empowering citizen scientists to invent medicine" further reveals the suggested relationship between the users and the game itself. It is noticeable that users are already referred to as "citizen scientists" before their participation has even begun. According to the project's slogan, they will not *become* "citizen scientists" *after* participating in the project, but they are *already* citizen scientists themselves, even if they have not described themselves as such beforehand. EteRNA

[2] The following analysis was performed together with Andreas Wenninger and Marcel Woznica.

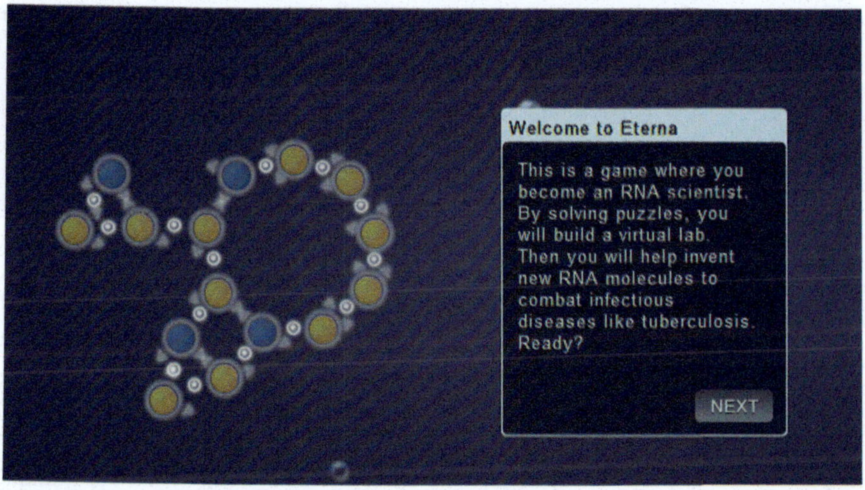

Fig. 14.3 ETERNA, welcome message

claims to offer these citizen scientists the necessary tools to "invent medicine". The game positions itself as an instrument of "empowerment", which gives players the opportunity to realize their latent potentials as citizen scientists.

The homepage also contains an introductory video. In this video, the speaker explains the function of RNA and its relation to EteRNA as follows: "Smart RNA has the potential to transform medicine, tackling some of the most prevalent and deadly diseases on the planet like tuberculosis, malaria, and cancer. But scientists don't fully understand its behavior. [...] That's where EteRNA comes in". Thus, EteRNA informs users, that by engaging in playful puzzle solving, they can contribute to address these serious problems.

Before one starts the game, the following box appears (Fig. 14.3):

The box suggests that citizen scientists, given the appropriate tools, can become actual "RNA scientists" by participating in the "game". Apparently, this game does not know any losers, because the mere participation in it guarantees success ("you become an RNA scientist", "you will build a virtual lab [...] will help invent new RNA molecules" – without any restriction). This choice of words gives the impression that failure is impossible. Becoming an "RNA scientist" seems to be guaranteed.

According to the website, participating in this game without losers will generate technoscientific products that "combat infectious diseases". The technoscientific products themselves, "new RNA molecules", are depicted as agents that will engage in combat. It is the responsibility of the citizen scientists to "invent" these technoscientific agents.

EteRNA suggests that by taking part in a game, citizen scientists may contribute to technoscientific inventions. In doing so, EteRNA interfaces two distinct spheres of action: playing games and engaging in technoscientific research. Its message: you can do technoscience as a playful leisure activity. If doing technoscience can be fun *and* may combat diseases, it offers something worthy for you *and* society. By taking part in EteRNA, civic responsibility can be performed without the need to leave your private life and home.

14.2.3 *Interfaced Participation – Responsibilized Crowds*

The promise of citizen science is to make research accessible to anybody. This 'crowdification' of research is often performed by designs of user interfaces that make participation possible for laypeople. In the citizen science projects analyzed above, the design of these participatory interfaces appears as a *condition* for translating the activities of a crowd into acceptable scientific outcomes. Thus, the social machines of Cell Slider and EteRNA are cases of "involvement made easy" (Marres 2012: 62). The first reason for this attributed 'easiness' is the fact that participation is made possible via an online platform which is accessible for every user – as long as he or she possesses the technological means of access. Participation in Cell Slider and EteRNA is mediated by websites. Everyone possessing an appropriate device like a PC, tablet, or smartphone can access a website. Participation via websites usually does not discriminate by legal status or location, but by technologies of access. Thus, online participation regulated by websites is, on the one hand, very inclusive in the sense that anyone can take part in principle. On the other hand, it is exclusive, because only users equipped with the proper technology can enter the realm of digital participation.

The user interfaces of Cell Slider and EteRNA are designed in such a way that the amount of biomedical knowledge needed for participation is considerably low. In this way, the interfaces construct a specific relationship between science and the public:[3] They configure users as contributors to science by involving them in a problem-solving crowd. This is accomplished by a predefinition of the task that users are expected to perform.

The large amount of implicit knowledge that people involved in research normally possess should be made superfluous by such a design. Thus, participatory interfaces are not only able to lower the barriers of inclusion through technologies of access (a commonplace of narratives of the digital age), but by making the rules of engagement as explicit as possible and by reducing possible forms of action (see also Collins and Evans 2007).

[3] User interfaces can be understood as contemporary ways in which humans and machines are related to each other. They are "a relation with technology rather than […] a technology in itself. In this relation the interface describes a boundary condition that is at the same time encountered and worked through toward some specific end" (Hookway 2014). User interfaces can be interpreted as the slash between the techno/social distinction, co-constituting and bridging both sides of the distinction and thus creating new technosocial realities. Interfaces are borders that allow interpenetration, regulate the flow of information and set the conditions for irritation (Karafillidis 2012: 53). Unlike other linkages between humans and machines (which can also happen by accident and coincidence), interfaces are *designed* to make a specific relation happen.

By decreasing the amount of expertise needed to participate, certain research tasks that are necessary for biomedical research can be outsourced to the crowd (Candido Dos Reis et al. 2015), and participants can generate information that could be used in scientific knowledge production. Cell Slider did not require its users to have expert knowledge. Their contribution was instead explicitly framed as an offering of *time*, because "[a]nalysing our data manually is time-consuming" (Cancer Research UK 2015b). In the case of EteRNA, users may acquire a considerable amount of specialized expertise. The evolution of the project demonstrates that players have indeed made new technoscientific discoveries that have led to academic papers (which list participants as collective co-authors). However, as in Cell Slider, no prior biomedical knowledge is required for participation.

My analysis reveals a discourse of *responsible citizenship* (see also Sutter 2004; Henkel 2015) in both cases. Users are requested to contribute to the fight against cancer and to invent medicine. The projects position participants as 'good citizens' who are expected to take responsibility for research. Cell Slider addressed them as heroes in a fight against cancer who perform their heroic deed by analyzing images. In the case of EteRNA, participants are addressed as players who can invent medicine by taking part in a game. Thus, responsibility for technoscientific innovation in service of the greater good is redistributed and extended.

This redistribution of responsibility, however, should not be understood as 'just' a suggestive discourse, decoupled from technology. This discourse is rather part of the participatory apparatus itself. Digital technology designed to harvest the contributions of volunteers has been put in place to *allow* such participation in the first place. The examples show that participatory interfaces not only make public involvement easy. They also *make responsibilization for a common good* easy. At Cell Slider "[a]nyone can be part of the fight". In EteRNA, everyone is already addressed as a latent "citizen scientist" waiting for the appropriate tools to realize his or her potential. In the case of Cell Slider, citizens were positioned as "heroes" in war. They can be heroic without leaving their homes or risking their lives; *the public becomes private*. In the case of EteRNA, the participants are addressed as "players" who contribute to technoscientific inventions by engaging in a leisure activity. Here, *the private becomes public*.

14.3 Technoscientific Citizenship as Functional Participation

Citizen science allows for a *functional participation* in science in the mode of "prosuming" – without the need to occupy a formal position in the organizations and/or professional communities that are usually deemed responsible for producing technoscientific knowledge and innovation. As citizen scientists, individuals are expected to imitate and complement the performance of professional scientists. Hence, participation in science is becoming more independent of organizational and professional membership. At the same time, scientific practice is becoming more dependent on information technologies.

The cases show that citizen science may not just *extend* participation, but also (re-)define the conditions and the meaning of participation and citizenship: Participating in research becomes part of the rights and obligations of 'good' citizenship. I suggest that this techno-logic of responsibilization is not restricted to Cell Slider or EteRNA. For example, the "Citizen Cyberlab" (Citizen Cyberlab 2016), which aims to create infrastructures for citizen science, used the following picture in its promotional materials (Fig. 14.4).

The picture communicates the normative expectation that science should be "for all" (open, accessible, and useful for the public). On the other hand, "all" should also do something "for science". A network of lines, symbolizing the infrastructures of digital citizen science, links these two slogans. The lines appear as predefined pathways, interfacing science and the public, enabling participation while weaving it into an infrastructured technosocial order.

Citizen science, in the form of crowdsourcing, is clearly a case of invited participation (Wynne 2007; Wehling 2012). It differs, however, from other forms of invited participation that STS usually deals with. In citizen science, citizens are not assembled because of their attributed ability for political judgement (Barney 2007). Participation in these cases is also not designed to tackle the problem of scientific legitimacy in the context of critical innovations. It does not conform to an ideal of deliberative democracy. Participants are rather assembled as a "performing audience" (Andersen and Knudsen 2016: 448) contributing to research.

Whereas Woolly et al. suggest that citizen science might just be a "populist rhetoric" (Woolley et al. 2016: 11) that conceals the real purpose of recruiting volunteers to provide data for large research projects, I argue that citizen science may actually redefine participatory citizenship in science, crossing the boundaries of public and private spheres. It reveals a new notion of technoscientific citizenship: After the paradigms of popularization (shaped by the deficit model) and deliberation (shaped by discourses of risk and uncertainty), the paradigm of "prosuming" positions citizens as contributors to scientific research and technological innovation. In contrast to notions of "amateur science" and related terms, citizen science imagines a public of participants who are responsible for the common good. Citizen science – at least in its dominant mode of "crowd science" (Franzoni and Sauermann 2014) – reconfigures citizens as functional parts of a social machine assembled to attain specific research goals. In their social position as citizen scientists, individuals do not primarily enact citizenship by taking part in political discourses but by becoming responsible members of a productive crowd of volunteers in a predefined technoscientific regime of knowledge production.

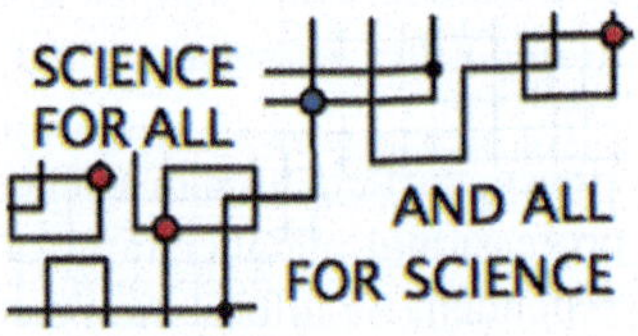

Fig. 14.4 Citizen Cyberscience

References

Andersen, Å.N., and H. Knudsen. 2016. Playful hyper-responsibility and the making of a performing audience. *Soziale Systeme* 19 (2): 433-455.

Anderson-Lee, J., E. Fisker, V. Kosaraju, M. Wu, J. Kong, J. Lee, and M. Lee. 2016. Principles for Predicting RNA Secondary Structure Design Difficulty. *Journal of Molecular Biology* 428 (5): 748–757.

Barney, D. 2007. *One Nation Under Google: Citizenship in the Technological Republic*. Toronto: Hart House Lecture Committee.

Beer, D. 2009. Power Through the Algorithm? Participatory Web Cultures and the Technological Unconscious. *New Media & Society* 11 (6): 985–1002.

Benkler, Y. 2006. *The Wealth of Networks: How Social Production Transforms Markets and Freedom*. New Haven: Yale University Press.

Bogner, A. 2012. The Paradox of Participation Experiments. *Science, Technology & Human Values* 37 (5): 506–527.

Bonney, R. 1996. Citizen Science: A Lab Tradition. *Living Bird* 15 (4): 7–15.

Bonney, R., T.B. Phillips, H.L. Ballard, J.W. Enck. 2015. Can citizen science enhance public understanding of science?. *Public Understanding of Science* 25 (1):2–16.

Bonney, R., J.L. Shirk, T.B. Phillips, A. Wiggins, H.L. Ballard, A.J. Miller-Rushing, and J.K. Parrish. 2014. Next Steps for Citizen Science. *Science* 343 (6178): 1436–1437.

Bora, Alfons, and H. Hausendorf. 2006. Communicating Citizenship – Social Positioning: Theoretical Concepts. In *Analysing Citizenship Talk: Social Positioning in Political and Legal Decision-Making Processes*, ed. H. Hausendorf and A. Bora, 23–49. Amsterdam: Benjamins Publishing.

Cancer Research UK. 2015a. *Cell Slider*. http://www.cellslider.net/. Accessed on 10 Aug 2018.

———. 2015b. *About Cell Slider*. http://www.cellslider.net/#/info. Accessed on 17 July 2016.

Candido Dos Reis, F.J., S. Lynn, H.R. Ali, D. Eccles, A. Hanby, E. Provenzano, and C. Caldas. 2015. Crowdsourcing the General Public for Large Scale Molecular Pathology Studies in Cancer. *eBioMedicine* 2 (7): 681–689.

Citizen CyberLab. 2016. *About the CCL*. http://www.citizencyberlab.org/about.html. Accessed on 10 Aug 2018.

Collins, H., and R. Evans. 2002. The Third Wave of Science Studies: Studies of Expertise and Experience. *Social Studies of Science* 32 (2): 235–296.

———. 2007. *Rethinking Expertise*. Chicago: University of Chicago Press.

Dickel, S., and M. Franzen. 2015. Digitale Inklusion: Zur sozialen Öffnung des Wissenschaftssystems. *Zeitschrift für Soziologie* 44 (5): 330–347.

———. 2016. The "Problem of Extension" Revisited: New Modes of Digital Participation in Science. *Journal of Science Communication* 15 (1): 1–17.

Dickel, S., and J.-F. Schrape. 2017. The Logic of Digital Utopianism. *NanoEthics* 11 (1): 47–58.

Edwards, R. 2013. The 'Citizens' in Citizen Science Projects: Educational and Conceptual Issues. *International Journal of Science Education* 4 (4): 376–391.

Eitzel, M.V., J.L. Cappadonna, C. Santos-Lang, R.E. Duerr, A. Virapongse, S.E. West, and C.M. Kyba. 2017. Citizen Science Terminology Matters: Exploring Key Terms. *Citizen Science: Theory and Practice* 2 (1): 1.

Epstein, S. 1995. The Construction of Lay Expertise: AIDS Activism and the Forging of Credibility in the Reform of Clinical Trials. *Science, Technology & Human Values* 20 (4): 408–437.

EteRNA. 2018. *Empowering Citizen Scientists to Invent Medicine*. http://eternagame.org. Accessed on 10 Aug 2018.

EteRNA Wiki. 2014. *EteRNA*. http://eternawiki.org/wiki/index.php5/What_is_EteRNA. Accessed on 10 Aug 2018.

Feenberg, A. 2011. *Agency and Citizenship in a Technological Society, Digital Citizenship*. Copenhagen: University of Copenhagen.

Felt, U. 2003. Scientific Citizenship: Schlaglichter einer Diskussion. *Gegenworte* 11.

Frankenfeld, P.J. 1992. Technological Citizenship: A Normative Framework for Risk Studies. *Science Technology Human Values* 17 (4): 459–484.
Franzoni, C., and H. Sauermann. 2014. Crowd Science: The Organization of Scientific Research in Open Collaborative Projects. *Research Policy* 43 (1): 1–20.
Hellmann, K.-U. 2010. Prosumer Revisited: Zur Aktualität einer Debatte: Eine Einführung. In *Prosumer revisited: Zur Aktualität einer Debatte*, ed. B. Blättel-Mink and K.-U. Hellmann, 13–48. Wiesbaden: VS Verlag für Sozialwissenschaften.
Henkel, A. 2015. Gesellschaftstheorie der Verantwortung. Funktion und Folgen eines Mechanismus der Reduktion sozialer Komplexität. *Soziale Systeme* 19 (2).
Hookway, B. 2014. *Interface*. Cambridge, MA/London: MIT Press.
Howe, J. 2010. *Crowdsourcing: Why the Power of the Crowd Is Driving the Future of Business*. http://crowdsourcing.typepad.com. Accessed on 10 Aug 2018.
Irwin, A. 1995. *Citizen Science: A Study of People, Expertise, and Sustainable Development*. London/New York: Routledge.
———. 2001. Constructing the Scientific Citizen: Science and Democracy in the Biosciences. *Public Understanding of Science* 10 (1): 1–18.
Jasanoff, S. 2004. Science and Citizenship: a New Synergy. *Science and Public Policy* 31 (2): 90–94.
Karafillidis, A. 2012. Die Recodierung der Soziologie: Zu Harrison C. Whites Interfaces. *Revue für postheroisches Management* 10: 46–59.
Kasperowski, D., and T. Hillman. 2018. The Epistemic Culture in an Online Citizen Science Project: Programs, Antiprograms and Epistemic Subjects. *Social Studies of Sciences* 44 (1).
Kelly, K. 2005. We Are the Web. *Wired*. https://www.wired.com/2005/08/tech/. Accessed 10 Aug 2018.
Lee, J., W. Kladwang, M. Lee, D. Cantu, M. Azizyan, H. Kim, and A. Limpaecher. 2014. RNA Design Rules from a Massive Open Laboratory. *Proceedings of the National Academy of Sciences of the United States of America* 111 (6): 2122–2127.
Marres, N. 2012. *Material Participation: Technology, the Environment and Everyday Publics*. Houndmills/Basingstoke/Hampshire/New York: Palgrave Macmillan.
Open Science Policy Platform. 2018. *Integrated Advice of the Open Science Policy Platform*. https://ec.europa.eu/research/openscience/pdf/integrated_advice_opspp_recommendations.pdf#view=fit&pagemode=none. Accessed on 10 Aug 2018.
Prainsack, B. 2014. Understanding Participation: The 'Citizen Science' of Genetics. In *Genetics as Social Practice: Transdisciplinary Views on Science and Culture (Theory, Technology and Society)*, ed. B. Prainsack, S. Schicktanz, and G. Werner-Felmayer, 147–164. Farnham: Ashgate.
Reichwald, R., and F.T. Piller. 2006. *Interaktive Wertschöpfung: Open Innovation, Individualisierung und neue Formen der Arbeitsteilung*. Wiesbaden: Gabler.
Ribble, M.S., G.D. Bailey, and T.W. Ross. 2004. Digital Citizenship: Addressing Appropriate Technology Behavior. *Learning and Leading with Technology* 39 (1).
Riesch, H., and C. Potter. 2014. Citizen Science as Seen by Scientists: Methodological, Epistemological and Ethical Dimensions. *Public Understanding of Science* 23 (1): 107–120.
Rose, N. 2000. Community, Citizenship, and the Third Way. *American Behavioral Scientist* 43 (9): 1395–1411.
Serrano. 2013. This website: https://sciencenode.org/feature/engaging-citizens-science-research-excellence.php.
Shadbolt, N. R., D. A. Smith, E. Simperl, M. V. Kleek, Y. Yang, and W. Hall. 2013. Towards a Classification Framework for Social Machines. *WWW 2013 Companion*. https://dl.acm.org/citation.cfm?id=2488078. Accessed on 10 Aug 2018.
Smart, P.R., and N.R. Shadbolt. 2014. Social Machines. In *Encyclopedia of Information Science and Technology*, ed. M. Khosrowpour, 6855–6862. Hershey: Information Science Reference.
Strasser, B. J., J. Baudry, D. Mahr, G. Sanchez, and Elise Tancoigne. 2018. 'Citizen Science"? Rethinking Science and Public Participation. *Science and Technology Studies*.

Surowiecki, J. 2004. *The Wisdom of Crowds: Why the Many Are Smarter Than the Few and How Collective Wisdom Shapes Business, Economies, Societies, and Nations*. New York: Doubleday.

Sutter, B. 2004. Governing by Civil Society: Citizenship Within a New Social Contract. In *Reflexive Representations: Discourse, Power, and Hegemony in Global Capitalism*, ed. J. Angermüller, D. Wiemann, R. Kollmorgen, and J. Meyer, 155–168. Münster/Piscataway: LIT.

Tapscott, D., and A.D. Williams. 2007. *Wikinomics: How Mass Collaboration Changes Everything*. New York: Portfolio.

Toffler, A. 1989. *The Third Wave*. New York: Bantam Books.

Wehling, P. 2012. From Invited to Uninvited Participation (and Back?): Rethinking Civil Society Engagement in Technology Assessment and Development. *Poiesis & Praxis* 9 (1–2): 43–60.

Weingart, Peter. 1982. The Social Assessment of Science, or the De-Institutionalization of the Scientific Profession. *Science, Technology, & Human Values* 7 (1):53–55.

Wellington-Oguri, R., E. Fisker, M. Wiley, and M. Zada. 2018. Evidence of an Unusual Poly(A) RNA Signature Detected by High-Throughput Chemical Mapping. *bioRxiv*.

Wenninger, A. 2015. Hermeneutische Analysen neuer Kommunikationsformen im Internet. In *Die qualitative Analyse internetbasierter Daten: Methodische Herausforderungen und Potenziale von Online-Medien*, ed. D. Schirmer, N. Sander, and A. Wenninger, 51–87. Wiesbaden: Springer.

Woolley, J.P., M.L. McGowan, H.J.A. Teare, V. Coathup, J.R. Fishman, R.A. Settersten Jr., S. Sterckx, J. Kaye, and E.T. Juengst. 2016. Citizen Science or Scientific Citizenship? Disentangling the Uses of Public Engagement Rhetoric in National Research Initiatives. *BMC Medical Ethics* 17 (33): 1–17.

Wynne, B. 2007. Public Participation in Science and Technology: Performing and Obscuring a Political–Conceptual Category Mistake. *Easts* 1 (1): 99–110.

Zooniverse. 2018. *About*. https://www.zooniverse.org/about. Accessed 10 Aug 2018.

Sascha Dickel is Assistant Professor for Sociology of Media. His research concerns social theories of science and technology, digital participation, and technoscientific futures. The edited volume *Personal Health Science* (with Nils B. Heyen and Anne Brüninghaus) came out with Springer (2019).

Chapter 15
Anti-modern Techno-Science: Cybernetics, Ontology, Practice in the European Hacker Tradition

maxigas

15.1 Introduction: The Case for Deep Hacker Studies

Hacker artefacts, knowledges, practices and scenes constitute an alternative engineering culture that can be characterised as anti-modern techno-science. Constructed in the wake of the 1960s cultural shock (Wallerstein 2004: 16), hackers' attitude to technology is reminiscent of the cybernetic ontology identified by Pickering (2010) and articulated in the sciences by cyberneticians. It privileges practice over representation, performance over cognition, and adaptive encounters with the unknown over the domination of Nature. In short, it is informed and forged by an experience in the ontology of becoming.

While it has been demonstrated that even engineers and scientists working in the heart of modern institutions grapple with the mangle of practice in their daily routines (Pickering 1995), it has also been pointed out that neither official accounts of engineering or scientific work, nor the institutions that frame them are necessarily constructed according to such realisations. Simply put, the role of material agency in techno-scientific research is not taken into account adequately by modern institutional frameworks. Pickering (2010) brought a historiography of scientists – the cyberneticians – who attempted to reformulate scientific tools, theories, practices and institutions to match their experiences in the ontology of becoming. Starting from the figurations of cybernetics in European experiences of hacking, I argue here that an anti-modern techno-science can be shown to be at work in our contemporary historical horizon. Some hacker practices do more than dealing with the contingencies of material resistance found everywhere in research: these practices are designed to invite such encounters and encourage resistance to shape processes and results. Furthermore, organisational forms of hackers take seriously such encounters rather than incorporating them to fit institutional expectations. Once again,

maxigas (✉)
Lancaster University, Lancaster, UK
e-mail: maxigas@anargeek.net

S. Maasen et al. (eds.), *TechnoScienceSociety*, Sociology of the Sciences Yearbook 30, https://doi.org/10.1007/978-3-030-43965-1_15

rather than claiming that hackers work with an ontologically different technology than mainstream engineers, what I seek to demonstrate is that *ontology matters* as part of culture.

Pickering shows that the classic cyberneticians' non-modern approach was so transversal that it did not fit into the grid of modern institutions, and therefore they failed to reproduce their social basis. What I intend to point out is that hackers' practices are articulated in opposition to best practices in the academic-military-industrial complex to the point where they set up their own organisational infrastructures to sustain their scenes and reproduce their social basis. In a more nuanced analysis below, I seek to capture the difference between how hackers and how cyberneticians enact a cybernetic ontology in the contrast between *non-modern* and *anti-modern* techno-science.

As Pickering when researching English cyberneticians, I am more interested in how to live and how to engineer in a different way, than to capture the essence or the statistical veracity of hackerdom. The wealth of the empirical material is drawn from the European hacker tradition, but I do not claim that it represents the mainstream of hacking there or elsewhere. I would rather like to chart an undercurrent of motifs that is theoretically revealing. Everything that is laid down here is taken from my decade long anthropological and historical research into hackerspaces, but this time the ambition is not ethnological documentation.[1] Studying hackers offers lessons on how artefacts, knowledges, practices and even organisations can be different when the mangle of practice is taken into account in a more consistent and coherent manner.

Asking such questions is essential in a historical moment when late modernity seem to have exhausted its potential, so that we have to look at other ways of dealing with the world in order to find adequate responses to social, environmental, economical, political and institutional crises. While hackers' encounters with the unknown are certainly not enough to address these issues, they may possibly offer lessons about alternative pathways to take. After all, as Pickering phrases it, alternative engineering cultures can be seen as a "collective exploration of the nonmodern selfhood", where new subjects in conjunction with new artefacts and, in tandem, new kinds of "odd institutions" are constructed (Pickering 2010: 200, 400). The advantage of inquiring into hacker culture rather than cyberneticians is that it is possible to analyse collective practices rather than individual contributions, and ones that are formative of our current historical horizon. *Notwithstanding the "social marginality" of hackers and the "hobbyist origins" of hacking* (Pickering 2010 *about cybernetics, 388), in the penultimate section I argue that hacker practices ground an anti-modern ethical relation to material agency framed by a symmetric ontology.* However, first it is necessary to situate hacker practices historically in relation to the cybernetics movement (Fig. 15.1).

[1] For an ethnological account see maxigas (2015).

Fig. 15.1 Logo of the Brussels Hackerspace

15.2 Figurations of Cybernetics in the European Hacker Tradition

Ross (1990) identifies a shift in the expression of countercultural dissent in the decades following the 1960s. He describes the ecologically-inspired anti-technology attitude of the hippies as the *technology of folklore*, contrasting it with the *folklore of technology* that inspired cyberculture – revolving around perseverance in a dystopian and oppressive world through technological expertise. He does acknowledge that it is harder to read such new dissent politically. Developing these doubts, Markoff (2005) and Turner (2006) focus on the cross-breeding of counterculturalists in decline and the ascendant Silicon Valley entrepreneurship – what Barbrook and Cameron (1996) called the *Californian ideology* or what Curtis (2011) blames directly on cybernetics.

Reactionary or not, as Leary (1994) vividly expresses, the translation happened through novel artefacts, knowledges and practices of the brain, which in pop culture meant mainly psychedelics. Essentially, subjects became hollow objects to be crudely manipulated in a determinist fashion, while objects expressed a richer subjectivity in their non-determinism. *Cyberpunk* – first as a sub-genre of science fiction and soon as a fully fledged subcultural phenomena – expressed such ontological ambiguity in the language of the gothic uncanny (Cavallaro 2000; Grant 1998; Rapatzikou 2004; Whatley 2014). Despite the dystopian bent of cyberpunk imaginaries, hackers drew the conclusion that in case the underdogs can grasp new technologies faster than the powers that be, they can at least go out with a bang.[2]

[2]These ideas are not far from the arguments put forward by proponents of Left Accelerationism today (Srnicek and Williams 2015).

As the "cyber-" prefix suggests, these advances were partly inspired by cybernetics, a transdisciplinary development in the sciences that was also organised around the ontological ambiguity of the brain as the instrumental locus of subjectivity. Classic cybernetics was expressed in books such as *The Computer and the Brain* (von Neumann 1958) or *The Brain of the Firm* (Beer 1972), and analysed in works such as the aforementioned *The Cybernetic Brain* (Pickering 2010) or *The Mechanization of the Mind* (Dupuy 2000). Cybernetic theories gave a good basis for technological determinist arguments ("solutionism"): e.g. that social problems can be transformed *to* and eventually solved *as* technological problems. In other words, that revolution can be engineered (Söderberg 2014). Most commentators, such as Tiqqun (2012) or Dammbeck (2003) draw dire conclusions from these notions, arguing that cybernetics grew into an overarching ideology justifying a more oppressive and exploitative society. Such analysis is both profoundly true and necessary to address political problems of our times. However, it takes for granted that there is a single unified essence of cybernetics, and all the originality of the cybernetic world view have been absorbed into the capitalist system.

On the other hand, Pickering single-handedly advanced a more positive outlook based on an alternative reading of British cybernetics, picturing it as an antidote to modernity's blind paths. I suggest that the two readings – symmetric and asymmetric cybernetics (Pickering 2013) – share an experience of subjectivity, which is first and foremost an encounter with objectivity: human intentionality with the material unknown. Their paths bifurcate in the moment of an ethical decision about how to handle the encounter. To rephrase Ross Ashby's remark on Clausewitz: "Who will adapt to who?" (Pickering 2010: 140). Engineering in general, and computer science in particular, serve as fertile grounds for experimenting with the ethics and epistemology of such an encounter, in light of its ontological implications. It makes sense that hackers, – as participants in an *alternative* engineering *subculture* – explored a different aspect of cybernetics than what has been absorbed into mainstream institutions.

The ambiguity of cybernetics and its close connection to computerised subcultures is aptly described by Christopher Kelty (2010) in an introduction to a special issue of Anthropological Quarterly on the topic:

> The "cybernetic-informatic consciousness" that he uses to link journalists and anthropologists together opens a door to understanding why computers and cultures are equally everywhere today – and maybe for some surprising reasons that have a lot to do with the mid-century successes of both cybernetics and anthropology. Cybernetics' fortunes look a lot like culture's – overused, diffuse, heavily critiqued, yet nonetheless compelling in their most rigorous forms. Cybernetics' dissemination looks a lot like culture's – taken up across nearly every discipline, attenuated by circulation, unpoliced by classic disciplinary modes of ownership and exclusion.

Indeed, there is consensus between scholars that cybernetics started as a scientific paradigm, but quickly disseminated into other spheres such as psychiatric practice (Pickering 2010), the arts (MacGregor 2002; Reichardt 1968) and its own subsequent branches such as chaos theory. Perhaps more than other cultural movements, the European hacker tradition soaked up motifs of cybernetics in its fabric, with a special emphasis on *chaos theory*. The latter became the organising principle for an engineering subculture that – especially in Europe – grew up on the fringes.[3]

Since autonomist struggles ran out of steam in the 1970s (Cuninghame 2005; Geronimo 2012; Schultze and Gross 1997; Wright 2002), the advent of chaos theory was perhaps the only and last hope for revolutionary change. If a very small, but incredibly smart move can produce systemic change, than a hacker coupled with a computer can save the world. Another – less radical – interpretation invested its faith in the instability of systems where small changes can propel into phase changes pretty quickly. The latter interpretation put the apparent stability of "the system"[4] in perspective, and allowed for revolution almost by accident. What is essential, however, is that both interpretations empowered marginal social actors[5] with much more freedom than what contemporary political ideologies (activist or not) or current systems theory (scientific or not) allowed. In the era branded as the *Information Age* (Kline 2015) ruled by Shannon's statistical approach to communication, *noise* became the rallying cry of outsiders (Krapp 2011 i–xx; Menkman 2011) – and as a corollary, positive feedback[6] the analogue to the revolutionary process (Larsen 2011). In a chaotic world, therefore, *adaptation* was the key for survival, resiliency and perseverance.

No other person represented such tendencies better than Wau Holland (Herwart Holland-Moritz). In 1980s Germany he was as much a public intellectual as a public engineer, as much an educator as an inventor. Underpinning his ideology about universal access to computers and networks, as well as the protection of private information and the free flow of public knowledge, were devices and practices as much as discourses (Kulla 2003). He founded the Chaos Computer Club to advance these causes and legitimise hacker practices. In an incident that is now part of "German computer science folklore" (Wikipedia contributors 2015a) members wired 134,000 DMs to themselves and presented the pile of money to the press after the German postal service (which held a monopoly on home banking) refused to fix the security issues they reported in its BTX system (Denker 2014). They saw the broken implementation of encryption for online banking as a social problem, but authorities did not heed their advice. Therefore, they *demonstrated* the problem in the long tradition of the *public experiment* (Born and Barry 2013): creating an ontological theatre which not only represented or described, but actually *performed* the vulnerability of informatic systems. The small action by these peripheral actors lead to the rehauling of the BTX as a national information infrastructure and helped to introduce competition to home banking. At the same time, it earned a good reputation for hackers for acting in the public interest. I argue later on that such formational experiences cemented hackers' trust in practice and performance over theory

[3] Incidentally, European cybernetics likewise developed in relative social marginality compared to its US counterpart (Pickering 2010: 388).

[4] Whether economic, political, technical or even world system.

[5] Whether single individuals, small groups or subcultures.

[6] The self-amplification of noise.

and representation. Soon, chaos theory figured in the name of what is now the largest and most influential hacker organisation on the continent (Brooke 2011):

> The Club was so named not because it set out to cause chaos but rather because one of the founders, Wau Holland, felt chaos theory offered the best explanation for how the world actually worked. Dutch hacker and entrepreneur Rop Gonggrijp says the club is about "adapting to a world which is (and always has been) much more chaotic and non-deterministic than is often believed".

The Club was founded on December 12, 1981, at a table belonging to Kommune 1, an emblematic site of the autonomist movement, suggesting both continuity with, and departure from the ideas and practices of the extra-parliamentary left (Chaos Computer Club 1981, 2009; Wunschick 2015). Chaos theory came to be officially embraced in the hacker scene as an explanation of *how the world actually worked.* In other words, an ontology, and one hinged on performance. Wau Holland explained that the goal was "causing Chaos as a means to display asymmetrical practices of Surveillance and control" (Kulla 2003: 14).[7] The organisation of hackers was itself to function as an ontological theatre: a public display of a chaotic world. "Hence the most personal manifestation of Chaos was the computer club of the same name, whose speakers persisted with nearly missionary fervor that there is no security, that it is nothing but a nice yet deceiving illusion" (Kulla 2003: 30).[8]

Following the establishment of regular hacker meetings such as the yearly Chaos Communication Congress (now in its 33rd edition), the proliferation of shared machine shops ushered in a subsequent period of hacking. These served as club houses for the hackers who could now organise publicly. One of the first was c-base in 1995, continuing the ***c***ybernetics, ***c***yberpunk, ***c***omputer associations. However, the mainstream success of hackerspaces started with Metalab in Vienna, which was initially proposed as KyBase ("Cybernetic Base") in 2005.[9] This was not an accident because the hacker group behind the idea was called Wiener Kybernetischer Kreis: a word play on the name of cybernetics founder Norbert Wiener, and the interdisciplinary Vienna Circle ("Wiener Kreis") that played a decisive role in the intellectual trajectory of the twentieth century (Stadler 2001; Wikipedia contributors 2015b). However, hackers decided that references to cybernetics were too obscure to draw a more diverse audience, and called their base Metalab instead. Indeed, Metalab then provided the blueprint and inspiration for many subsequent hackerspaces (Bre and Astera 2008: 7, 68, 87).

Hackerspace Brussels (HSBXL) materialised in 2008, while hackerspaces were exploding in numbers. Its identity also plays on Norbert Wiener, who is pictured in the logo wearing *flicker glasses* (Hackerspace Brussels Wiki Contributors 2010). Flicker glasses are essentially sunglasses fitted with LEDs on the inside, which

---

[7] "Chaos stiften als Mittel als Mittel der Sichtbarmachung von asymmetrischen Praktiken der Überwachung und Kontrolle."

[8] "Der erste Hackangriff rief den Schutzpatron der Sicherheit auf den Plan, die Ordnung. Chaos as a counterpole to the patronsaint of security, order/system."

[9] Metalab was formally founded the next year (2006).

simulate the brain by flashing in specific patterns. Pickering notes that they are descendants of British cybernetic artist and inventor Brion Gysin's Dreamachine: a revolving tube dotted with holes that produced a similar effect thanks to internal illumination (419). It was built in collaboration with Ian Sommerville, a computer programmer and companion of Burroughs and other Beats. The invention was inspired by The Living Brain, (Walter 1963), the book of a seminal British cybernetician. Walter was the first to describe the psychological effects of flicker ("The Central Effects of Rhythmic Sensory Stimulation"; 1949). He later commented on the commercial success of stroboscopes in discos to the effect that he should be paid a royalty (Pickering 2010: 78–83). Much like other cybernetic motifs, flicker arrived to European hackers through American mediation. It was the San Francisco-based Mitch Altman who popularised remakes of the device in hacker circles (Cornfield Electronics, Inc 2013). He soon acquired a standing in the hackerspaces scene comparable to Wau Holland's role earlier, spending much of the year travelling between hacker clubs, teaching soldering and electronics while spreading a DIY[10] anti-military message. The emblem of his teachings came to be the Tripglass, which participants could assemble at soldering workshops.[11] One could go on recounting other instances of cybernetic figurations in European hacking (such as maxigas 2015: 272–273), but let three examples suffice for suggesting a pattern.

To summarise, in the 1950s and 1960s both hacking and cybernetics emerged from the socialisation and systematisation of Second World War techno-scientific advances, at a specific moment characterised by ample funds for base research (at least in the United States). Notwithstanding their origins in the academic-military-industrial complex, they were closely intertwined with the countercultural movements that flourished in their wake (Turner 2006). As both tendencies show, the relationship between science and technology drew ever more close, yet it is fair to argue that cybernetics was articulated primarily as a scientific culture, while hacking as an engineering culture.

Throughout the 1970s and 1980s cybernetics – often through chaos theory – provided a much needed explanation of how the world actually worked to hackers who acquired an identity as a movement and a self-image as the underdogs of engineering. Semi-underground scenes formed, which often clashed with larger social structures like the authorities or corporations (Sterling 1992). Hackers started to reflect on their socio-political roles and responsibilities as renegade engineers, especially in the European context, where links with social movements were stronger (maxigas 2012).

Since the 1990s, prominent personalities and organisations in the European hacker tradition were aware of some classic cybernetic tropes, artefacts and theories. Even if the translation of cybernetics into hacker culture has been very superficial, partial and fragmented, it remains that cybernetic motifs were woven into its

[10] Do It Yourself.

[11] Later the Tripglass was replaced by the TV-B-Gone universal remote control that can turn off any television. Mitch's latest product is the Neurodreamer, a return to Gysin's Dreamachine.

fabric. This is not to suggest that the structural similarities between the ontology enacted by Pickering's cyberneticians and the ontology performed through hackers' practices can be explained by a shared genealogy. The use of cybernetic imagery in hacker culture is not internally coherent, even if its presence is consistent. For instance, where hackers champion chaos, their idol Norbert Wiener may champion order. Kline (2015: 96) shows that Wiener appeared to the average American as a humanist enemy of chaos:

> In his masterwork, America as a Civilization (1957), journalist and educator Max Lerner quoted Wiener's idea that "communication is a dialogue between people united against the common enemy, whether we call it entropy or chaos." It was one of the few popular references to Wiener's extensive discussion of entropy in his books.

Instead of a history of ideas, figurations of cybernetics (especially chaos theory) in European hacking amount to a *reconfiguration* of cultural historical imaginaries "that hold particular relations of persons and things – with land, or with information – in place" through a simultaneous reflexive and generative movement, including leakages and entanglements (Suchman 2012: 52, 58). Cybernetic motifs *worked* for hackers, generating theoretical insights and practical results that both guided their understanding of themselves in the world, and helped to orient their technological endeavours. As I proceed to present in the next sections, everyday enactments of a cybernetic ontology are more deeply ingrained in hacker culture than a reflection on the scientific content or motives of cybernetics.

15.3 Practices Favouring Practice

Before introducing *engineering* practices, I would like to recount a scene that everyone can relate to. An unspecified francophone hackerspace got flooded in June 2015 because the landlord repeatedly refused to fix the roof of the old factory. Equipment was damaged and water covered a large part of the premises. A member proposed to drill a hole in the floor and sweep water down to the workshop of the landlord below the club. The idea was adopted on a whim, and since machine tools were stocked in the hackerspace, the hole was ready in a matter of minutes. The hackers did not bring it up with the landlord, but he fixed the roof the next week.

What I find revealing about the incident is that members were "thinking" directly in material agencies and how to mobilise them, sidestepping institutions and symbolic communication between humans. Just like in the case of the BTX, it was not enough to report vulnerabilities in the infrastructure: *hackers had to demonstrate them with a performance*. Instead of talking, then, the hackers had to do something to get their point across. Moreover, instead of mopping up the water, they accommodated its resistance by letting it continue its flow through the space. They entered a risky situation in the course of the exercise, which could have gone in different unexpected ways, and they were ready to accept the consequences. The most striking aspect for me, observing the scene, was how little discussion and hesitation went into it. I was clearly immersed in a culture that favours practice over representation, and performance over cognition, ready to go where material agency takes the engineer.

15.3.1 Exploratory Programming

According to computer science professor Keunwoo Lee (2004), **exploratory programming** is about "programming by' trying stuff out' and seeing what happens". It is a programming methodology taught in computer science and recommended for tackling problems that are not well understood. For the same reason, it is ideal for learning by doing. Since hackerspace members are not required to produce anything in particular, but often eager to learn, it is not surprising that exploratory programming is popular amongst them. Compared to their counterparts with monolithic institutional backgrounds, programmers in hackerspaces tend to meddle into a higher number of programming languages, paradigms and problems. While CTOs and computer science teachers often enforce a homogeneous technological repertoire in their institutions, experience with a particular technology or problem domain is not a criteria for hackerspace membership – only technical curiosity is. Just to follow the intricate jokes that Coleman (2012) pinpoints as the pinnacle of hacker culture, one has to know the basics of the dozen most popular programming languages. Of course, once members actually start to build things in their favourite languages, some of those things become part of the infrastructure for running the space, at which point – like it or not – more members will have to know at least how to repair them.

Conversely – like in any organisation – there *are* in fact favourite (if not so exclusive) languages in hackerspaces: Python and LISP. Python, which allows for a high level of reflection; and LISP, the ultimate "programmable programming language". Reflection and programmability both support the exploratory programming paradigm, because they allow programmers more opportunity to understand and change programming language behaviour. Exploratory programming is most interesting when it is used to understand a particular property of the programming language itself, or even when it is loosely concentrated on a problem domain, rather than a particular real-world problem. The methodology and its technical repertoire is designed to iterate over possible solutions and hopefully find new ones through the process.

Therefore, exploratory programming is a methodology to engage with the temporal emergence of material agency in order to find a way to approach a problem that was not obvious before. Material agency in this context simply means that algorithms, languages and environments exhibit resistance in an analogous way to the physical reality that physicists – or mechanical engineers – face. Pickering (1995) already recognises the materiality of human language and abstract theories, so this is but a small extension to the theory of resistance and accommodation. The temporal emergence of material agency simply means that there is no way to know if something will work or not, that is until it has been tried out in practice. Even then, in case of failures it is impossible to know if there *is* a way to make it work. Practices that respect the temporal emergence of material agency are distinguished by the possible range of consequences that the resistance of the material may have on the overall endeavour.

Exploratory programming allows the programmer to engage with the unknown encountered during programming by changing the entire orientation of the activity, to the point where even the problem set out to be explored may shift. Similar conclusions have been reached by scholars of Human-Computer Interaction. Lindell (2014) is interested in the materiality of programming and craftsmanship in software development in the sense of Sennett (2009). His claim is that exploratory programming "mainly describes a conversation with the material". I return to the idea of the conversation as the central metaphor for ontological ethics in the penultimate section.

One Autumn afternoon I sat around the table with several members in the hackerspace of Budapest.[12] One of the best programmers of the club happily announced that he is ready with paid company work, so now it is time to start another side project. After a few hours of intense coding, I asked him what will the new program solve? He answered that it was not possible to know yet. I was astonished, because he already wrote so much (Python) code. It was as if he walked out in the wilderness without a destination. He was building a general system to programmatically parse web pages, but it was not clear yet what are the real world tasks where it could outperform existing solutions. This way of approaching a problem was very different from most average programming jobs, where coders are paid to implement a prewritten specification, or work with clients on improving a product incrementally.

It was close to blue skies research where ideas are tried out and practical applications can wait. Nonetheless, even academic researchers have to write research proposals, acquire funding and produce an institutionalised timeline of deliverables, not to mention answering to superiors. In this case, there was no supervisor or institutional framework, only a club created by hackers themselves to support tinkering with materials. Conversely, the result would not be directly remunerated by anything other than the experience itself, and perhaps the respect of peers.

While in most cases exploratory programming may not be so efficient as a less spontaneous software development methodology, it allows programmers to perform the part of their job closest to engagement with materiality: coding. Negotiating the task, designing the software architecture, writing the documentation – and in most cases implementing a ready-made specification – are all part of the programmers' job that exploratory programming puts less emphasis on. While these activities arguably present their own resistances, it is not the resistance of the program code itself. Hacker culture values engaging with the resistance of the program code itself over other tasks involved in programming. As the hacker saying goes, "two hours of planning can save you two days of coding – but then again, two days of coding can

[12] The Hungarian Autonomous Center for Knowledge, alias H.A.C.K., located in Hungary.

save you from two hours of planning" (personal communication, Ralf Bendrath, 2007-11-09).

15.3.2 Penetration Testing

Penetration testing is about probing a technological system in order to find faults in it. It can be done with malicious intent, but it can also be done under contract, in order to improve security by reporting problems. Penetration testing can be thought of as an antonym of the scientific experiment. While in a classic scientific experiment the point is to create an isolated environment where a repeatable operation is carried out in order to prove a hypothesis, the point of penetration testing is that it takes place outside of the laboratory, carried out on actual infrastructure through means that are difficult to foresee, in order to find problems in practice. For instance, banks are generally required to hire security companies to make sure that their infrastructure conforms to the relevant industry standards (such as ISO/IEC 27002:2007, 21188:2006, etc.). Many hackers run such companies. Others practice penetration testing in teams at *Capture the Flag* events, which typically last for a few sleepless days, and became a staple of hacker conventions. Online platforms set up for the pleasure and pain of playing penetration testing are called *wargames*.

Penetration testing can include anything from breaking into a computer network, through physical security like stealing paper documents, to social engineering where the attacker talks their way to sensitive information. However, these are related practices requiring a similar mindset. Here, the focus is on software security, because it is the most widespread area of expertise in hackerspaces.[13]

My claim is that in order to even begin working on penetration testing, one first has to accept that there is a gaping field to be explored between representation and performance. Most programmers and administrators *must* assume that the actual performance of a program is more or less accurately described in its specification, because it is but a building block of their systems. On the contrary, penetration testers could not even start their work without assuming that the program actually behaves differently under certain circumstances. It is *those* circumstances that the penetration tester is interested in, and most often they are to be found by trying out applications in a wide variety of ways, while looking for typical integration mistakes and known flaws in the individual components. However, the real prize of a penetration tester is a grave but unforeseen implementation mistake, one with interesting unintended consequences. Those latter can be developed into new ways to compromise the system: *vulnerabilities* [^CVE]. Previously unknown vulnerabilities are called *zero days*, because vendors, developers and administrators have had

[13] Hackerspaces often host lock picking sport clubs, and sometimes social engineering workshops too.

no opportunity to fix them yet. They play an important part in hackers' imaginaries, reputation and economies.

Vulnerabilities are effectively the (most lucrative) products of penetration testers. Hackers construct vulnerabilities by engaging with the material resistance of an infrastructure, often through long time periods. Therefore, vulnerabilities are a good example of temporal emergence. A vulnerability report is often a mere description of a weakness, but eminent vulnerability reports also include *exploits*. An exploit is a short piece of code that demonstrates the vulnerability by triggering it in practice. Therefore, exploits can be used to compromise systems even by people who have only basic informatics knowledge. [^kiddies] Exploits are also recognised for their exchange value and consequently brought and sold on the black market, or acquired legally by vendors through bug bounty programs, or even reported on public mailing lists by independent researchers (similarly to academic findings in scientific journals). In the latter cases, exploits and vulnerabilities are meticulously credited, attesting to their status as products.

Exploits can be conceptualised as ontological theatres, which extort systems into strange performances. "No system is secure" is a mantra of hackers and each vulnerability is an additional proof of this point. Vulnerabilities are studied by hackers both for "fun and profit". Their entertainment value comes from the fact that they achieve some unusual, often seemingly impossible task – like gaining administration rights to a system without knowledge of the appropriate credentials. The roots of the vulnerability lay in the unintended consequences of software written by developers, or the actions of the system administrators who run the vulnerable systems. Since the consequences are unintended, most vulnerabilities are discovered by examining the performance of the systems in practice, rather than the underlying theories, even though the vulnerability may prove the theory behind the system flawed. [^backdoors] Therefore, a vulnerability is often theoretically impossible, and by definition unknown before its discovery. The penetration tester have to be open to the unknown and do things that invite it to manifest itself. There are penetration testing methodologies taught at universities, but searching for vulnerabilities remains a craft that requires a calling, rather than a routine procedure that can be described in manuals.

An ontological theatre does not only take advantage of a particular ontological outlook in its design, but also showcases it through its performance. As a living proof that a system does not only work as advertised, but also in other "strange" ways, an exploit points to the distance between representation and performance. Such a performance is "strange" because it goes beyond – and very often against – the way given system is supposed to work. As such it also showcases and legitimises the epistemological advantages of working in close consideration of the cybernetic ontology.

15.3.3 Hardware Hacking

Hardware hacking is arguably the *raison d'être* of hackerspaces (Kostakis et al. 2014; Dafermos 2014; maxigas 2014). On the one hand, the rise of hardware hacking accentuated a need for hackers to join forces in clubs commanding their own real estate, because sharing the necessary manufacturing resources and tacit knowledges solely online is difficult. On the other hand, once various strands of hackers came together in the same premises, working on tangible artefacts of *physical computing* (Igoe and O'Sullivan 2004) provided an opportunity for common engagement between enthusiasts of software, hardware, mechanics and more.[14] Electronics classes – especially introduction to rapid prototyping tools such as the Arduino line of microcontroller boards – are usually amongst the first organised activities in newly founded hackerspaces. Similarly, "Learn to solder" and "Soldering is easy" is the signature workshop and fanzine of the aforementioned hackerspaces personality Mitch Altman, who likes to give it in budding hackerspaces around the world.

Hardware hacking can be conceptualised as the appropriation of more widely used rapid prototyping practices. Prototyping in an industry is typically practiced with an eye on developing a final product, or sometimes used in an academia to learn something specific. Rapidity, when it is applied to prototyping, refers to the fact that desktop manufacturing technologies such as 3D printing and development tools such as programmable microcontrollers can be used to speed up design and implementation of hardware.

I argue that the appropriation of rapid prototyping practices in hackerspaces may be driven by a desire for engagement with material resistances and strange performances. At the same time, it is enacted under specific material constraints answering to the wider structural constraints on hackerspaces. Of course, both desires and constraints can be similar in certain institutional settings. However, institutions adopt rapid prototyping as a means to an end, whether learning by doing, research and development, or straightforward product development. Even if some workers are motivated by the encounter with the material resistances and strange performances during product development, most companies are constructed as an organisation geared towards making products. Such is the modern relationship to technology – reflected in the discourse of *innovation*: instrumental rationality frames most organised engagements with technology. As a counterpoint, it seems that in the organisational environment of hackerspaces, encountering material resistances and strange performances is not a means to an end but an end in itself.

Hardware hacking is highly dependent on the components and manufacturing equipment that developers can get their hands on. Ordering new components or acquiring funding is cumbersome and boring. Instead, hackerspaces maintain a junkyard of sorts for scavenging components. Any project or product – assembled or not – can fall prey to rapid prototyping in a hackerspace. Therefore, both

[14] Building hardware always already includes software engineering (for interfaces and especially the firmware) and mechanics (for the tangible components, especially robotics actuators).

purpose-made stickers ("Do Not Hack") and ad-hoc notes ("Don't Touch") are used to distinguish *loot* from *work in progress* projects and *infrastructures* that are essential for operating the space. Members have lockers or boxes to keep their stuff away from prying hands.[15]

Hackerspaces are filled with the technological refuse of latter generations, to the point where several of them collect typewriters (used by typewriting clubs at Noisebridge, San Francisco or at Technologia Incognita, Amsterdam that meet regularly at the premises) and operate museums of working computers (for instance at Arnhem, The Netherlands or in Catania, Sicily). In turn, obsolete machine tools from factories and legacy laboratory equipment are sometimes dropped at hackerspaces. Technical solutions, design insights and salvaged components from these machines may become building blocks of prototypes. Constraints such as using readily available components and manufacturing tools are another form of material agency that plays its part in hardware hacking and emerges both historically and stochastically. Like the ontological theatres of the classic cyberneticians, a prototype made in a hackerspace often showcases an idea or an enigma, an intervention whose actual engineering could be deceptively simple or primitive – or its practical value questionable.

Returning to Coleman's point that sophisticated in-jokes are the heart and soul of hacker culture, it can be argued that prototypes are often not more than elaborate practical jokes, a running commentary on the absurdity of rational instrumentalisation in modern technologies. A case in point is the common trick implemented in hackerspaces, where step-motors (of a 3D printer, laser cutter or even a stack of old CD-ROM drives) are used to generate evocative of popular classical music, such as the Empire theme from Star Wars. Rather than accomplishing a practical task, such contraptions function as ontological theatres showcasing strange performances of otherwise familiar machines.

When Latour (2008) writes about the virtues of design thinking as an amodern form of thought, he emphasises that design always starts with something which is already there. Therefore, design is always redesign. For Latour, this is in opposition to the Promethean hubris of modernity where the inventor is supposed to create ex nihilo. Latour praises the careful humility of design which never starts from scratch, but shapes things attentive to the circumstances. However, it is curious that it never occurred to the author of *We Have Never Been Modern* that Prometheus actually stole the fire rather than invented it. The fire was already there, what happened is that it was made public. Nonetheless, Latour's description sits well with rapid prototyping, which starts from the material available at hand, strives to create a functioning thing as soon as possible, and tinker on from there, rejecting "'destroy it all and build something completely different' modernism" (Flynn 2014).

[15] From the Code of Conduct draft of the Xinchejian Hackerspace in Shanghai: "DO NOT HACK means DO NOT HACK If you're leaving something and you would rather it was not hacked, please label it with a DO NOT HACK notice. If you see something has a DO NOT HACK notice on, don't hack it" (Xinchejian Hackerspace Wiki Contributors 2013). For other examples, see Juul (2010).

What is carried on in the translation of rapid prototyping into hardware hacking is the conviction that practice, not theory – materials, not blueprints – should guide development. The ontological commitment necessary for pursuing development in this way is to a world that is not infinitely knowable through what Pickering calls "positive knowledge" (a staple of modernity, see 2010, 401). In other words, there is no way to know how a prototype will turn out. Therefore, prototypes are built not only rapidly but in quick succession, so that every iteration can build on the practical experience gained. In the connected world of hackerspaces, this often means reproducing something that was done elsewhere, sometimes with personal help from the previous designer. Due to the haphazard source materials and other local factors, exact copies are downright impossible. Moreover, it is widely acknowledged that as long as the experience is new to a participant, rebuilding something old can be as rewarding as building something new.

Education, research and production in hackerspaces are mere corollaries to encounters with material agency that produce unpredictable effects. The irrelevance of the very categories that define modern institutions is demonstrated in how seamlessly they can blend into each other in the quotidian life of the hackerspace. In particular, prototypes made in hackerspaces often evade these modern categories of instrumentalisation. It may not be clear whether they are prototypes of personal projects, learning exercises, artworks, commercial products or activist interventions. Indeed, the very hackers working on the project are often altogether uninterested in such questions. At some point the project may be abandoned, nailed to the wall of the hackerspace as an example of how to (or how *not* to) connect two specific electronic components – or it could become an activist intervention in a political debate, but just as well provide the basis for a lucrative startup. These destinies matter little when the point of the process is to encounter the unknown on a common ground.

15.4 Odd Institutions Hosting Strange Performances

The "lack of material and social discipline" is the double thread that ties together the trio of ontological theatres emblematic of UK cybernetics: Grey Walter's tortoises, Ross Ashby's homeostats and Gordon Pask's Musicolour (Pickering 2010: 324).

Materially, they were constructed from the junk left over after the Second World War: surplus bomb relays from the RAF and obsolete telephone relays from the Post Office. "Elizabeth Pask (n.d.) recalled that Gordon and Harry Moore built Musicolour from'old relays and uniselectors junked from post office telephone exchanges' – the same components that Walter and Ashby used in their model brains" (Ibid.).[16] In fact, cybernetics itself was largely a product of synthesising the

[16] "The only other reference I have found to Ashby's source of parts for the homeostat is to the detritus of World War II: 'It has four ex RAF bomb control switch gear kits as its base, with four cubical aluminium boxes' (3 March 1948, p. 2341)." (Pickering 2010: 426)

theoretical and technical legacy of the war effort coming "out of the detritus of war and a technological society" (Ibid.). But what is more crucial for the present argument is that the design and implementation of these artefacts was shaped by the technological debris of the time as a material force in its own right.

I have demonstrated at lengths that hardware hacking relies on a similarly haphazard ensemble of refuse – a refuse that is only structured by the fact that these components and machines spilled over to the hackerspace from modern institutions such as companies and laboratories where they were deemed obsolete or superfluous. In this sense it is not only the practices of hackers that fall outside modern categories but also the materials on which they act, let it be archaic programming languages such as LISP, or old machinery such as typewriters. It is no wonder that the results can be sometimes interpreted as a subversive commentary on the ideals of modern engineering.

Socially, UK cyberneticians pursued cybernetics as a side project barely tolerated by their respective institutions and hardly fitting into their respective disciplines. "I am struck, first, by the profound amateurism of British cybernetics. Key contributions often had an almost hobbyist character: Walter built his first tortoises at home in his spare time; so did Ashby his homeostat (at least, in the apocryphal version of the story); likewise Beer and Pask's experimentation [...]. Cybernetics welled up outside the usual channels, and it found little support within those channels" – writes Pickering (2010: 10). Even if the results of British cybernetics have been distilled into (pseudo-)scientific books and articles, how the substance of the work was carried out was nothing short of tinkering with engineering in a largely peripheral setting.

Hackerspaces are likewise marginal engineering institutions, with the crucial difference that they were explicitly constructed to inhabit marginality as a countercultural position. For some of them, social marginality was a desirable position – for others it became a necessity. Either way, Coleman and Golub (2008) notes that repression as an outside pressure was instrumental for the organisation of the hacker scene, forging a compact set of ideas and practices that hackers held their own. Realising that their exploits and out of the box engineering culture goes beyond the confines of modern institutions, they banded together to pursue an alternative trajectory. Consequently, hacking is today a popular folk tradition that developed a relatively wide social basis, while cybernetics disseminated because it failed to reproduce its social basis.

15.5 From Cybernetic Ontologies to Hacker Ethics

Ethics mediates between ontology and its performance in practice. Therefore, eliciting the ethical consequences of the view of the world distilled in the cybernetic ontology can further illuminate the practices described in the previous section, while preparing the ground for the political conclusions in the next section. Departing from Pickering (2010), I theorise the ethical consequences of the cybernetic ontology by way of the ontological ethics of Emmanuel Lévinas (1969).

The starting point of Lévinas' ontological ethics is a Hegelian understanding of the Other as an existential threat to the phenomenological subject. The threat comes from the ambiguity of the Other between being another agent like me, but also fundamentally different. The proposed eminent ethical relationship with the Other is modelled on the conversation. In short, the conversation allows for keeping a distance and therefore recognising difference, while at the same time exposing oneself, and therefore be effected by the relationship. The conversation does not seek to destroy or assimilate the Other, but to learn and adapt to it. There are obvious parallels here with the Conversation Theory developed by the classic cybernetician Gordon Pask (1976). Moreover, neither authors conceptualised the conversation as necessarily verbal, leaving theories open for performative interpretations (such as Lindell 2014). Lévinas uses further timely metaphors such as hosting the other in my home, but also falling hostage to the Other in a foreign land.

My suggestion is that this formulation of ontological ethics is descriptive of hackers' relationship to material agency found in their interactions with machines, just like Pask's experience with the Musicolor. In line with the unstable ontological divide between subjects and objects expressed in cyberpunk works, the Other takes an objective form in these experiences. While certainly a departure from Lévinas original thought, I hope that the recognition of the Other as the material resistance performed by objects explains many aspects of symmetric cybernetics as well as hacker culture.

Pickering recounts how Ashby's prolonged struggle with prototyping DAMS made him abandon "the modern engineering paradigm of knowledge-based design in favor of evolutionary tinkering" (2010: 128) that "necessarily entailed a degree of *respect* for the other" (2010: 32). His desperation marked the ethical moment in cybernetics, and resonates well with hackers' difficulties in exploratory programming, penetration testing or hardware hacking (described in Kleif and Faulkner 2003). I argued that hackers such as penetration testers *construct* bugs through enacting an anti-modern ontology - yet vulnerabilities are always both unexpected and unsettling even to them. Technical expertise allows hackers to recognise the face of specific technical problems as strikingly familiar yet unsettlingly different: hence the fascination with security.

Echoing Lévinas' claims about the construction of subjectivity in relation to the Other, I would explore the notion that hacker subjectivity is primarily ethical and constituted through subjection to the resistance of materials encountered in practice. On closer examination, each of the practices above share a few characteristics that point to such a realisation. All have to do with the recognition of material agency in the form of bugs, or glitches in computer science.

The resistance of the material manifests itself as a bug or glitch in a complex technological system. First, the creation of situations in which bugs can occur and unfold. Since bugs are the products of penetration testing, it is no wonder that security experts have mastered the art of summoning errors through unexpected inputs into the system. Second, accommodation but not recuperation of bugs into the development process. Exploratory programming is about meeting the full force of the problem in the very spontaneity of practice rather than solving it first theoretically. In hardware hacking each iteration of a prototype has serious design flaws – since this is exactly what distinguishes it from a finished product.[17] Third, bugs resulting in unforeseen, and thus potentially unexpected and potentially dangerous results. I would like to emphasise once again the exploratory nature of drilling a whole on your floor (which is the ceiling of your landlord). It is hard to argue that it is a safe approach to conflict resolution. The strength of the gesture is not in finding a safe solution but on the contrary: opening up the possibility of an unexpected, if dangerous solution to work out or not.

As ontological notions of ethics should, such formulation of hackers' ethical experience logically precede Levy's (1984) list of concrete commandments. In light of ontological exegesis, the spirit of the commandments is to remove particular barriers – remnants of modernity – that could interfere with the existential encounter between subject and object, user and computer, and give free reign to performativity. The first and last commandments – "All information should be free" and "Computers can change your life for the better" are explicitly written in the performative idiom that recognises material agency. The intermediate ones target modern social institutions such as bureaucracies, prejudices (sexism, racism, etc.) and instrumental rationality (in opposition to art and beauty).

The ontological interpretation puts connections with classic liberalism tackled by Coleman and Golub (2008) in a new perspective: anti-racism, for instance, is not a commitment of solidarity with the downtrodden, but a precondition for the materiality of the code to speak for itself. As Coleman puts it, "Hacking, even if tethered to liberal ideologies, spills beyond and exceeds liberal tenets or liberal notions of personhood" (2012: 4). Liberalism might be merely a precondition for including non-humans in the cosmopolitics of hackers.

The moral of the story is that bugs function analogously to Lévinas' Other in the diagram of the hackers' engineering culture. Once this assertion is secured, the consistency of the exploratory practices and ethical commitments above is apparent, as well as the common ground of politics that grows out of them. The unknown is sought out for the experience of the encounter and material agency is accommodated, but not recuperated into the engineering enterprises of hackers. Furthermore, as I argue in the previous section, the organisational infrastructures of the scene are

[17] Indeed, debugging the prototype to get an actual product is most often more difficult and definitely more tiresome than developing an almost perfect prototype.

built to accommodate such an ethical relation with the unknown – unlike modern institutions.

15.6 Conclusion: Anti-modern TechnoScienceSociety

Finally, we are in a position to ask the crucial question: how does hackers' enactment of a particular ontology – through practices, organisations and forms of life that take into account that ontology – challenge modernity? Latour (1993) criticised modernity as a series of purification processes that seek to counter a parallel process of hybridisation: an ontological horror that the cyberpunk imaginary captures suggestively. I suggest that hacking as an alternative engineering culture features some practices that are articulated against modern institutional best practices: against specifications that can be translated into working software, information infrastructures that can be secured, prototypes that can be turned into perfect products. As Latour advocated in his works since, culture and institutions have to be reinvented and rebuilt so that they are not conditioned by such disastrous modern myths. Hacker scenes include some organisational forms such as the hackerspaces that provide the social basis for the sort of hybridisation that modern institutions are meant to prevent – for instance between education, research and production. Finally, I theorise that hackers' diverse and sometimes contradictory ethical commitments and political interventions can be, perhaps, understood on the basis of an ontological ethics. Lévinas' work on the encounter with the Other is a good entry point, even if hacker subjectivity is constructed in relation to the non-human, material agency found in technical problems.

The most significant contribution of hackers to the project of surpassing modernity may well be establishing popular genres of engineering and setting up organisations that are constructed in a way that points beyond the modern view of the world. A kaleidoscopic overview such as this article can identify patterns that are worth exploring but does little in way of arguing concrete cases. Further studies are needed to understand the historical trajectory, ontological sensibilities, and social content of these practices and how they come together in innovative organisational forms such as the hackerspaces.

In the meantime, hacking addresses new audiences (such as women and children) and incorporates new techniques (such as synthetic biology) – but it has also been absorbed into modern institutions such as political parties, university campuses and international corporations. Delfanti (2013) notes that these tendencies testify to the versatility and complexity of the hacking phenomena, but also contribute to its diffusion. Ironically, if cyberneticians failed to reproduce their social basis, the success of hacker culture could have a similar effect on its ontological dimensions, as much as it is recuperated by contemporary pop culture and late modern capital accumulation (Söderberg and Delfanti 2015).

15.6.1 Cybernetics vs. Hackerdom

Notwithstanding the connections established so far, hacker practices differ from those explored by Pickering in three major ways. First, **hacking is anti-modern** since it is deliberately articulated in opposition to modern practices. Hackers – even powerful ones – have always positioned themselves on the periphery of modern society and regarded their movement as an anomaly – even if an inevitable one. The practices above have each been developed in diametrical opposition to industry best practices of mainstream engineering. A single example would be penetration testing which is part of the *offensive security* tradition, which challenged assumptions that building a secure system according to specifications is sufficient for resilience, advocating that all security experts have to be familiar with the mindsets, skill sets and tool sets of attackers.[18]

Second, ***hackers can reproduce their social basis*** through an expanding network of hackerspaces, periodic meetings and online communities. While classic cybernetics revolved around the cyberneticians themselves, hacking is organised around practices which can be appropriated by anybody. Hacking, like popular culture, is contained in and communicated through genres. Classic cybernetics did practice and propose organisational innovations from the dinner societies of the UK cyberneticians through the Kingsley Hall and Archway communities to the Fun Palace. However, scenes of hackers have been arguably more innovative, more concentrated and more successful in organisational innovations. From underground phreaking groups that convened on telephone networks, through security conferences in festive settings, to hackerspaces accessible to the general public, hackers found different ways to reproduce their culture and pass it on to the subsequent generations. Indeed, one may say that hackers are more focused on collectively perpetuating their folk culture than classic cyberneticians were.

Third, ***hacker culture is organised around engineering***, while the cybernetic ontology was originally articulated as a scientific culture. Of course, since both point beyond modern categories, both disseminated beyond their surface of emergence, even if retaining a heartland in science or engineering, respectively. In fact, it could be argued that the decline of classic cybernetics and the rise of hacker culture are both embedded in a larger process, where engineering itself is becoming hegemonic not just in its relation to science but many more areas of social life. The popularisation and recuperation of hacking are both happening through the appropriation of engineering practices and metaphors.

[18]Another evaluation of exploratory programming by scientists is from Bundy (2009) 37: "This exploratory programming methodology has been the target of much criticism of AI, especially from the theoretical computer science community. There are elements of both justice and unfairness in this criticism. Such an *ad hoc* methodology is no way to develop a robust program. It is the antithesis of the structured programming and formal methods advocated by software engineers."

15.7 Summary

Throughout the article I pick up practices strategically where the cybernetic ontology manifests in hacker practices as ontological theatres (showcasing another way of approaching material agency). Notwithstanding such selective bias, I believe that a similar outlook is present, albeit in a much more diluted form, in many areas of hacker culture. The deep hacker studies research program is grounded in the hypothesis that the sociological implications of enacting such an ontology – perhaps mediated by an ethical regime – can help to understand specific areas where hackers made contributions. For instance, the disruptive effects of practices as disparate as hacktivism, online piracy, digital fabrication or free software may find a common vector of explanation in ontological analysis. On a final note, I argued that hackers' efforts to research and organise in close consideration of an anti-modern view of the world – the socialisation of cybernetics, if you like – spawned a vibrant popular culture with wide implications for technology, science and society.[19]

References

Barbrook, R., and A. Cameron. 1996. The Californian Ideology. *Science as Culture* 26: 44–72. Available at: http://www.imaginaryfutures.net/2007/04/17/the-californian-ideology-2.

Beer, S. 1972. *The Brain of the Firm*. 1st ed. London: Allen Lane, The Penguin Press.

Born, G., and A. Barry. 2013. Art-Science: From Public Understanding to Public Experiment. In *Interdisciplinarity: Reconfigurations of the Social and Natural Sciences*, Culture, Economy and the Social, ed. A. Barry and G. Born, 1st ed., 247–272. Routledge. Available at: http://libgen.io/book/index.php?md5=7617b674bdf31db99c5e264348cfb073.

Bre, and Astera, eds. 2008. *Hackerspaces: The Beginning*. Ebook. Available at: http://blog.hackerspaces.org/2011/08/31/hackerspaces-the-beginning-the-book/.

Brooke, H. 2011. Inside the Secret World of Hackers. Article in *The Guardian*. Available at: http://www.theguardian.com/technology/2011/aug/24/inside-secret-world-of-hackers.

Bundy, A. 2009. Prospects of Artificial Intelligence. Reissue Edition. In *Computing Tomorrow: Future Research Directions in Computer Science*, ed. I. Wand and R. Milner, 33–48. Cambridge: Cambridge University Press.

Cavallaro, D. 2000. *Cyberpunk & Cyberculture: Science Fiction and the Work of William Gibson*. 1st ed. London/New Brunswick: The Athlone Press.

Chaos Computer Club. 1981. Founding Document. Scanned Typescript. Available at: http://berlin.ccc.de/~tim/tmp/tuwat-protokoll.pdf.

———. 2009. Chaos Computer Club. Page on the Official Website of the Chaos Computer Club. Available at: http://www.ccc.de/de/club.

Coleman, G. 2012. *Coding Freedom: The Ethics and Aesthetics of Hacking*. Princeton: Princeton University Press.

Coleman, G., and A. Golub. 2008. Hacker Practice: Moral Genres and the Cultural Articulation of Liberalism. *Anthropological Theory* 8 (3): 255–277.

[19] With the support of a postdoctoral grant from the Universitat Oberta de Catalunya (UOC) and the sponsorship of the Central European University Foundation, Budapest (CEUBPF) for a fellowship at the Center for Media, Data and Society in the School of Public Policy.

Cornfield Electronics, Inc. 2013. Tripglasses Product Page. Product Page on Vendor Website. Available at: http://cornfieldelectronics.com/tripglasses/trip.home.php.

Cuninghame, P. 2005. Autonomia in the 1970s: The Refusal of Work, the Party and Power. *Cultural Studies Review* 11 (2): 77–94. Available at: http://epress.lib.uts.edu.au/journals/index.php/csrj/article/view/3660.

Curtis, A. 2011. *All Watched Over by Machines of Loving Grace*. Documentary Series, BBC. Available at: https://thepiratebay.se/torrent/6457676/BBC_All_Watched_Over_By_Machines_Of_Loving_Grace.

Dafermos, G. 2014. *Distributed Manufacturing: Commons-Oriented Productive Capacities*. Policy Paper by the FLOK Society. Available at: http://floksociety.org/docs/Ingles/2/2.4.pdf.

Dammbeck, L. 2003. *Das netz*. Documentary.

Delfanti, A. 2013. *Biohackers: The Politics of Open Science*. London: Pluto Press.

Denker, K. 2014. Heroes Yet Criminals of the German Computer Revolution. In *Hacking Europe: From Computer Cultures to Demoscenes*, History of Computing, ed. G. Alberts and R. Oldenziel, 1st ed., 167–188. London/Heidelberg/New York/Dordrecht: Springer.

Dupuy, J.-P. 2000. *The Mechanization of the Mind: On the Origins of Cognitive Science*. Princeton/Oxford: Princeton University Press.

Flynn, A. 2014. *Solarpunk: Notes Toward a Manifesto*. Blog Post. Available at: http://hieroglyph.asu.edu/2014/09/solarpunk-notes-toward-a-manifesto/.

Geronimo. 2012. *Fire and Flames: A History of the German Autonomist Movement*. Oakland: PM Press.

Grant, I.H. 1998. Cyberpunk. In *The Handbook to Gothic Literature*, ed. M. Mulvey-Roberts, 263–264. New York: NYU Press.

Hackerspace Brussels Wiki Contributors. 2010. *Who's That Dude?: NorbertWiener*. Entry on Hackerspace Brussels Wiki. Available at: https://hackerspace.be/smw/index.php?title=NorbertWiener&oldid=4696.

Igoe, T., and D. O'Sullivan. 2004. *Physical Computing: Sensing and Controlling the Physical World with Computers*. London: Premier Press.

Juul, M. 2010. *Don't Hack My Tools!* Email to the Hackerspaces.org Discussion List. Available at: http://lists.hackerspaces.org/pipermail/discuss/2010-April/002121.html.

Kelty, C. 2010. Culture in, Culture Out. *Anthropological Quarterly* 83 (1): 7–16.

Kleif, T., and W. Faulkner. 2003. 'I'm No Athlete [But] I Can Make This Thing Dance!': Men's Pleasures in Technology. *Science, Technology, & Human Values* 28 (2): 296–325. Available at: http://www.studioincite.com/PNT/athlete_dance.pdf.

Kline, R.R. 2015. *The Cybernetics Moment: Or Why We Call Our Age the Information Age*, New Studies in American Intellectual and Cultural History. 1st ed. New York: Johns Hopkins University Press. Available at: http://gen.lib.rus.ec/book/index.php?md5=254527E594A49CE50D59528C18556C7F.

Kostakis, V., V. Niaros, and C. Giotitsas. 2014. Production and Governance in Hackerspaces: A Manifestation of Commons-Based Peer Production in the Physical Realm? *International Journal of Cultural Studies*. Available at: http://libgen.io/scimag/?s=10.1177/1367877913519310.

Krapp, P. 2011. *Noise Channels: Glitch and Error in Digital Culture*, Electronic Mediations. Minneapolis/London: University of Minnesota Press.

Kulla, D. 2003. *Der phrasenprüfer: Szenen aus dem leben von Wau Holland, mitbegründer des chaos-computer-clubs* [The Voltage Tester – Scenes from the Life of Wau Holland, Co-Founder of the Chaos Computer Clubs]. Werner Pieper & The Grüne Kraft.

Larsen, L. B. 2011. *Anti-disciplinary Feedback and the Will to Effect. Mute Magazine: We Gladly Feast on Those Who Would Subdue Us* 3(1). Available at: http://www.metamute.org/editorial/articles/anti-disciplinary-feedback-and-will-to-effect.

Latour, B. 1993. *We Have Never Been Modern*. Cambridge: Harvard University Press.

———. 2008. *A Cautious Prometheus? A Few Steps Toward a Philosophy of Design (with Special Attention to Peter Sloterdijk)*. Keynote Lecture for the Networks of Design Meeting of the

Design History Society, Falmouth, Cornwall. Available at: http://www.bruno-latour.fr/sites/default/files/112-DESIGN-CORNWALL-GB.pdf.

Leary, T. 1994. *Chaos and Cyber Culture*. Berkeley: Ronin Publishing.

Lee, K. 2004. *Exploratory Programming Using Squeak and Morphic*. Course Slides. Available at: http://courses.cs.washington.edu/courses/cse341/04wi/lectures/19-morphic-exploratory-programming/html/slide_2.html.

Lévinas, E. 1969. *Totality and Infinity: An Essay on Exteriority*. Pittsburgh: Duquesne University Press.

Levy, S. 1984. *Hackers: Heroes of the Computer Revolution*. Doubleday: Anchor Press.

Lindell, R. 2014. Crafting Interaction: The Epistemology of Modern Programming. *Personal and Ubiquitous Computing, Springer-Verlag* 18 (3). Available at: http://libgen.io/scimag/index.php?s=10.1007/s00779-013-0687-6.

MacGregor, B. 2002. Cybernetic Serendipity Revisited. In *C&C'02 Proceedings of the 4th Conference on Creativity & Cognition*, 11–13. New York: ACM.

Markoff, J. 2005. *What the Dormouse Said: How the Sixties Counter Culture Shaped the Personal Computer Industry*. London: Penguin. Available at: http://libgen.io/book/index.php?md5=631E8A3A67AB0EA19ECD3DB0E689ADA1.

maxigas. 2012. Hacklabs and Hackerspaces: Tracing Two Genealogies. *Journal of Peer Production* 2. Available at: http://peerproduction.net/issues/issue-2/peer-reviewed-papers/hacklabs-and-hackerspaces/.

———. 2014. *Hackerspaces: The Infrastructure of Open Hardware Production*. Workshop at Backbone409, Calafou. Available at: http://backbone409.calafou.org/.

———. 2015. *Peer Production of Open Hardware: Unfinished Artefacts and Architectures in the Hackerspaces*. Doctoral Dissertation, Barcelona: Universitat Oberta de Catalunya, Internet Interdisciplinary Institute. Available at: https://relay70.metatron.ai/maxigas_dissertation.pdf. Accessed 18 Sept 2017.

Menkman, R. 2011. *The Glitch Moment(um)*, Network notebooks. Amsterdam: Institute of Network Cultures.

Pask, G. 1976. *Conversation Theory: Applications in Education and Epistemology*. New York: Elsevier.

Pickering, A. 1995. *The Mangle of Practice: Time, Agency, and Science*. Chicago: University of Chicago Press. Available at: http://libgen.io/book/index.php?md5=44ad4df824c04f1e8160cc16f61d2bec.

———. 2010. *The Cybernetic Brain: Sketches of Another Future*. Chicago/London: University of Chicago Press.

———. 2013. *Cybernetics and the Politics of the Dark Universe*. Paper Presented at Sonic Acts XV: Festival for Exploration in Art, Music & Science, Amsterdam.

Rapatzikou, T.G. 2004. *Gothic Motifs in the Fiction of William Gibson*, Postmodern studies. Amsterdam/New York: Rodopi.

Reichardt, J., ed. 1968. *Cybernetic Serendipity*. London: W. & J. Mackay. Available at: http://cyberneticserendipity.com/cybernetic_serendipity.pdf.

Ross, A. 1990. Hacking Away at the Counterculture. *Postmodern Culture* 1 (1).

Schultze, T., and A. Gross. 1997. *Die autonomen: Ursprünge, entwicklung und profil der autonomen bewegung broschiert* [The Autonomists: Origins, Development and Profile of the Autonomous Movement]. Hamburg: Konkret Literatur Verlag.

Sennett, R. 2009. *The Craftsman*. New Haven/London: Yale University Press. Available at: http://libgen.io/book/index.php?md5=DB85DF3B46F2E519A111FD5EB961084E.

Söderberg, J. 2014. Reproducing Wealth Without Money, one 3D Printer at a Time: The Cunning of Instrumental Reason. *Journal of Peer Production* 4. Available at: http://peerproduction.net/issues/issue-4-value-and-currency/peer-reviewed-articles/reproducing-wealth-without-money/.

Söderberg, J., and A. Delfanti. 2015. Hacking Hacked! The Life Cycles of Digital Innovation. *Science, Technology and Human Values*: 1–6.

Srnicek, N., and A. Williams. 2015. *Inventing the Future: Postcapitalism and a World Without Work*. 1st ed. London/New York: Verso.
Stadler, F. 2001. *The Vienna Circle: Studies in the Origins, Development, and Influence of Logical Empiricism*. Vienna/New York: Springer.
Sterling, B. 1992. *The Hacker Crackdown: Law and Disorder on the Electronic Frontier*. New York: Bantam.
Suchman, L. 2012. Configuration. In *Inventive Methods: The Happening of the Social*, Culture, Economy and the Social, ed. C. Lury and N. Wakeford, 1st ed., 48–60. London/New York: Routledge. Available at: http://libgen.io/book/index.php?md5=712B065634D6563D3D70051FB275602A.
Tiqqun. 2012. *The Cybernetic Hypothesis*. The Anarchist Library. Available at: http://theanarchistlibrary.org/library/tiqqun-the-cybernetic-hypothesis .
Turner, F. 2006. *From Counterculture to Cyberculture: Stewart Brand, the Whole Earth Network, and the Rise of Digital Utopianism*. 1st ed. Chicago: University of Chicago Press. Available at: http://libgen.io/book/index.php?md5=6916B53A2F276602174090943602E3F2.
von Neumann, J. 1958. *The Computer and the Brain*. 3rd ed. New Haven: Yale University Press.
Wallerstein, I. 2004. *World-Systems Analysis: An Introduction*. Durham/London: Duke University Press.
Walter, W.G. 1963. *The Living Brain*. London: W. W. Norton & Company.
Walter, V.J., and W.G. Walter. 1949. The Central Effects of Rhythmic Sensory Stimulation. *Electroencephalography and Clinical Neurophysiology* 1 (57–86): 84.
Whatley, J. 2014. Gothic Self-Fashioning in Gibson's Novels: Nature, Culture, Identity, Improvisation, and Cyberspace. In *A Companion to American Gothic*, Blackwell Companions to Literature and Culture, ed. C.L. Crow, 1st ed., 418–433. Wiley Blackwell: Chichester.
Wikipedia Contributors. 2015a. Bildschirmtext. *Entry on Wikipedia, The Free Encyclopedia*. Available at: https://en.wikipedia.org/w/index.php?title=Bildschirmtext&oldid=686449225.
———. 2015b. Metalab. *Entry on Wikipedia, The Free Encyclopedia*. Available at: https://en.wikipedia.org/w/index.php?title=Metalab&oldid=679800154.
Wright, S. 2002. *Storming Heaven: Class Composition and Struggle in Italian Autonomist marxism*. London: Pluto Press.
Wunschick, F. 2015, 12. September 1981: Gründung des chaos computer club. *Deutsche Welle Kalenderblatt*. Available at: http://www.kalenderblatt.de/index.php?what=thmanu&manu_id=1794.
Xinchejian Hackerspace Wiki Contributors. 2013. Weekly Staff Meetings, 2013-05-21, Code of Conduct Draft. Wiki page. Available at: https://wiki.xinchejian.com/wiki/Weekly_Staff_Meetings/2013-05-21/code_of_conduct_draft#DO_NOT_HACK_means_DO_NOT_HACK.

maxigas seeks answers to philosophical questions through empirical investigations from a broad science and technology studies perspective. His research interests include the refusal of technology (the Luddite aspects of hackerdom), critical approaches to cybernetics (in contemporary techno-optimist discourses), and the history of hacking (in Europe). He works as a Lecturer in Critical Digital Media Practice in Lancaster University and lives in Calafou, the post-capitalist eco-industrial colony.

Chapter 16
An Unfolding TechKnowledgy: Technology and Knowledge in Open Digital Fabrication

Christoph Schneider

The production, exploration and valuation of novel technical capabilities is of central concern in TechnoScienceSociety.[1] New technologies are often the centre of gravity of unruly technosocial assemblages that increasingly define this society. This is the case not only for how new technologies are used but also for how they are produced. In this text, I examine an example of one of the novel technical capabilities that has been produced during the past 15 years and which constitutes a window into the novel assemblages of expertise, technologies, and organisations that transgress older and more established institutions. Through a focus on 'open digital fabrication', where technologies such as 3D printing have become increasingly accessible and produced in a more public and 'open' manner, I argue that it is necessary to ask how novel technical capabilities are differently constituted, since this tells us much about the new ways that TechnoScienceSociety constitutes itself. The ways in which technical creativity is taking shape and being contested have multiplied. The invention, design, production, and circulation of technologies no longer only follow established paths, for example, between institutionalised science and industry. However, the new paths that emerge are only beginning to be understood, since they are often highly dynamic and classic concepts may not necessarily be easily transferrable.

Open digital fabrication is a particularly striking example of these new paths. Starting from technoscientific research activities into computer numerically controlled (CNC) machinery and nanotechnology, new organisations such as FabLabs

[1] For inspiring discussions and comments on this text, I want to thank Sascha Dickel, Bettina-Johanna Krings, Andreas Lösch, Sabine Maasen, Anne Seidenstücker, and the contributors of this book, who commented on a presentation of the argument.

C. Schneider (✉)
Institute for Technology Assessment and Systems Analysis, Karlsruhe Institute of Technology, Karlsruhe, Germany
e-mail: christoph.schneider@kit.edu

S. Maasen et al. (eds.), *TechnoScienceSociety*, Sociology of the Sciences Yearbook 30, https://doi.org/10.1007/978-3-030-43965-1_16

and collaborative and public open source platforms have come into being and aim to make the technical capabilities of such machines more widely accessible. Such 'open digital fabrication' merges particular cultures, forms of organisation, and subjectivities that were established in the digital realm with the cultures of material object fabrication. I will discuss and analyse the relevant networks of organisations such as FabLabs and the platforms and object-centred sociality of open digital fabrication below. For a start, it will suffice to note that thousands of people across the globe have been developing, producing, and using technologies such as 3D printing in highly collaborative ways, drawing on open source approaches and other ways of sharing these technologies and the knowledge about them. Open digital fabrication takes place at the intersection of different societal spheres, with participants working professionally or voluntarily within this field to explore the potentials of these technologies or to simply enjoy tinkering with them.

Since we can no longer simply locate the becoming of technology within the professional institutions of science and industry, but also need to consider such novel assemblages, we must ask what brings processes such as open digital fabrication to life in TechnoScienceSociety. However, they haven't replaced older and more deeply embedded processes that organise technical creativity, such as industrial complexes. Instead, they have emerged alongside them and in their interstices. It might even be that the multiplicity of pathways of technical becoming, their complex intersections and contestations, is a defining feature of TechnoScienceSociety. However, this means we need to get to grips with the diverse realities of technical creativity and to understand their different qualities. This is what this text sets out to do in two steps. First, I introduce the concept of 'TechKnowledgies' to focus on the procedures that produce and organise technology and knowledge.[2] This allows us to bring different qualities of social organisation and their entanglement with the production of technologies into focus. Second, an empirical case study of open digital fabrication uncovers the particular qualities that enable and produce open digital fabrication and highlights the usefulness of 'TechKnowledgy' for thinking about emerging assemblages of technical creativity. The chapter moves from a more theoretical approach to open digital fabrication towards more empirical detail. The empirical 'intensity' of my argument will increase towards the second half of the text, and it weaves analyses drawn from several years of participant observation and action research in open digital fabrication[3] into the argument.

---

[2] For complex power/knowledge processes that shape subjects and technology see Tyfield, Chap. 13, this volume. For an in-depth discussion of TechKnowledgy see Schneider 2018.

[3] The two case studies from which I draw from took place between 2013 and 2015 and involved expert interviews with practitioners in an open source development project for a laser cutter, the Lasersaur project, and participant observations and action research in which I was part of a group that set up a public and non-profit workshop to share digital fabrication machines, FabLab Karlsruhe. Both cases are analysed in detail in Schneider 2018.

16.1 TechKnowledgies

Open digital fabrication is filled with technical objects that have received a lot of attention, have been admired and desired, and have astonished observers by the technical procedures that they can perform. During my research,[4] I encountered many instances where, for example, a 3D printer was used to create a sense of wonder amongst visitors to the FabLab that I have set up as an action researcher together with a group of citizens. 'See, by combining digital templates, a sophisticated mechanical design, and some plastic, this machine creates tangible, three-dimensional objects for use', was the typical statement implied in such performances. And while many other machines in digital fabrication, such as laser cutters, equally combine digital and material technology to produce objects, from the perspective of these technical objects, open digital fabrication is actually not that spectacular. Industry uses much more expensive and sophisticated versions of such machinery and, furthermore, we now encounter combinations of digital and material technology almost everywhere. Open digital fabrication starts to become interesting and unique, however, when technical objects and the ways in which they are being designed and used are considered together. Much of the wonder people have when seeing a 3D printer operate also relates to the fact that the machines have not been seen in action before, since they are literally locked away in companies that tightly regulate access to and use of the machines. How can we combine a perspective on these novel technical objects and capabilities and the specific ways that they are organised in order to understand the processes that have been unfolding both?

Technology can mean many things, and it is a pervasive notion in everyday life and in STS.

'The word technology is as capacious as it is unspecific. It covers an astonishing diversity of tools and instruments, products, processes and systems. A composite of Greek techne (skill) and logos (study of), "technology" in its earliest usage, back in the seventeenth century, meant the study of skilled craft. Only in the 1930s did the word begin to refer to objects produced through the application of techne' (Jasanoff 2016: 8).

Precisely because phenomena such as open digital fabrication appear in TechnoScienceSociety we have to rediscover and update the older meaning of technology that is closely linked to procedures, rationalities, and knowledge processes that constitute the production and usage of technological artefacts. The wisdom gained from established 'modern' institutions such as universities and companies, which have been dominant in these processes, is not enough.

With the concept 'TechKnowledgy', I want to rediscover and update this older meaning of technology that foregrounds technological processes instead of their

[4] I draw on extensive qualitative empirical research including interviews with participants, participant observation, and action research in starting a FabLab. For an extensive analysis of the cases that feature in this text, FabLabs as a dynamic network of organisations, and the Lasersaur open source development project, see Schneider 2018.

products. When we study technical skills, which are a form of knowledge, we also need to ask what the processes are that regulate, produce, and organise such skills and their acquisition. What are the processes through which technical skills are studied and passed on? And in what ways are particular technical objects an inextricable part of such processes and transformed through them? In short, how are the becomings of technology and knowledge produced and organised together? TechKnowledgy thus extends one of the main concerns of classical STS literature, namely, how objects and materials are part of complex social practices of knowledge production (e.g., Latour 2013; Knorr-Cetina 2007; Rheinberger 1997).

However, TechKnowledgy also extends the long tradition of seeing human labour and technology as constitutive of each other. In this tradition, historian David Noble, following Marx, writes the following in his analysis of the emergence of modern engineering: 'Like every other social process, technology is alive. People – particular people in particular places, times, and social contexts – are both the creators of modern technology and the living material of which it is made. Designers and builders of an ever more sophisticated productive apparatus, they are at the same time the critical constituents of that apparatus, without which it could not function' (Noble 1977: 167). Writing in the Marxist and machinic tradition, Noble's argument is similar to Deleuze and Guattari's (2004) conception of 'machinic assemblages' that weave heterogeneous elements into a productive process.[5] The concept of assemblage, however, is often narrowed down in STS research to a sole focus on heterogeneity. However, the qualities of the processes that weave heterogeneous elements together are also central in conceptualising TechKnowledgies – processes that Deleuze and Guattari call the 'abstract machines', the procedures within the machinic assemblages. An inquiry into TechKnowledgies uncovers what it means to act and to become in particular techno-*logical* manners like, for example, open digital fabrication. The shifts in perspective that go along with this theoretical concept are twofold. On the one hand, this is a shift towards the temporal and unfolding character of the production and organisation processes of technology and knowledge. And on the other hand, this is a shift that moves beyond the diagnosis that heterogeneous elements are related in sociotechnical processes – the defining ontological statement of much STS – towards an understanding of *how* particular heterogeneous elements work together to constitute specific becomings of knowledge and technology.

A look back at the word's history already provides some important insights. The term 'TechKnowledgy'[6] was coined in the literature on 'knowledge management' in management and organisation studies. It has been used there, however, to designate ICT applications that store and transmit knowledge within an organization, such as a wiki or a mailing list (Davenport and Prusak 1998). It is significant that the term

[5] In fact, Noble and Deleuze and Guattari have a common intellectual inspiration in the work of Lewis Mumford (1970) and his analysis of the 'mega machine'.

[6] The term has also been used to refer to universities by Böhm (2002), a usage closer to mine. Also, an edited book with texts by humanities scholars is entitled 'TechKnowledgies' (Yablonsky 2007), which designates interdisciplinary interpretations of novel technical dynamics.

was coined at a time when digital information technologies made the constructed nature of knowledge evident again and turned it into a digitally malleable relation. My use of the term transcends this narrow meaning. Yet 'knowledge management' also hints at an important aspect, namely that knowledge and technology are both 'managed', consciously produced, and organised. I define a TechKnowledgy *as a historically emergent and dynamic set of collective procedures that produce and organise technology and knowledge*. A TechKnowledgy uses and transforms knowledge of technologies by using and transforming technologies of knowledge.[7] The concept is not restricted to open digital fabrication. Instead, there are many different TechKnowledgies in the contemporary world, and phenomena such as open digital fabrication help us see the diversity and contingency of assemblages that produce and organise technology and knowledge. In the following, I focus on three entangled procedures that are central to open digital fabrication and that will further enrich its theoretical conceptualisation as well as its empirical grounding: desiring openness, publicising technical objects, and organising collaborations.[8]

16.2 Desiring 'Openness'

One of the key cultural sources for open digital fabrication is the open source culture. From its beginnings as Free Software in the 1980s, this culture has been directed against the corporate monopolisation of technological knowledge, i.e. 'proprietary' knowledge, while fostering a scientifically inspired culture of knowledge sharing. Its emphasis on sharing the source code of software to make it freely usable and transformable aimed to make technological knowledge a public instead of a private affair. The cluster of open source practices that this involved had a huge influence on the early Internet, and they started to spread beyond software development beginning in the late 1990s (Kelty 2008). What is key here is that this spread is not simply an extension of these practices but their entanglement and transformation with other practices and contexts. Along with its more popular promoters such as Wikipedia, 'openness must therefore be understood as a powerful new form of political desire in network cultures' (Tkacz 2015: 28) that give way to such fields of practice as open access, open data, open education, open government, and open source hardware. In particular, the open source 3D printing project RepRap (www.reprap.org), which started in 2004, formed a rapidly growing sphere of 3D printing enthusiasts who build their own printers and it greatly influenced the imaginaries of individuals who use and share this technology, which had previously only had industrial and scientific uses.

[7] For an in-depth discussion of TechKnowledgies and why it is analytically useful to focus on the relationships of subjects, objects, organisational forms, and desires in the processes under investigation, see Schneider 2018, especially Chap. 2.

[8] Open digital fabrication is clearly inspired by the practices of Free Software that have been thoroughly analysed by Kelty 2008 – a classic study that was also inspirational for my work.

Openness cannot be reduced to one particular aspect since it is found in very different practical settings, all of which enact their particular form of openness. The history of openness in digital cultures has given it a certain contour, however. The discourse of openness is filled with calls for participation, transparency, and sharing through digital technology. It alludes to collective agency and ideas of publicness (Tkacz 2015). Practically, many technosocial assemblages have been created that foster the publication of information and cooperation, which are both linked. As Kelty put it: '[open source software developers] argue about technology, but they also argue through it. They express ideas, but they also express infrastructures through which ideas can be expressed (and circulated) in new ways' (2008: 29). Many technosocial assemblages are by now shaping the discourse and the practices of openness (Benkler 2013; Coleman 2012). In them, it is particularly evident that digitised information is vastly reproducible and distributable, which creates practical tensions with dominant conceptions of intellectual property and exchange value (Gorz 2010). In different ways, we can see a wider vision of Internet-utopian colours shades that imaginatively links the Internet with the emergence of a less hierarchical society. This vision has been emerging from the US counterculture of the 1960s and has been stabilised through various discourses, organisations, and technosocial arrangements (cf. Dickel and Schrape 2017; McCray 2012; Turner 2006). The politics of the desires for openness become evident when considered in contrast to the large emphasis on intellectual property rights in neoliberalism, which created many technoscientific endeavours whose goal was to make private profit instead of to create public knowledge (Mirowski 2011). The main TechKnowledgies of the contemporary 'knowledge economy' are thus in a paradoxical situation: their processes, practices, and norms are geared towards constraining the techosocial potential of knowledge production and circulation – fostered by digitisation and novel organisational forms (Gorz 2010; Hardt and Negri 2009; Mason 2015). And 'openness' is the discursive strategy to critique this.

Instead of following the lens that the normative discourse of openness engenders,[9] however, I approach openness from a meta-perspective, as a desire. This way, openness appears as part of the moral economy[10] of open digital fabrication – any TechKnowledgy takes part in a moral economy and vice versa, designating what is 'good' and 'bad' about particular assemblages. As an ethos, openness shapes collective and individual evaluations. It is something that can feature differently in what is and what is not yet. 'We live between what is and what could or should be; ethical ideas themselves are related both to the kinds of beings we (think we) are and the

[9] Whether something is more or less 'open' is a question made possible by the ethos of openness. Doing research on this question remains within the frame of this ethos; seeing openness as an ethos that structures practices, however, is a meta-perspective.

[10] '[Moral economy] examines and assesses the moral justifications of basic features of economic organisation, in particular property relations and what institutions and individuals are allowed and required to do. It also examines and assesses the moral influences on, and implications of, economic activities, and how economic practices and relations are evaluated as fair, unfair, good or bad by those involved in them.' (Sayer 2015: 2)

kind of beings we (think we) should become through our actions' (Sayer 2011: 145). Such evaluations are not simply cognitive, argumentative, and reflexive endeavours, as most academic and philosophical ethics would have it, but are often practical engagements in the world, including objects (Thévenot 2002). Accordingly, 'openness' is a practical engagement that highlights the (possible) differences with more 'closed' and proprietary assemblages. In the next section of this chapter, I show how 'openness' is entangled with digitised infrastructures and how 'open objects' are objects of desire and practical enablers in open digital fabrication.

16.3 Publicising Technical Objects

Open source practices have received a lot of attention because of their forms of organisation, the motivations of people engaging in them, or their critique of intellectual property regimes. Little research has so far addressed the technical objects at the centre of these knowledge practices. In this section, I focus on the technical objects that mediate open digital fabrication technically, socially, and imaginatively (Knorr-Cetina 1997, for object relations in 'online' communities: Mackenzie 2005). Objects are crucial for the production and circulation of knowledge (Knorr-Cetina 1997; Rheinberger 1997; Simondon 2009).

As an empirical entry into these objects, I use the Lasersaur (www.lasersaur.com) open source laser cutter project, which started in 2010. Like 3D printers, laser cutters are digitally operated to produce material things, and these machines are widely used in open digital fabrication. The project's founders, Addie and Stefan, two artists/technologists with a background in US art, new media, and hacking culture, wanted a laser cutter as a tool, yet industrial machines with prices in the tens of thousands of euros were out of their financial reach. They thus shared their plan to build one in open source online and, to their own surprise, a group of volunteers quickly formed to support this plan financially, socially, and technically. Clearly, this is the desire for 'openness' in action. To analyse the important role of technical objects for the TechKnowledgy of open digital fabrication, I first discuss the ecology of technical objects in which the Lasersaur project happened. Second, I discuss how efforts to publicise this object and knowledge about it created an unfolding 'open object' (Simondon 2009) that enables the developer community to collaborate and to produce knowledge.

The last 10–15 years have seen the massive spread of a digitised knowledge infrastructure (on infrastructures: Barry, this volume) in wealthier and technologised societies: PCs, broadband Internet, digital photography and video, and online platforms and services for sharing data and content. Furthermore, 'the Internet' has also contributed to the emergence and distribution of organisational forms that are built around digital objects such as mailing lists, wikis, different forms of social media, and so on, which are often available for use at little to no financial cost. Thus whilst 'the Internet' is certainly a key force in contemporary transformations, it is itself dramatically changing and looks quite different from, for example, 'the

Internet' in 2000, and its technosocial assemblages keep on changing at a fast pace. This infrastructure is key to 'objectifying' knowledge through creating digitised objects that store and transmit information and which make knowledge shareable with others. In particular, digital texts (via mailing lists) are important here, as are digital images, videos, and CAD (computer aided design) drawings. Many such objects can be found publicly circulating on the Lasersaur project's homepage, designed, commented upon, and digitally shared by the group of developers who organise through a mailing list. Technical objects are the media that form and shape public knowledge of the project as the product of interactions between those who participate in these publics. This circulation of objects in particular creates the publicness of the knowledge (cf. Warner 2002).[11] Such a circulation of digital objects is also key to platforms where digital design files that instruct the machines to produce particular shapes are shared and sometimes sold. Many online platforms have emerged on which either professionals or hobbyists can find thousands of such files, often as 'open access', for designing forms for material objects on their PCs using software programmes. These platforms and the people engaged in them have contributed to the recontextualisation of 'computer aided design', usually found in the design and research departments of companies or universities. The production of knowledge in the Lasersaur project is largely due to people creating and transforming relations to these objects and transmitting explicit knowledge based on them.

In the actual process of building a Lasersaur, two other sorts of objects come into play. When you've decided that you want to build such a machine based on its publicly available open source instructions, you need to first engage with many standardised industrial objects, such as aluminium extrusions, screws, cables, etc. These are the mechanical building blocks of the actual machine and can – as in the case of a laser cutter – be rather expensive.[12] With their standardised sizes and wide commercial availability, they play an enormous role in making the building process repeatable. Selecting and suggesting suitable objects and their suppliers represents a large part of the labour in the Lasersaur project, even though globalised online commerce has afforded relatively easy access to these objects and to information about them. Furthermore, this task was greatly supported by other open source projects that had already used parts and created experiences. Secondly, open source projects and their technical objects provide valuable information and building blocks for each other. The Lasersaur, for example, uses open source software, micro controllers, and 3D-printed parts. While there is still a great need for industrial objects, one can see how modularity is emerging amongst the increasing number of open source objects available for use and modification. Even the Lasersaur project's knowledge – in the form of documentation or software (an object) – has already been used for other projects.

[11] Warner's 'ontology' of publics is, however, too narrowly focused on the circulation of texts to capture the publics of open source.

[12] Although, the Lasersaur's materials cost between 5000 and 7000 €, this is still much cheaper than a similar commercial machine.

An increasing number of digital fabrication machines like the Lasersaur are 'open source', which basically means there are building instructions that are freely accessible online – the open 'source code' – and often a 'community' that produces and maintains this source code. The objects, such as 3D printers or laser cutters, form the centre of the 'communities' that assemble around them and they are the medium for their 'communication'. A recursive logic is at play in this form of sociality (cf. Kelty 2008), which is thoroughly entangled with technical objects: 'the effects and products are necessary to the process that creates them. The product is producer of that which produces it' (Morin 2008: 61). Knorr-Cetina have argued that, in contemporary societies, new forms of 'post-social' relations appear, at the centre of which are object-relations to unfolding objects that are desired by people (Knorr-Cetina 1997). Thus these desires and the objects of desire are a key element of the forms of technosociality that are building up here.

Furthermore, these relationships between objects, subjects, and desire have a quality that is at the centre of these assemblages. The philosopher of technology, Gilbert Simondon, placed technical objects at the centre of this thought and argued several decades ago that a new class of technical objects had appeared: 'open objects' that corresponded with what he calls the 'technical mentality'. '[I]n order for an object to allow for the development of the technical mentality and to be chosen by it, the object itself needs to be of a reticular structure [...] an *open* object that can be completed, improved, maintained in the state of perpetual actuality' (Simondon 2009: 24, italics in original). Open objects need connectivity, the possibility to be connected to other objects and, through this, to change their properties. A central aspect of open objects is *indeterminacy*, the malleability and the multifunctionality of these objects that is achieved by linking them with other objects. And this process of linking and changing technical objects is what Simondon sees as the effect of the technical mentality. The above analysis of the roles of diverse technical objects in enabling and forming the assemblages for producing and organising technology and knowledge show how such connecting and changing of objects is key to open digital fabrication. These desires of open digital fabrication are not limited to creating things using the existing sets of machines and trying out uncommon uses for them. Desires reveal tendencies in objects that are framed as becoming more and more malleable, customisable, networked, and informed. Seen in this light, '3D printing' is a mere metaphor for this larger process of the entanglement of material and digital technologies. Here, the powerful imaginaries of the technosciences come into play as well, imaginaries that promise to make matter more malleable and controllable through technology (Nordmann, Chap. 2, this volume; Schmidt, Chap. 3, this volume).[13] Additionally, however – and this is a transformation of institutionalised technoscience – there is also a desire that the control over the technologies will become distributed throughout society.

[13] In particular, ideas in nanotechnology, such as the 'molecular assembler' (Eric Drexler) and a 'next industrial revolution', spread widely and were incorporated into the early discursive framing of digital fabrication (Gershenfeld 2005) and nowadays feature in a broad discourse about a digitised industrial revolution (e.g., Rifkin 2014).

The TechKnowledgy of open digital fabrication in development projects like the Lasersaur is fundamentally the configuration and connection of relatively indeterminate 'open objects' and people into a process of object-mediated communication and knowledge transmission. Desires for 'social' openness deeply correspond with the open objects and desires to create 'open' relations to them. The Lasersaur project thus not only had to develop a technical object but also a way of communicating about and with this object in a digitised public that the project itself brought into existence. It was, however, no 'one-click wonderland' but involved several years of hard labour by dozens of people who interacted online and partly offline to foster mutual learning and a reworking of the documentation to improve the machine's design. In this way, they created and circulated socially robust knowledge that can be used to relatively easily repeat the building process, which can be done by following the online instructions. In the next section I turn in more detail to how such collaborations are organised.

16.4 Organising Collaborations

I now swivel my perspective and analyse digital fabrication from its 'organisational' side. What used to be 'online' and 'offline' forms of organisation become entangled in digital fabrication, and this is the case in both development projects and FabLabs. The latter are a global movement of organisations that aim to make digital fabrication machines locally accessible. I briefly turn to the initial phase of the Lasersaur project to illustrate how digital platforms help create public forms of technosocial organisation. Then, I show why FabLabs in particular play an important part in defining and spreading digital fabrication and why their history is one of being co-defined in public. Besides the online collectives, FabLabs are particularly important in making the development and use of digital fabrication technologies more public.

Addie and Stefan not only designed a technical machine, they co-designed a social development and design project centred on a technical object, the Lasersaur. As Stefan put it: 'I guess there are two things we did. It's, you know, we developed the technical thing, the reference design, and the other thing is sort of, you know, develop the community'. When the Lasersaur project was conceived in 2010, one of the first 'crowdfunding' websites went online, something that is now a hugely popular format amongst open source projects and start-up companies. The two put their plan to build an open source laser cutter on the crowdfunding website and were curious what would happen. The goal in crowdfunding is to let the online 'crowd' fund the realisation of diverse projects with small to medium amounts of money (starting from around one dollar). The crowd then buys an incentive if the project is successful. For the Lasersaur, 260 people signed up to support the project and pledged 20,000 dollars to it. The project had become a social project with mutual expectations and a shared vision. This vision entailed discursive references to 'open source' and 'making' and to the vision of digital fabrication in the hands of individuals, which is also promoted by Neil Gershenfeld, the founder of FabLabs

(Gershenfeld 2005, cf. 2012). Crowdfunding also enabled the collection of money as a form of support in economic and symbolic terms. People literally made an 'investment' in the object, which was still just the vision of something that would emerge in the future. Addie and Stefan told me that the money that accumulated was important, but that the most important outcome was the publicity for the project and the group that had formed. The platform enabled an initial form of inter-subjectivity centred on the (vision) of a technical object, entangled with a desire for the object's unfolding based on the streams of data, texts, images, and money. Eventually the crowdfunding experiment turned into a form of public decision making about how to realise the envisioned technical object, which was mediated by the technologies of attention that the crowdfunding platform affords. Many more such platforms that enable interactions, the sharing of ideas, and the assembling of crowds and groups have since emerged, and there is an increasing digitised support structure for such novel organisational forms that can be mobilised in open digital fabrication and similar fields of practice.

This strong linkage of organisational form and technological structures can also be found in FabLabs. The first FabLab started in 2002 and there are now, in 2018, around 1700 FabLabs across the globe.[14] These FabLabs have a relatively diverse array of organisational structures, for example, they are part of universities, community run non-profits, or companies (Troxler and maxigas 2014). What they all share, however, is the centrality of digital fabrication machinery, such as laser cutters and 3D printers, which defines what a FabLab is. Furthermore, FabLabs have been modelling themselves using the ontologies and processes that machines for digital fabrication afford. Each FabLab, in theory, should have a similar technical infrastructure to allow for an exchange of data between the labs and to materially produce these shared digitised objects and ideas. Thus, FabLabs, like the machines, facilitate the materialisation of the digital flows of technical data and forms on the Internet. The global 'movement' of these labs is largely held together by a set of similar technical objects that are even recommended in a FabLab inventory list circulated by MIT. These objects function as the locus of most of the practices in FabLabs. Other than this, there are only loose regulation structures in the form of guidelines that function as a kind of code of conduct. Furthermore, open source projects regarding digital fabrication machines, such as the Lasersaur and 3D printing, have emerged. With their co-definition of open source hardware culture, they have co-defined FabLabs as well and have drastically lowered the cost of setting up a FabLab with a basic digital fabrication infrastructure. Digital fabrication enthusiasts, who can often be found in FabLabs, often build the machines themselves. Technically, culturally, and economically, many FabLabs build strong relations with open source technology. Here we also find that the collaborations are organised around the technologies and, again, object-mediated communication plays an important role in holding this global assemblage together.

[14] They are unevenly global, however, with the densely technologised zones of rich countries dominating the global landscape of FabLabs, although they can be found on every continent by now. See also www.fablabs.io for an overview of the labs.

FabLabs were initially conceived at an elitist and exclusive place, MIT's Media Lab, which is also a prominent place for institutionalised digital technoscience. Yet the labs have transformed into a more inclusive concept (Schneider and Lösch 2019). Since its inception, the Media Lab has been seen as a place where the future of digital technology was already here and where it was made. This is based on a digital utopianism that grew out of parts of the US counterculture and which was influential at the Media Lab (Turner 2006: 180). Around the year 2000, Neil Gershenfeld, director of the lab's Center for Bits and Atoms, saw that, in a massively overcrowded course on digital machines such as CNC mills, students from different disciplines started teaching each other and built stuff instead of doing research. From this experience, along with a government funded programme, a prototype of a FabLab was set up in 2002, making just a few small CNC machines available to the public in a part of Boston. After a handful of FabLabs had been established, Gershenfeld thought the experiment was over. Yet, other organisations adopted the concept and a massive growth started around 2006 and eventually there were around 150 labs in 2013 and around 400 in 2014 and growing. Notably, from 2010 onwards, a wave of community-run grassroots FabLabs began. These were formally independent from MIT, which organisationally and culturally diversified FabLabs (Troxler 2014).

The imaginary of FabLabs, of experimenting with new digital technology, decentralisation, and heterarchy, resonated with other universities, in the start-up scenes of 'creative cities', and in the cultures of hackers, DIYers, open source enthusiasts and activists for alternative economies – groups which are not entirely exclusive to each other but which also do not necessarily overlap (Troxler and Maxigas 2014). Like the documentation of technical open source projects, the FabLab idea spread online through the images and texts that the first FabLabs produced. Furthermore, from 2004 onwards, open source 3D printing started to capture the practices and imaginations of a growing number of hackers, students, and tinkerers who recognised in FabLabs a suitable concept for sharing and fostering their desired objects. Thus, while escaping the dominance of MIT – although Gershenfeld and the FabLab organisations set up at MIT are powerful within this global 'movement' – FabLabs have been co-defined in public through imaginaries, objects, practices, and their creative entanglement with other organisations and concepts. FabLabs have become the 'objects' at the centre of a distributed collaboration of many people and organisations that have tried to foster the idea of open digital fabrication, to build supportive infrastructures – also by 'opening' the organisational form of the FabLab itself – and to link digital technosociality based on platforms to local and materialised organisational forms.

Practices of making knowledge public and shared are common in FabLabs: First, the development, construction, and use of technical objects – which are often directly related to digital fabrication but not always – is taking place as a shared practice that is visible to others and sometimes involves mutual learning and help. Secondly, whilst not all FabLabs are open to the public on most days, the labs are asked by MIT's guidelines to host open days that are open for everyone and which include free usage of the lab. Such 'open access' to the lab is also fostered by the desires for openness prominent in many of the cultures that are typical of FabLabs. Making the labs accessible involves, amongst other things, giving courses and teaching others to use the machines. Thirdly, in addition to such technical practices, another important practice involves talking about the machines and the labs, to visitors for example. This is a kind of public presentation of FabLabs. Similar to the projects that are coordinated online, this part of open source digital fabrication is based heavily on organising technosocial entanglements between technical objects and people, and between people, to create a form of publicness for the knowledge that is enabled through this. Yet, to facilitate the actual process of digital fabrication, a lot of labour, coordination, learning, and communication amongst people is necessary: collaborations need to be organised and there needs to be the relevant knowledge for such organising. There need to be technosocial translations of digital information into the material sphere and vice versa. Particular subject skills are needed to enable the publicness of digital fabrication. This includes are how to use digital technologies and communication with others. Social skills are needed for communication, exchange, and learning, often involving movements between various languages and areas of expertise and of organising such technosocial exchanges and participating in these. Therefore, beyond being an important sphere for further development and creative use of digital fabrication machines, FabLabs are important places for people to become the subjects to 'publicise' digital fabrication. As Noble writes, and which I quoted above, any TechKnowledgy needs the 'materials' involving technical objects and people that arrange themselves or are being arranged into productive assemblages, and FabLabs are important in facilitating this.

16.5 The TechKnowledgy of Open Digital Fabrication and Beyond

Considered on their own, many practices and technical objects that I have analysed here are not absolutely new. Crucial to the TechKnowledgy of open digital fabrication are the re-arrangements of these practices and the corresponding technical objects as well as the cultures and technosocial possibilities for knowledge sharing engendered by an increasingly digitised and technologised society within which this takes place and shape. What is crucial to open digital fabrication is its arrangement and not the individual elements of it, such as 3D printers. The TechKnowledgy of open digital fabrication is a process of producing and organising technology and knowledge through the entanglements of desires for openness, technical objects for digital fabrication, and organisational forms that enable object-mediated and digital, information-rich collaborations between people. This equally affects objects and subjects of digital fabrication, both of which are enabled by this arrangement and both of which are involved in propelling and transforming its trajectories. The procedures that I have analysed, desiring openness, publicising objects, and organising collaborations, are of course not exclusive to open digital fabrication, but neither are they necessarily the only procedures within it. However, they give a sense of the qualities that make a technological practice part of open digital fabrication and not of other endeavours. It is these qualities that we need to attend to in TechnoScienceSociety when technologies in different areas are multiplying and the ways that they are embedded within society are multiplying, as well. Conceptions about the 'hybrid' nature of society and its 'sociotechnical' character are all valid. However, we need to take the next step and understand what different qualities such entanglements can entail and how they can be shaped in ways that are conducive to flourishing within TechnoScienceSociety. TechKnowledgies is a concept that makes this possible.

Seen from the perspective of TechKnowledgies, Responsible Research and Innovation (see Fisher, Chap. 10, this volume), for example, can be read as an effort to re-arrange the procedures that organise 'innovation', which is to say, procedures that produce and organise knowledge and technology. However, from the concept of TechKnowledgies and the above analysis, we can also draw the insight that such re-arranging needs to include subjects, objects, organisational forms, and desires in order to enable different becomings for all of them. Appropriate procedures must be found, or re-arranged, or even be invented. It is here, especially imaginatively and experimentally, that phenomena like open digital fabrication may have their biggest impact in TechnoScienceSociety. We already have practical examples showing that technological change may increasingly be shaped by assemblages that transcend the 'modern' scientific-industrial complexes or that extend it into new areas. Our conceptions of what 'technology' is and who is in charge of it may fundamentally change, and options for shaping different embeddings of technological processes within society are emerging. Open digital fabrication is one of them, and the recent and enormous interest that research funders, participation practitioners, and researchers have had in 'hackathons', 'maker spaces', and the use of digital technologies is an indicator of this. New TechKnowledgies show us that technology and knowledge can be differently organised and produced and that we can try to re-arrange educational, research, and economic institutions to shape this. As the expectations about who could and should be included in processes of technosocial creativity widen, this also constitutes a possibility for STS to bring its reflexivity regarding the technosociality of 'technology' to such emerging fields of practice (see also Tyfield, Chap. 13, this volume). STS can more thoroughly engage in the unfolding multiplicities of TechKnowledgies that are central to TechnoScienceSociety. Some of these multiplicities invite us to experiment with their uncertain potential for re-defining what the technologies and knowledges of TechnoScienceSociety are and could become.

References

Benkler, Y. 2013. Practical Anarchism Peer Mutualism, Market Power, and the Fallible State. *Politics and Society* 41 (2): 213–251.

Böhm, Steffen. 2002. Movements of theory and practice. *Ephemera: critical dialogues on organization* 2 (4): 328–351.

Coleman, E.G. 2012. *Coding Freedom: The Ethics and Aesthetics of Hacking*. Princeton: Princeton University Press.

Davenport, T.H., and L. Prusak. 1998. *Working Knowledge: How Organizations Manage What They Know*. Cambridge, MA: Harvard Business Press.

Deleuze, G. and F. Guattari. 2004. *A Thousand Plateaus: Capitalism and Schizophrenia* (B. Massumi, trans.). London: Continuum.

Dickel, S., and J.-F. Schrape. 2017. The Logic of Digital Utopianism. *NanoEthics* 11 (1): 47–58.

Gershenfeld, N. 2005. *Fab: The Coming Revolution on Your Desktop–from Personal Computers to Personal Fabrication*. New York: Basic Books.

———. 2012. How to Make Almost Anything: The Digital Fabrication Revolution. *Foreign Affairs* 91: 43.

Gorz, A. 2010. The Exit from Capitalism Has Already Begun. *Cultural Politics: An International Journal* 6 (1): 5–14.

Hardt, M., and A. Negri. 2009. *Commonwealth*. Cambridge, MA: Harvard University Press.

Jasanoff, S. 2016. *The Ethics of Invention: Technology and the Human Future*, The Norton Global Ethics Series. New York: W.W. Norton.

Kelty, C.M. 2008. *Two Bits: The Cultural Significance of Free Software*. Durham: Duke University Press.

Knorr-Cetina, K. 1997. 'Sociality with Objects: Social Relations in Postsocial Knowledge Societies. *Theory, Culture and Society* 14 (4): 1–30.

Knorr-Cetina, Karin. 2007. Culture in global knowledge societies: Knowledge cultures and epistemic cultures. *Interdisciplinary Science Reviews* 32 (4): 361–375.

Latour, Bruno. 2013. *An inquiry into modes of existence*. Cambridge MA: Harvard University Press.

Mackenzie, A. 2005. The Performativity of Code: Software and Cultures of Circulation. *Theory, Culture & Society* 22 (1): 71–92.

Mason, P. 2015. *PostCapitalism: A Guide to Our Future*. London: Allen Lane.

McCray, W.P. 2012. *The Visioneers: How a Group of Elite Scientists Pursued Space Colonies, Nanotechnologies, and a Limitless Future*. Princeton: Princeton University Press.

Mirowski, Philip. 2011. *Science-mart: Privatizing American science*. Cambridge, Mass: Harvard University Press.

Morin, E. 2008. *On Complexity*. Cresskill: Hampton Press.

Mumford, L. 1970. *The Myth of the Machine: The Pentagon of Power*. New York: Harcourt Brace Jovanovich.

Noble, D.F. 1977. *America by Design: Science, Technology, and the Rise of Corporate Capitalism*. Oxford: Oxford university press.

Rheinberger, H.-J. 1997 *Toward a History of Epistemic Things: Synthesizing Proteins in the Test Tube*, Writing Science. Stanford: Stanford University Press.

Rifkin, J. 2014. *The Zero Marginal Cost Society: The Internet of Things, the Collaborative Commons, and the Eclipse of Capitalism*. New York: Palgrave Macmillan.

Sayer, A. 2011. *Why Things Matter to People: Social Science, Values and Ethical Life*. Cambridge: Cambridge University Press.

———. 2015. Time for Moral Economy? *Geoforum* 65: 291–293.

Schneider, C. 2018. *Opening Digital Fabrication: Transforming TechKnowledgies*. Karlsruhe: KIT Scientific Publishing.

Schneider, Christoph, and Andreas Lösch. 2019. Visions in assemblages: Future-making and governance in FabLabs. *Futures* 109 (Mai): 203–212. https://doi.org/10.1016/j.futures.2018.08.003.
Simondon, G. 2009. Technical Mentality. *Parrhesia* 7: 17–27.
Thévenot, L. 2002. Which Road to Follow? The Moral Complexity of an "equipped" Humanity. In *Complexities: Social Studies of Knowledge Practices*, ed. J. Law and A. Mol, 53–87. Durham/London: Duke University Press.
Tkacz, N. 2015. *Wikipedia and the Politics of Openness*. Chicago/London: University of Chicago Press.
Troxler, P. 2014. Fab Labs Forked: A Grassroots Insurgency inside the Next Industrial Revolution. *Journal of Peer Production* 5: 1–3.
Troxler, P., and Maxigas. 2014. Editorial Note: We Now Have the Means of Production, but Where Is My Revolution? *Journal of Peer Production* 5.
Turner, F. 2006. *From Counterculture to Cyberculture: Stewart Brand, the Whole Earth Network, and the Rise of Digital Utopianism*. Chicago: University of Chicago Press.
Warner, M. 2002. Publics and Counterpublics. *Public Culture* 14 (1): 49–90.
Yablonsky, Paula, Mary, Valentis, and Tara P. Monastero. 2007. *TechKnowledgies: New imaginaries in the humanities, arts, and TechnoSciences*. Newcastle upon-Tyne: Cambridge Scholars Publishing.

Christoph Schneider works as a Sociologist at the Institute for Technology Assessment and Systems Analysis of KIT, Karlsruhe. His research focuses on the possibilities of democratizing technoscientific research and innovation. As Public Sociologist, he is active in progressive technopolitics and consults organizations in search of more humane forms of digitalization.

The manufacturer's authorised representative in the EU is Springer Nature Customer Service Centre GmbH, Europaplatz 3, 69115 Heidelberg, Germany. If you have any concerns regarding our products, please contact ProductSafety@springernature.com

Printed and bound by CPI Group (UK) Ltd, Croydon, CR0 4YY
30/09/2026
02225445-0008